Applied Signal Processing

Concepts, Circuits, and Systems

Applied Signal Processing

Concepts, Circuits, and Systems

Nadder Hamdy

AAST
Miami, Alexandria, Egypt

CRC Press
Taylor & Francis Group
Boca Raton   London   New York

CRC Press is an imprint of the
Taylor & Francis Group, an **informa** business

CRC Press
Taylor & Francis Group
6000 Broken Sound Parkway NW, Suite 300
Boca Raton, FL 33487-2742

© 2009 by Taylor & Francis Group, LLC
CRC Press is an imprint of Taylor & Francis Group, an Informa business

No claim to original U.S. Government works
Printed in the United States of America on acid-free paper
10 9 8 7 6 5 4 3 2 1

International Standard Book Number-13: 978-1-4200-6702-6 (Hardcover)

Library of Congress Cataloging-in-Publication Data

Hamdy, Nadder A.
    Applied signal processing : concepts, circuits, and systems / author, Nadder Hamdy.
        p. cm.
    Includes bibliographical references and index.
    "A CRC title."
    ISBN 978-1-4200-6702-6 (hardback : alk. paper) 1. Signal processing. 2. Signal processing--Digital techniques. 3. Signal processing--Equipment and supplies. 4. Electric filters. 5. Electric filters, Digital. 6. Fourier transformations. I. Title.

TK5102.9.H354 2008
621.382'2--dc22   2008009173

Visit the Taylor & Francis Web site at
http://www.taylorandfrancis.com

and the CRC Press Web site at
http://www.crcpress.com

Dedication

To my beloved family: wife, sons, daughter, and grandchildren, with love and gratitude

Contents

Preface

Applied Signal Processing is a textbook for electrical, computer, and mechatronics engineering students at the senior and graduate levels. It can also help practicing engineers enhance and upgrade their knowledge and skills in the field of signal processing. It provides material that can be covered in two semesters. The book is the outcome of more than 25 years of experience in teaching signal processing. It introduces underlying principles and basic concepts, definitions, classic and up-to-date designs, and implementations of signal processing techniques and signal processing systems. Since analog and digital signal processing are closely related, it seems advantageous to treat them together in a single book. This should help students get a global overview of the whole subject and establish links between them. Moreover, becoming familiar with the classical analog processing techniques should make it easy to understand their digital processing counterpart. The book presents the topics in clear language that should be accessible for native English speakers as well as international students. Several illustrative solved examples are found throughout the text to expose students to real-design problems and enhance their grasp of the subjects. In addition, a number of review questions and problems are given at the end of each chapter. The use of MATLAB® is recommended as a tool for verifying and visualizing solutions to the given problems. Many examples are solved employing MATLAB instructions.

The book is divided into ten chapters that cover most signal processing fields. The chapters are written to be mostly independent so that instructors can select the teaching sequence they prefer to suit their curriculum. After the introduction given in Chapter 1, some basic analog processing operations are presented in Chapter 2. Analog filters, being an integral component in any signal processing system, are discussed in detail in Chapter 3. Their types, describing equations, and design techniques are provided. Chapter 4 presents a comprehensive and detailed description of the necessary interfacing devices between the analog and digital domains, namely, analog-to-digital and digital-to-analog converters, where basic concepts as well as new trends in designing them are introduced. Some basic digital signal processing operations such as correlation, convolution, and the Z-transform are described in Chapter 5, with an introduction to digital filters. In Chapter 6, detailed descriptions of digital filters design and realization techniques are provided. Chapter 7 introduces the concept of multirate signal processing and its applications. Transforming signals from the time domain into any frequency domain is the subject of Chapter 8 in which several discrete-time transforms such as Fourier, cosine, and wavelet transforms are covered in an illustrative and simple way. In Chapter 9 basic digital signal processing hardware is described giving some practical architectures such as those developed by Texas Instruments, Motorola, and Analog Devices. Finally, Chapter 10 provides an overview of several practical applications of DSP in many fields such as forensic and biomedical applications, remote sensing, active noise control, pattern recognition using neural networks classifiers, biometrics, and data compression. The book has a CD solutions manual for instructors. Readers may also visit the interactive web site, http://book.aast.edu/eng/asprocessing.

Acknowledgments

A number of people have contributed, either directly or indirectly, to this book. I have been fortunate to have them. In this respect, I appreciate the thorough review of Chapter 10 of Professor S. El Kahamy, Alexandria University, and the significant contributions of Professor M. Sharkas, AAST, in reviewing parts of the manuscript and preparing the solutions manual. Special thanks should go to both of them. I am equally grateful to Professor H. Baher, Dublin Institute of Technology, for his review of Chapters 3 and 6. I am highly indebted to Professor M. Saeb, Professor W. Fakhr, and Dr. I. Badran, AAST, for their comments and careful review of some chapters. I am also grateful to my students in different universities who have helped over the years, through their feedback, in bringing the course and hence the book to the form that matches their needs. Among many people at Taylor & Francis, two have especially contributed much through their limitless help and cooperation; namely, Nora Konopka, publisher of engineering and environmental sciences, and Catherine Giacari, project coordinator, I am really grateful to them. I am also grateful to Macmillan's team for their valuable help. My appreciation also goes to Iman Eid for the wonderful cover design and Mona Shahin for designing the book's Web site.

Lots of thanks and gratitude should go to my daughter Noha for improving the quality of the language, and to my brother Naeil and my sons Mohamed and Ahmed for their moral support. Last but not least, I wish to thank my wife Nagat for her continuous encouragement and many lost weekends.

Nadder Hamdy

Author

Nadder Hamdy is a professor of electronics and signal processing at the Arab Academy of Science and Technology (AAST), Alexandria, Egypt. He is currently chairman of the Electronics and Communication Engineering Department. He holds a PhD in electronics and signal processing from the University of Erlangen, Germany. His research interests are high-speed analog-to-digital converters, image and speech signal processing, watermarking, and other related topics.

Dr. Hamdy coauthored *Electronic Analog to Digital Converters*, published in 1983 by John Wiley & Sons. He cochaired the IEEE International Midwest Symposium on Circuits and Systems (MWSCAS) held in 2003 in Cairo. He has published 44 papers and supervised several MS and PhD theses. He has been teaching signal processing for more than 25 years. Dr. Hamdy is a member of the steering committee of the MWSCAS and is an IEEE senior member where he is currently the secretary of the Alexandria subsection.

A Note to the Instructor

Teaching signal processing is a unique experience. A good textbook should make it a pleasant undertaking. This book is the outcome of almost 25 years of teaching the subject. Selection of a suitable book to accompany the lectures was always hard. There were always many signal processing books. However, no single title covered the whole curriculum. Therefore, I felt compelled to change the text several times over the years due to either the students' feedback or the publication of a new promising title. As the lecture notes grew over the years, I came to the idea of compiling them into one book. I, therefore, have tried to include every subject that might be part of a signal processing curriculum.

Applied Signal Processing can cover the span of two semesters. The chapters are written to be independent (stand alone) so that each instructor can select a teaching sequence according to his or her needs. The following illustration is a suggestion for three different courses. The sequence of presentation is a matter of individual choice and is left to the judgment of the instructor. The many solved examples should boost the students' assimilation of the subjects; the review questions at the end of each section should serve as self-testing for the students. A solutions manual for the problems can be posted to interested instructors. The author welcomes any feedback, comments, and corrections.

Suggestions for three course plans

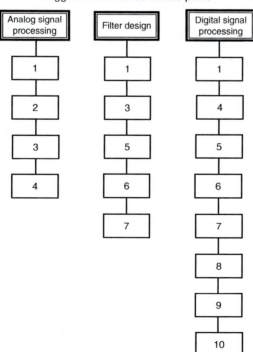

A Note to the Student

WHY THIS BOOK?

Applied Signal Processing is a student-oriented textbook. It presents up-to-date subjects together with classical ones clearly written to be accessible for native speakers of English as well as international students. Involved mathematical derivations, which are off-putting for the majority of students, are avoided wherever possible without affecting the in-depth treatment of the matter. Few books on the market navigate through the areas of both analog and digital signal processing the way *Applied Signal Processing* does. Also, inclusion of a detailed description of data converters, which are an interface between the real world of analog signals and the artificial world of digital signals, gives the book an extra edge. Many students are surely curious to know the theory underlying modern devices such as MP3 players, cell phones, DVD players, etc. and how some complicated systems such as e-banking, security systems in airports, etc. operate. Therefore, some current signal processing systems are discussed. The solved examples scattered throughout the text should help expose the reader to real-design problems and enhance a grasp of the subjects. The summary and a number of review questions and drill problems found at the end of each chapter are included to ensure mastery of the presented techniques. The book recommends the use of MATLAB® as a tool for verifying and visualizing the solutions to the given problems. In addition, a MATLAB project closes each chapter when applicable.

1 Introduction

1.1 WHAT ARE SIGNALS?

A *signal* is a time-varying function that carries information representing a certain phenomenon. Thus, it is analog by nature. A typical example of a signal is a telephone call, where speech (audio signal), in the form of acoustical (mechanical pressure) waves, is changed into electrical variations through the handset-microphone. These propagate via a cable to the receiver, where they are turned back into their original (acoustical) form through the built-in loudspeaker in the receiver's handset. Figure 1.1 illustrates examples of some signals in practice. The signal shown in Figure 1.1a is part of a speech signal, where low-amplitude variations are the periods of silence. In Figure 1.1b, a radio frequency signal is modulated by an audio frequency one. Such types of signals are common in wireless communications and radio and TV broadcasting. The composite TV signal depicted in Figure 1.1c represents one line out of the 625 lines that are used to represent one frame. Each line carries information representing variations in the illumination together with other control pulses (synchronizing and blanking pulses), as we shall see in Chapter 10.

Digital processing techniques are now replacing analog processing in several applications, due to the many advantages of digital signal processing. Such advantages can justify the added delay and cost of converting analog signals to the digital format. Data converters (analog-to-digital converters [ADC] and digital-to-analog converters [DAC]) will be described in Chapter 4. Representative examples of practical digital systems are advanced communication systems, such as satellite communications, mobile (cellular) phones, computer communications through local-area and wide-area networks (LANs and WANs, Wi-Fi and WiMAX), high-speed Internet communications through asymmetric digital subscriber line (ADSL, a high-speed Internet connection), software radio, digital television (as in HDTV), and so on. All these applications implement different digital processing techniques aiming at, for example, reducing the huge amount of data that would otherwise be transmitted, stored, and so on. Modern data compression techniques (an important signal processing field) such as JPEG and MPEG have recently been developed and became mature industry standards for audio and video compression. Compression techniques will be described in Chapter 10. The number of Internet users is exponentially increasing annually, and so is the amount of information to be exchanged. Special data transfer protocols and techniques such as packet switching, and modern modulation techniques such as Code Division Multiple Accesses (CDMA) and Orthogonal Frequency Division Multiplex (OFDM) have been developed to help reduce communication time and to improve the quality of service (QoS).

1

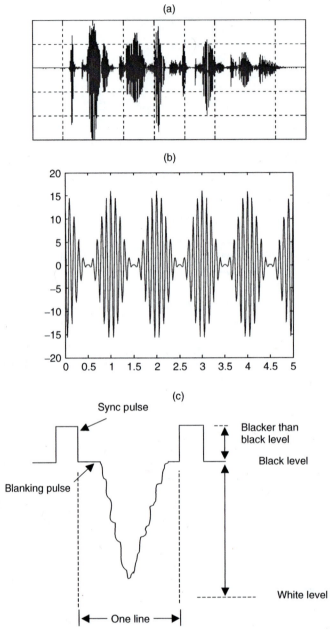

FIGURE 1.1   Example of signals. (a) An oscillogram representing a speech signal, (b) an RF signal, (c) part of a composite TV signal.

Several operations are usually performed on signals in either format. Although analog signals can be amplified, filtered, rectified, modulated/demodulated, recorded, and converted to digital format, digital signals can be stored, compressed, modulated/demodulated, transformed into/from the frequency domain, encoded/decoded, and converted back to analog format. Processing of signals is sometimes mandatory in order to make them useful or intelligible; for example, by "de-noising" them. In such cases, special types of digital algorithms are employed to extract useful data from noisy signals (corrupted by noise). Also, without data compression it would practically be impossible to fit, for example, a whole encyclopedia (several volumes) including illustrations onto one or two compact discs (CDs), or to fit a complete movie on a single DVD. Due to its superior quality, high reproduction quality coupled with digital recording of either audio or video information is such that the term "CD quality" has been coined to indicate top reproduction quality, and has become synonymous with studio quality.

1.2   SIGNAL PARAMETERS

Signals are usually described by one or more of their parameters, such as amplitude (envelop), frequency, and phase spectra. They carry information in form of changes in one or more of these parameters. The signal-to-noise ratio (SNR) is rather an important parameter of signals, especially in the field of communications, where a certain minimum SNR value should be provided to guarantee good communication between the transmitting and receiving ends. It is defined as

$$\text{SNR} = 10 \log P_S/P_N \text{ dB} \tag{1.1}$$

where P_S is the signal power consumed in a certain load and P_N is the noise power consumed in the same load.

The magnitude and phase spectra of a signal are important parameters, as they describe the signal in the frequency domain, thereby providing information about the contributions of the different frequency components to its magnitude and phase. This information is quite important in several applications, as it defines the useful bandwidth (BW) of the signal. Such a parameter indicates how fast the signal is, and hence the phenomenon that it represents changes. The BW of a signal is essential for better understanding its nature. It enables the designer to select proper circuitry for processing, proper transmitting or storage media, suitable processor speed, and so on. Furthermore, the BW of a signal is a measure of its quality. Compare, for example, the sound quality reproduced from a compact cassette (BW = 10–15 kHz) with that obtained from a CD (BW = 20 kHz). Needless to say, the unbeatable quality of the CD is a direct result of occupying the full audio BW of 20 kHz as compared to the modest BW of the cassette, in addition to the high-quality sound provided by digital recording techniques. The same is true when comparing the low-quality sound of an AM radio receiver with the high-quality sound of a stereo FM receiver. This should be obvious, if we remember that the BW of an AM station is just 10 kHz (5 kHz audio BW) as compared to the 200 kHz BW (15 kHz audio BW) of an FM station. Moreover, although AM signals are vulnerable to amplitude modulating noise, FM signals have noise immunity, by nature.

1.3   WHY SIGNAL PROCESSING?

To be useful, raw signals should undergo one or more processing steps. Examples of such operations are amplification, de-noising, enhancing some features for better recognition, and correcting errors encountered due to transmission. Reshaping the magnitude and/or the phase spectrum of a signal through filtering is a rather important operation in several applications; for example, in telephone communications. In such applications, filters are employed to band-limit voice channels to 4 kHz. This process allows optimum utilization of the available BW of the communication channel (coaxial cable, optical fiber, microwave link, and so on). More audio calls can thus be accommodated while avoiding adjacent channel interference.

1.4   ANALOG VERSUS DIGITAL SIGNAL PROCESSING

Many processing techniques are now applied to digitized signals rather than analog ones. This is a direct result of the rapid development in digital circuitry that has resulted in high speed, low cost, and less power-hungry digital processors (some of them are designed exclusively for signal processing; see Chapter 9). However, this application implies that the signal should be first converted to the digital domain prior to processing, where it is then converted back to its analog format after processing. The power of digital signal processing, in spite of the added cost and time due to the necessary peripheral devices, can be explored by considering the following facts:

1. Digital components—for example, logic gates, memories, registers, processors, and so on—are becoming faster, cheaper, more efficient, and more reliable, while occupying less space.
2. Digital processing systems are more flexible, as they can easily be reprogrammed to perform different functions using the same hardware.
3. Time sharing of one signal processor among several signals has been possible due to the recently attainable high processing speeds.
4. Functions and properties that were almost impossible to achieve with analog components such as linear phase and/or constant group delay filters are possible.
5. Drifts and spurious interferences, if any, do not have any effects on performance.
6. Higher accuracy is achievable by increasing the word length of digitized signals.
7. Time varying operations are possible.

On the other hand, digital processing also has the following disadvantages:

1. Processing of high-speed signals is limited by the speed capabilities of the current technology.
2. Analog processing of signals is still more economical, in some applications, as compared to digital processing.

3. The required time to design and implement (time to market) of some systems is still excessive.
4. Due to the inherited finite word length, truncation errors can seriously degrade the performance and affect the specifications. However, such errors could be minimized by increasing the word length.
5. Possible instability of the clock source might impair the timing of the system, causing errors. However, a crystal-controlled clock source can provide stable timing.

1.5   PRACTICAL SIGNAL PROCESSING SYSTEMS

Signal processing techniques are applied extensively in many areas in practice. Among them are data acquisition systems that are used for monitoring and logging data in complex plants such as nuclear reactors. In such cases, several hundreds of variables are monitored and continuously compared with their nominal values 7 days a week, 24 hours a day. Any detected deviation should release an alarm, while corrective actions are immediately undertaken to avoid catastrophes. Similar systems can be adopted for patient's monitoring in intensive care units (ICUs), as well as in telemetry systems. In telecommunications, on the other hand, digital signal processing finds extensive use in applications such as mobile communications, data compression, ADSL, echo cancellation, spread spectrum, and secure communications.

Pattern recognition is a rather important signal processing technique that finds wide applications in many fields like e-banking, forensic, and medical applications. In such applications, one or more *biometric(s)** such as voice prints, signatures, ear prints, lip prints, iris prints, and so on are used for authentication and verification of a person's identity. In the biomedical field, electrocardiogram (EKG, a plot for the electrical activity of the heart) and electroencephalograph (EEG, a plot for the electrical activity of the brain) interpretations are helping doctors precisely diagnose heart and brain diseases, respectively. Many other indispensable diagnostic tools, such as computed tomography (CT) scans and magnetic resonance imaging (MRI) are digital signal processing–based.

1.6   SUMMARY

In this chapter, the field of signal processing has been introduced. Some basic definitions and examples of real-life signals, together with several practical applications, have been briefly described. Basic operations that are performed on signals were introduced. A comparison between analog and digital signal processing was made, where the advantages and disadvantages of each technique were highlighted. The chapter closed with a list of some modern practical signal processing applications.

---

* Biological features of a human being that do not repeat even among identical twins.

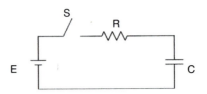

FIGURE 2.1   A simple *RC* circuit.

or more of the above-mentioned parameters. Examples of such operations are amplification, rectification, modulation, clipping, and clamping. In the following sections, we are going to discuss the most important of these operations, classified according to the property of the signal they deal with.

Because several amplitude operations rely on the charging and discharging of a capacitor, we shall start by describing the voltage/current-time relationship in a simple *RC* circuit. The term "time constant" will be introduced, along with its effect on the behavior of capacitor voltage and capacitor current.

2.1.1   SIMPLE *RC* CIRCUIT

Consider the simple circuit shown in Figure 2.1 with an initially uncharged capacitor; that is,

$$V_C(0^+) = 0$$

It is known that capacitor voltage does not change instantaneously. Thus as the switch S in Figure 2.1 is turned on (closed) at $t = 0$, the voltage at the capacitor, $V_C(0^+)$, will start increasing from ground potential. An initial current I_o will flow, whose magnitude is given by

$$i(0^+) = I_o = \frac{E}{R} \tag{2.2}$$

causing capacitor C to build up a charge $Q(t)$ and hence a voltage $V_C(t)$. As $V_C(t)$ increases, the current $i(t)$ will correspondingly decrease, as given by Equation 2.3:

$$i(t) = \frac{E - V_C(t)}{R} \tag{2.3}$$

$$V_C(t) = \frac{Q(t)}{C} = \frac{1}{C}\int_0^t i(t)\,dt \tag{2.4}$$

Substituting from Equation 2.3 in 2.4, we get

$$i(t) = \frac{E}{R} - \frac{1}{RC}\int_0^t i(t)\,dt \tag{2.5}$$

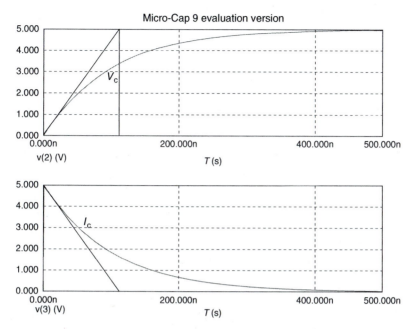

FIGURE 2.2 **(See color insert following page 262.)** Current and voltage waveform in a simple *RC* circuit.

Solving this differential equation under the assumed initial conditions gives

$$i(t) = \frac{E}{R}e^{-t/RC} = I_0 e^{-t/RC} \tag{2.6}$$

which indicates that the current decreases exponentially to reach a steady state value of zero current at $t = \infty$.

Substituting in Equation 2.4, we get an expression for the voltage $V_C(t)$:

$$V_C(t) = \frac{1}{C}\int_0^t I_0 e^{-t/RC} \cdot dt$$

$$= \frac{E}{RC}\left|-RCe^{-t/RC}\right|_0^t = E(1 - e^{-t/RC}) \tag{2.7}$$

from which we can easily conclude that the capacitor voltage reaches a steady state voltage that corresponds to the battery voltage E at $t = \infty$. The product $R \cdot C$ is called the circuit *time constant* τ. Equations 2.6 and 2.7 are illustrated in Figure 2.2.

Differentiating Equation 2.6 with respect to t and then substituting for $t = 0$, we get

$$\frac{d}{dt}i(t)\Big|_{t=0} = I_0\left(-\frac{1}{RC}\right)e^{-0/RC} = -\frac{I_0}{\tau} = \text{slope of the tangent at } t = 0$$

and similarly,

$$\frac{d}{dt}V_C(t)\Big|_{t=0} = -E\left(-\frac{1}{RC}\right)e^{-0/RC} = \frac{E}{\tau} = \text{slope of the tangent at } t = 0 \quad (2.8)$$

Substituting in Equation 2.7 for $t = \tau$, we get

$$V_C(t) = E(1 - e^{-1}) = 0.6321E \quad (2.9)$$

from which a definition for the time constant τ could be concluded as follows:

Time constant τ is the time required for the voltage on a charging capacitor to reach 63.21% of its steady state value (at $t = \infty$).

In the previous equations, we have assumed zero initial conditions; that is, the capacitor was initially uncharged. However, we can generalize the previous equations for the capacitor voltage and its current for any initial conditions other than zero ($V_i \neq 0$ and $i_i \neq 0$), giving

$$V_C(t) = V_{ss} + (V_{ss} - V_i)e^{-t/RC} \quad (2.10)$$

$$i(t) = i_{ss} + (i_{ss} - i_i)e^{-t/RC} \quad (2.11)$$

where V_{ss} and i_{ss} are the steady state voltage and current (i.e., at $t = \infty$), respectively. V_i and i_i are the initial (i.e., at $t = 0$) voltage and current, respectively.

To check the validity of Equations 2.10 and 2.11, the reader is asked to substitute zero initial conditions in both of them.

EXAMPLE 2.1

Calculate the capacitor voltage in Figure 2.1 after $t = 2\tau, 3\tau, 4\tau, 5\tau$. What conclusion could be drawn from the obtained results?

Solution

$$V_C(2\tau) = E(1 - e^{-2}) = 0.8647E$$

$$V_C(3\tau) = E(1 - e^{-3}) = 0.9502E$$

$$V_C(4\tau) = E(1 - e^{-4}) = 0.9817E$$

$$V_C(5\tau) = E(1 - e^{-5}) = 0.99E$$

From these results, it can be concluded that the charging rate of the capacitor diminishes drastically with time, and that after a time corresponding to 4τ, the capacitor is almost fully charged, with an error of about 1.83%.

2.2 AMPLITUDE OPERATIONS

Amplitude operations are intended to reshape the instantaneous amplitude of a signal. They can be divided, based on the components used, into linear and nonlinear operations.

2.2.1 LINEAR OPERATIONS

As the name implies, linear amplitude operations are performed using linear circuit components, such as resistors, inductors, capacitors, and linear amplifiers. They are "linear" in the sense that they obey Ohm's law; that is, their current/voltage relationships are linear. Most linear operations are based on charging and discharging of a capacitor through one or more resistors. Resistor/inductor circuits behave similarly, yet with interchanged roles of the currents and voltages. Therefore, we are going to study the most important amplitude operations as applied to *RC* circuits only.

2.2.1.1 Amplification or Attenuation

In many applications, it is necessary to magnify the amplitude of a weak signal in order to make it understandable, as in radio and television receivers, or to attenuate a large signal at the input of a certain device to avoid distortion of the signal. The process is described by Equation 2.12, which is a duplication of Equation 2.1, except for the multiplier A:

$$V_2(t) = A \cdot V_1(t) \tag{2.12}$$

Thus the signal is either amplified if $A > 1$ or attenuated if $A < 1$. The amplifier is said to be linear if the factor A remains constant over a wide range of amplitudes. This operation is depicted in Figure 2.3, where signal V_2 is the amplified version of signal V_1.

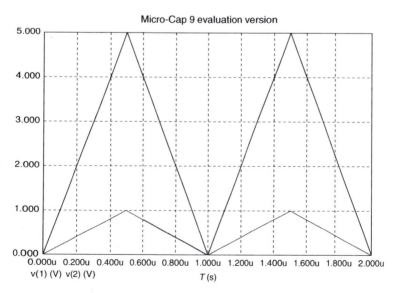

FIGURE 2.3 Amplification of a signal.

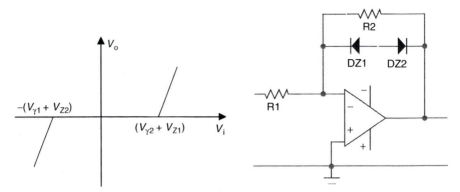

FIGURE 2.22   VTC for Example 2.5.   FIGURE 2.23   Circuit for Example 2.6.

2.2.2.3 Rectification

Bipolar signals have instantaneous amplitudes that can assume positive or negative values. *Rectification* is the process of changing such signals into unidirectional
signals that have either all positive amplitudes or all negative ones. In contrast to
clamping, no DC components are added; instead, the polarity of either negative parts
or positive parts are turned into the opposite polarity to get either an all-positive
signal or all-negative signal. A simple example of this is the process of changing
the AC line voltage into DC voltage. This operation takes place in almost all electrical equipments we use—for example, computers, radio receivers, chargers for
cell phones, handheld calculators, and laptops. A basic rectifier circuit is shown in
Figure 2.24. The polarity of the output voltage depends on the allowed direction of
current flow; that is, on the way the diode(s) are connected. Practical diodes, due to
their inherited "small" forward resistance, do have a forward voltage drop that would
delay the conduction of the diode, thereby affecting the precision of the circuit.

   The circuit shown in Figure 2.25 presents a solution to this problem. With the
help of an Op. Amp., a near-perfect rectifier could be obtained [2]. The resulting
circuit is commonly called the *super diode*. The circuit provides a nearly perfect
(linear) rectifier.

   Considering the equivalent circuit of the super diode given in Figure 2.25, we
can write

$$V_o = A(V_i - V_o) - 0.7$$
$$V_o + AV_o = AV_i - 0.7$$
$$V_o = \frac{A}{(1 + A)} V_i - \frac{0.7}{(1 + A)} \equiv mV_i - C$$

with $A \gg$, the open loop again of the Op. Amp. We get

$$V_o \approx V_i$$

with $C \ll$ indicating that it has an almost linear transfer characteristic with nearly
zero offset, as depicted in Figure 2.26. Moreover, better performance can be
obtained from the improved half-wave rectifier circuit shown in Figure 2.27. During

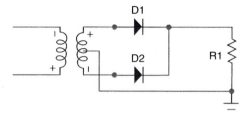

FIGURE 2.24 A simple diode rectifier circuit.

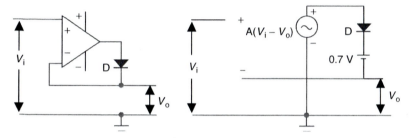

FIGURE 2.25 The super diode and its equivalent circuit.

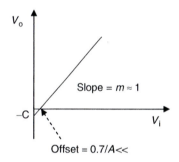

FIGURE 2.26 A linearized transfer characteristic of a rectifier.

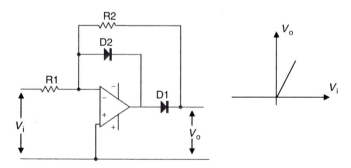

FIGURE 2.27 An improved super diode circuit and its transfer characteristics.

while one transistor, Q_1, is on (deeply saturated) over a certain period, the other, Q_2, remains off or relaxed. During the next half cycle, they exchange their states; that is, Q_1 relaxes while Q_2 goes into saturation. The output waveform is the collector (drain) voltage of either transistor, which can assume two distinct states—either $V_{CC}(V_{DD})$ or the saturation voltage $V_{CE}(V_{DS})$. The process continues indefinitely, producing a continuous rectangular wave. Based on the modes of operation of the inverters, there are three different types of relaxation oscillators:

1. Astable (free-running) MV: self-oscillating; that is, needs no triggering.
2. Monostable (one-shot) MV: needs to be triggered once to oscillate at the desired frequency.
3. Bistable (flip-flop) MV: should be triggered twice each cycle; that is, at double the required frequency.

Beside the classical transistorized realizations of multivibrators, other realizations that are based on a single Op. Amp. are known [3]. In the following sections, each one of the three basic types of multivibrators will be described, assuming Op. Amp. realization, together with its associated waveforms.

2.5.3.1 Astable Multivibrators

Astable multivibrators, as the name implies, are self-oscillatory; that is, they need no triggering to continue oscillation. In the classical transistorized version, two transistors are connected such that the collector of each transistor is coupled to the base of the other to form a closed loop, thereby providing a phase shift of 360°. A much simpler and more reliable circuit can be constructed employing an Op. Amp., three resistors, and a single capacitor (as illustrated in Figure 2.36). The operation of the circuit is based on the fact that for large signal operations, the output depends mainly on the sign of the potential difference at the input terminals of the Op. Amp. The timing and hence the frequency of oscillation is determined by the charging and discharging periods of capacitor C through resistor R_3, with the output switching between V_{CC} and $-V_{EE}$.

The periods T_1 and T_2 can be proven to be [3]

$$T_1 = R_3 C \ln\left[\frac{V_{CC} + \beta V_{EE}}{V_{CC} - \beta V_{CC}}\right]$$

(2.31)

and

$$T_2 = R_3 C \ln\left[\frac{V_{EE} + \beta V_{CC}}{V_{EE} - \beta V_{EE}}\right]$$

where $\beta = R_1/(R_1 + R_2)$.

For the special case of $|V_{CC}| = |V_{EE}|$, we get

$$T_1 = T_2 = R_3 C \ln\left[\frac{1 + \beta}{1 - \beta}\right] = R_3 C \ln\left[1 + \frac{2R_1}{R_2}\right]$$

(2.32)

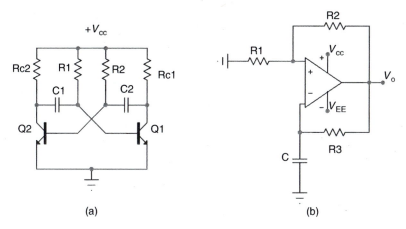

FIGURE 2.36 Two realizations for an astable MV and their associated waveforms. (a) A transistorized MV, (b) an Op. Amp. realization of an MV.

The frequency of oscillation of the transistorized version is determined by the components R_1, R_2, C_1, and C_2, as follows:

$$f_o = \frac{1}{T} = \frac{1}{T_1 + T_2}$$

where $T_1 = R_1 C_1 \ln 2$ is the period of relaxation (off) of transistor Q_1 while its collector voltage V_{C_1} is high. And $T_2 = R_2 C_2 \ln 2$ is the off period of transistor Q_2 while V_{C_2} is high.

The periods T_1 and T_2 each should have a minimum value of at least four time constants to insure complete charging (*recovery*) of the capacitors C_1 and C_2. Therefore, T_1 and T_2 should be checked for recovery. This step also insures that the output waveform is rectangular, with edges as steep as possible. The conditions for complete recovery are thus

$$T_1 \gg 4C_2 R_{C_1}$$

$$T_2 \gg 4C_1 R_{C_2}$$

2.5.3.2   Monostable Multivibrators

This type of relaxation oscillators has one stable state, which if attained, could be maintained indefinitely, as long as no external triggering is applied. If, on the other hand, it is triggered, it changes state spontaneously for a finite period, then returns to its stable state. The result is as a rectangular pulse output having a width T that is determined by the circuit components and a frequency that is given by the trigger frequency. In the single Op. Amp. realization illustrated in Figure 2.37, the output switches, as before, between the power supply voltages V_{CC} and $-V_{EE}$. The pulse

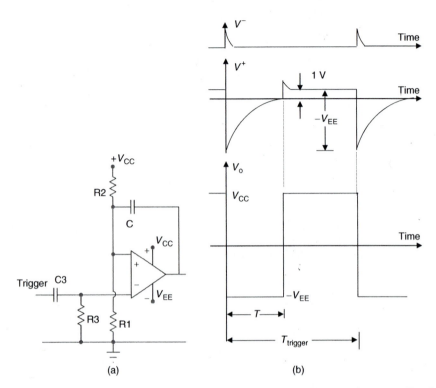

FIGURE 2.37   A single Op. Amp. monostable multivibrator and its waveforms. (a) Circuit diagram, (b) waveforms.

width (gate width) T is determined by capacitor C and the parallel combination of R_1 and R_2. The constant bias (about 1 V) that is developed at the noninverting input of the Op. Amp. through the potential divider R_1 and R_2 keeps the circuit at its stable state. A trigger signal whose amplitude is slightly greater than 1 V, if applied at the inverting input, would force the output instantaneously to switch to $-V_{EE}$, thereby causing capacitor C to charge towards V_{CC}. As seen from the waveforms of Figure 2.37, this transitional state is sustained until the input voltage V^+ reaches ground potential, where the output regains its stable level of $+V_{CC}$. The time taken to reach ground potential determines the period T. It is given by

$$T = C\,R_p \ln 2$$

where

$$R_p = \frac{R_1 \cdot R_2}{R_1 + R_2} \tag{2.33}$$

The circuit can be modified to produce a variable gate width without affecting its DC conditions, as shown in Figure 2.38. In this case, the gate width is given by

$$T = C \cdot (R + R_p) \cdot \ln 2 \tag{2.34}$$

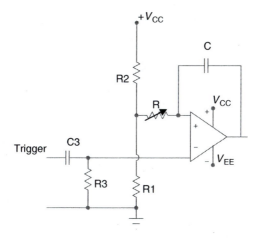

FIGURE 2.38 A variable gate–width monostable multivibrator.

2.5.3.3 Bistable Multivibrators (Flip-Flops)

A bistable circuit can remain indefinitely in any attained state as long as no external trigger is applied. Due to this property, *flip-flops*, as they are usually called, are used in the construction of computer memories and registers. They can also be used as frequency dividers, as the output frequency is half that of the input trigger. Cascading several stages can provide division of the input frequency by four, eight, sixteen, and so on. Schmitt triggers are important members of the bistable multivibrator family. They are used in several analog and digital applications. In the Op. Amp. realizations given in Figure 2.39, the output switches between V_{CC} and $-V_{EE}$ based on two set thresholds. The factors α and β of circuits (a) and (b), respectively, are given by

$$\alpha = \frac{R_1}{R_2} \qquad \beta = \frac{R_1}{R_1 + R_2}$$

The point of intersections with the horizontal axis are called the *trip points*. More specifically, in the noninverting Schmitt trigger (Figure 2.40), the point on the extreme right of the transfer characteristics is called the upper trip point (UTP), whereas the point on the left is called the lower trip point (LTP).

2.5.4 SAWTOOTH GENERATORS

Sawtooth generators, usually called *sweep generators*, are mostly needed in systems where an electron beam is required to scan a screen, as in cathode ray oscilloscopes, TV receivers, and video cameras. The principle of scanning is illustrated in Figure 2.41. The beam is deflected starting from the upper-left corner to the right side, while also being deflected slightly downward. As the rightmost point is reached, the beam is cut off and driven back to the left to begin a new cycle. Therefore, to scan a screen, two sawtooth signals (sweep signals) are needed: a low-frequency one

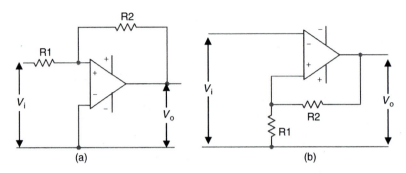

FIGURE 2.39 Two realizations for a Schmitt trigger. (a) Noninverting Schmitt trigger, (b) inverting Schmitt trigger.

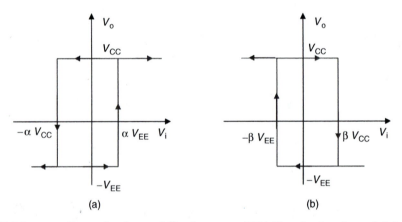

FIGURE 2.40 **(See color insert following page 262.)** Transfer functions of Schmitt trigger circuits of Figures 2.39a and 2.39b.

for vertical deflection and a relatively higher frequency one for horizontal deflection. The number of lines drawn depends on the respective system used (e.g., NTSC, PAL, SCEAM, HDTV).

An ideal sawtooth wave (shown dotted) and a practical wave are diagrammed in Figure 2.42, where it is evident that each cycle consists of two periods:

T_1 Trace period
T_2 Fly-back period

During T_1, a sawtooth wave should be as linear as possible to guarantee a linear trace (undistorted) for the displayed signal. It should also drop to zero in almost zero time to minimize the *retrace* or *fly-back* period. Main requirements on a saw-tooth wave are:

1. Linearity, as measured by the deviation δ
2. Periodicity
3. Minimum fly-back period T_2 (minimum duty cycle $= T_2/T_1 + T_2$)

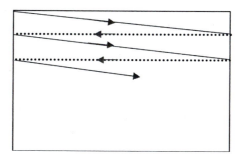

FIGURE 2.41 Principle of scanning a screen.

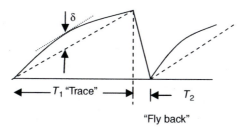

FIGURE 2.42 Ideal and practical sawtooth wave.

Sawtooth waves are either current or voltage signals, depending on the deflection system used. Electromagnetic deflection systems, as in TV receivers, are driven by sawtooth current signals. On the other hand, electrostatic deflection systems, as in oscilloscopes, are excited by voltage sawtooth signals. Because both types are more or less similar, only voltage sweep generators will be discussed here. Almost all voltage sweep generators are based on the charging and discharging of a capacitor, where the charging current decays exponentially as the capacitor builds up a charge resulting in an exponential capacitor voltage, as depicted in Figure 2.42. To get linear sawtooth waves, capacitors should be charged from constant-current sources rather than voltage sources, as we shall see.

2.5.4.1 Linear Sweep Generator

A circuit diagram for a sweep generator that is based on constant-current charging of a capacitor is given in Figure 2.43. During period T_1, transistor Q_1 is turned off, due to the applied control signal at its base. Transistor Q_2, together with diode D, acts as a current mirror that drives constant-charging current I_{ch} to capacitor C. At the end of period T_1, transistor Q_1 is turned on (actually, it saturates), thereby introducing a low-resistance path for C to discharge. Period T_2 should be enough to fully discharge C:

$$V_C = V_o = \frac{Q}{C} = \frac{\int_0^t i \cdot dt}{C} = \frac{I_{ch}}{C}t = \text{i.e., linear}$$

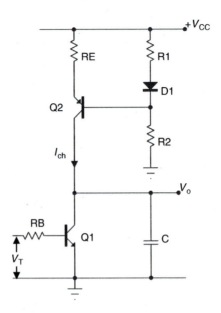

FIGURE 2.43   A linear sweep generator.

EXAMPLE 2.7

Design a linear sweep circuit to generate a ramp signal having a peak amplitude of 4 V at a frequency of 1 kHz using transistors having $\beta = 100$, $V_{CEsat} = 0.2$ V, $V_{BEsat} = 0.8$ V, $r_{CEsat} = 50$ Ω and a 12 V supply (Figure 2.44).

Solution

We start by choosing constant-current charging using a current mirror (D with Q_2) as a collector load. We further assume a 1:10 duty cycle for the control signal (Figure 2.45).

As agreed upon, the fly-back period should be long enough to fully discharge the capacitor; therefore, the condition

$$T_2 \geq 7\tau_{discharge} \geq 7 \cdot C \cdot r_{CEsat} = 100 \ \mu s$$

should be maintained, from which the capacitor C can be found to be

$$C = \frac{10^{-4}}{7 \times 50} = 285.7 \quad nF \approx 300nF$$

The required charging current can then be estimated from

$$V_o = \frac{I_{ch}}{C} \cdot t$$

$$4 = \frac{I_{ch}}{300 \times 10^{-9}} \times 900 \times 10^{-6}$$

$$I_{ch} = \frac{4}{3} mA$$

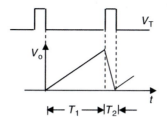

FIGURE 2.44   Waveforms of a constant-current sweep circuit.

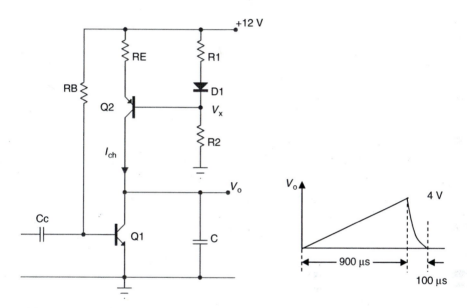

FIGURE 2.45   A circuit diagram for Example 2.7.

Assuming a current ratio of 1:1 in the current mirror, then $R_1 = R_E$, giving $I_{ch} = I_1$. Let $R_1 = 2.2$ kΩ. Hence,

$$V_x = 12 - V_{BE} - I_{ch} \times R_E$$

$$= 12 - 0.7 - \frac{4}{3} \times 10^{-3} \times 2.2 \times 10^3 = 8.366 \text{ V}$$

$$= I_{ch} \times R_2$$

giving

$$R_2 = \frac{8.388}{4/3} \times 10^3 = 6.275 \text{ k}\Omega$$

then

$$I_B = \frac{40}{3 \times 100} \text{mA} = \frac{12 - V_{BEsat}}{R_B}$$

$$R_B = \frac{(12 - 0.8) \times 300}{40} \times 10^3 = 84 \text{ k}\Omega$$

2.5.4.2 Miller Integrator

The concept of integrating waveforms was given previously in Section 2.2.1, together with the associated waveforms. It was concluded that

$$i = \frac{v_i}{R} = \text{constant}$$

$$V_o = \frac{Q}{C} = -\frac{\int i \cdot dt}{C} = -\frac{v_i}{R \cdot C} t = -\alpha \cdot t$$

which means that the resulting voltage waveform is linear with time, provided that the Op. Amp. is ideal. Therefore, to get a linear integrator using practical Op. Amps., the following modifications should be introduced (Figure 2.46):

1. Due to the nonidealities and the high gain of Op. Amps., an output (offset voltage) is measured, even if no input is applied, because the voltage difference between the two inputs V_d is nonzero. To minimize such an effect, the gain should be reduced greatly to a value around 10, by bridging C with a finite resistor R_1, as shown in Figure 2.47, such that

$$R_1 = 10\,R$$

2. Errors due to the input bias current of the Op. Amp. should be compensated for. This is possible by connecting a resistor R_2 to the noninverting input of the Op. Amp., as in Figure 2.47, such that

$$R_2 = R_1//R$$

2.5.4.3 Bootstrap Sweep Generators

Bootstrap generators also rely on constant-current charging of a capacitor. In the circuit given in Figure 2.48, transistor Q_1 acts as a switch that is controlled by the signal V_i. When Q_1 is off, capacitor C is charged through the current I_{ch}, resulting from the voltage drop across resistor R. Transistor Q_2 acts as a source follower; it thus provides unity gain and causes zero phase shift. The output voltage V_o is therefore an exact replica of the capacitor voltage V_C. The voltage V_K will have the shape of the voltage V_C, yet raised by the voltage on C_B. A constant voltage drop is thus established across R, causing a constant current I_{ch} to charge C. As the input control

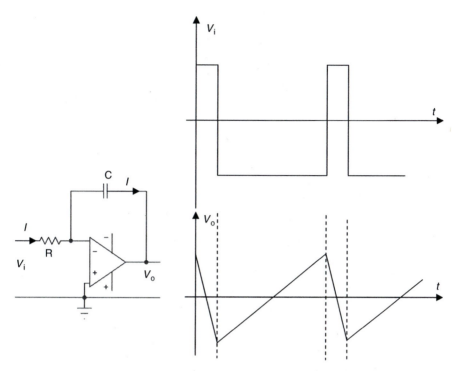

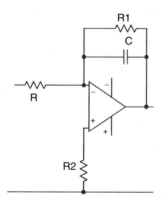

FIGURE 2.46   A Miller integrator and its associated waveforms.

FIGURE 2.47   Modifications for improving Miller integrators.

signal goes higher, Q_1 will saturate, forcing capacitor C to discharge. The associated waveforms are illustrated in Figure 2.49.

Capacitor C_B, being the source of I_{ch}, should be large enough compared to C to have an almost unaltered charge, even after charging capacitor C. Figure 2.50 illustrates an alternative realization for the bootstrap circuit using an Op. Amp. voltage follower.

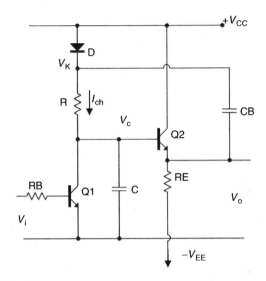

FIGURE 2.48 A bootstrap sweep generator.

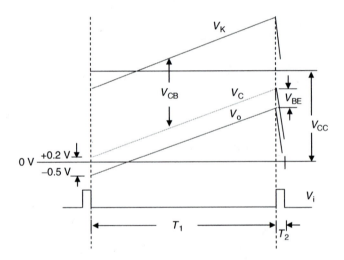

FIGURE 2.49 **(See color insert following page 262.)** Waveforms for a bootstrap sweep generator.

2.5.5 A PRACTICAL ANALOG SIGNAL PROCESSING SYSTEM

A good example of a practical system that contains several analog processing blocks is a TV receiver, of which a simplified block diagram is given in Figure 2.51. The blocks studied so far in this chapter are marked with a double frame.

The tuner at the front end consists of the RF amplifier, the mixer, and an *LC* oscillator. The received signal containing both the FM audio signal and the AM

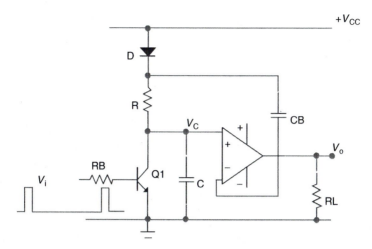

FIGURE 2.50   An alternative realization for the bootstrap sweep generator.

video signal is mixed with the local oscillator signal to produce the IF signal. The resulting IF signal, after being amplified, is then subjected to an AM detector, where the sound and the video signals are separated. The audio path is similar to that of an ordinary FM receiver, containing a noise limiter (a clipping circuit) to eliminate the noise usually superimposed on the amplitude of the FM audio signal. The video signal is applied to a DC restorer (a clamping circuit) that reinserts its lost DC in the video amplification stage (*RC* coupled amplifiers usually block any DC component in the input) to regain its proper level of brightness. The output of the AM detector is subjected to the sync separator—a "clipping circuit"—to extract the composite synchronization signal from the video information (see Figure 1.1c). Sync signals are necessary to lock the horizontal and vertical oscillators (sawtooth oscillators), used for the scanning of the screen tube, with those in the studio camera. In the parallel connected sync separators—namely, a "differentiating circuit" and an "integrating circuit"—the horizontal and the vertical sync train of pulses are further separated.

2.6   SUMMARY

In this chapter, several techniques for processing and generation of analog signals were described. They were classified into operation categories based on the amplitude, frequency, and phase of the signal. This includes operations like differentiation, integration, clamping, clipping, modulation, and delay equalization. The design and construction of waveform generators such as sinusoidal, square wave, and sweep generators were then considered. Finally, the block diagram of a TV receiver was given as an example of a practical system that contains almost all the studied analog signal processing circuits.

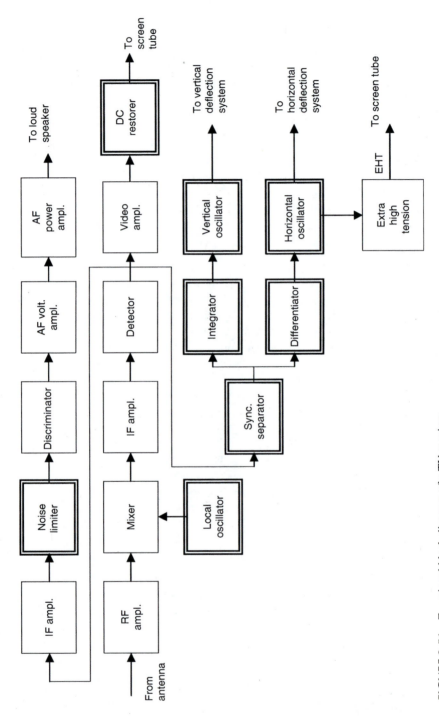

FIGURE 2.51 Functional block diagram of a TV receiver.

2.7 REVIEW QUESTIONS

1. Give some practical applications for the following circuits:
   - Integrating circuits
   - Differentiating circuits
   - Clamping circuits
   - Clipping circuits
2. Show how the proper choice of the time constant of an RC circuit for a specific input signal frequency can affect its function. Use an integrating circuit as an example.
3. What is the limiting factor for the frequency of oscillation of each of the following: astable, monostable, and bistable MVs?
4. Sketch a block diagram for an FM receiver, showing important signal processing blocks.
5. Prove that the frequency of oscillation of a three-section RC-phase shift oscillator is given by

$$f_o = \frac{1}{\sqrt{6}RC}$$

   and the minimum required gain of the amplifier is 29.
6. Show how horizontal and vertical sync pulses are separated from a composite TV signal and from each other.

2.8 PROBLEMS

1. Design a simple sync separator for a TV receiver, given that the black level is set at +1 V. Show how you can turn the circuit into an amplitude limiter to get the video signal out of the composite video signal. In both cases, draw the circuit diagram as well as the VTC, indicating amplitudes.
2. The signal shown in the following figure is applied to a differentiating circuit that is composed of a 0.1 μF capacitor and a 10 kΩ resistor. Sketch the output waveform, indicating amplitude.

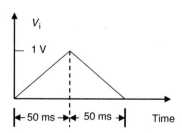

3. The two waveforms shown in the following figure were applied to a summing amplifier. Sketch the resulting output waveform, giving values.

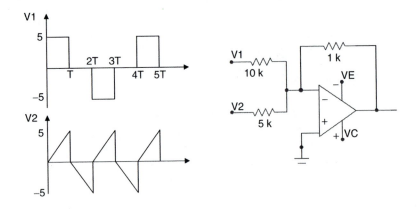

4. For the circuit shown in the following figure, sketch the capacitor voltage $V_c(t)$ as well as the voltages and currents through both resistors, with time, if the switch is first set to position 1 and then turned to position 2 after 10 μs.

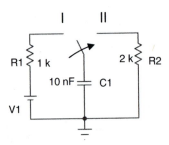

5. For the *RC* circuit shown in the following figure, calculate:
   (a) The initial slope $V_c(t)$
   (b) The rise time
   (c) Total charge time
   (d) The time taken to reach 63.2% of the steady state value

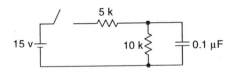

6. The switch shown in the following figure is closed at $t = 0$ and then reopened after 1 µs. Sketch $V_c(t)$, as well as the switch current i_{sw}, during both intervals.

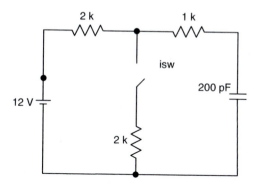

7. Sketch V_{01}, V_{02} as well as i for the circuit shown in the following figure.

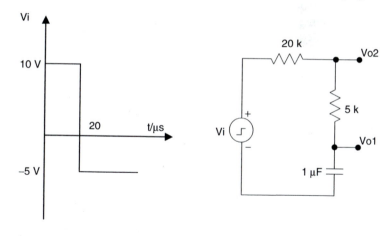

8. For the circuit shown in the following figure, calculate $i(t)$, $V_o(t)$, and $V_r(t)$.

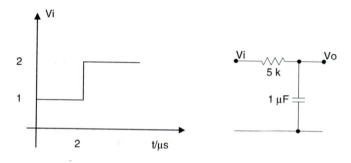

9. If at $t = 0$, SW_1 is closed and SW_2 is opened, and at $V_o = 10$ V, SW_2 is closed, calculate $V_o(t)$, $V_C(t)$, and $i(t)$ for the circuit shown in the following figure.

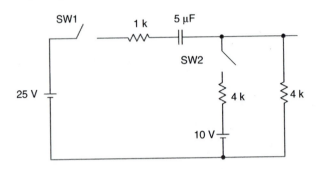

10. Sketch the transfer C/S as well as the DPI $(i_i - v_i)$ for the clipping circuits shown in the following figure, indicating the coordinates of the break points with their slopes.

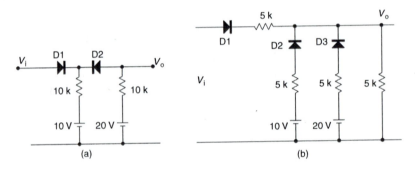

11. The circuit in the following figure is required to present a piece-wise approximation to the equation 0

$$i = 0.1v^2$$

The approximation should be exact at the voltages 2, 4, and 8 V. Design the circuit and calculate the error at the voltages 3, 5, 7, and 10 V.

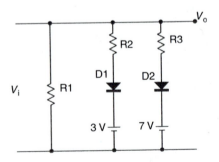

12. Sketch the transfer function as well as the DPC/S, indicating all coordinates, as well as all slopes for each of the circuits shown in the following figure, assuming ideal diodes.

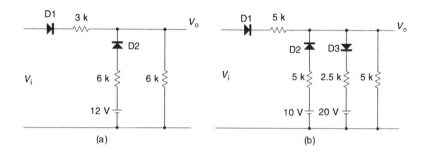

(a)   (b)

13. Find the circuit that has the transfer C\S shown in the following figure.

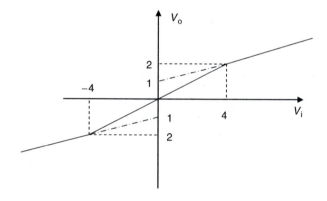

14. Calculate the repetition frequency and the duty cycle for the unsymmetrical MV illustrated in Figure 2.36 if the circuit has the following data:

$$R_{C_1} = R_{C_2} = 1 \text{ k}\Omega, \quad R_1 = R_2 = 100 \text{ k}\Omega, \quad C_1 = 20 \text{ pF}, \quad C_2 = 80 \text{ pF},$$

and $V_{CC} = 10 \text{ V}$

15. Design a symmetrical astable MV that oscillates at 10 kHz with a duty cycle of 1/4 using a single Op. Amp. Draw all waveforms, indicating values. (*Hint:* Take $V_{CC} = 10 \text{ V}$, $V_{EE} = -5$, $R_1 = 5 \text{ k}\Omega$, $R_2 = 15 \text{ k}\Omega$, and $R_3 = 2 \text{ k}\Omega$.) Make any necessary assumptions.

16. Design a single Op. Amp. monostable MV (like the one shown in Figure 2.27) to generate a pulse having a duty cycle of 10% at a frequency of 5 kHz. Assume any missing data.

17. Design a noninverting Schmitt trigger to have a trip point at 2.5 V and –2 V. Make any necessary assumptions.

18. Design an astable MV that is required to develop a 5 V, 10 kHz sawtooth signal having a trace/fly-back ratio of 9:1. Draw the circuit diagram and sketch all waveforms (use transistors that have $\beta = 50$ at $I_c = 1$ mA). Make any necessary assumptions.

19. Design a transistorized astable MV to generate 5 V oscillations at 10 kHz and a duty cycle of 25%, using transistors that have $\beta_{min} = 25$. Assume any missing data.

20. Design a three-section RC-phase shift oscillator to operate at 5 kHz using a JFET that has an amplification factor $\mu = 50$ at $I_d = 1$ mA, $V_{DD} = 10$ V and a drain resistance $r_d = 5$ kΩ. (*Hint:* Voltage gain of a JFET amplifier = $f_o = 1/\sqrt{6}RC$.)

21. Design a transistorized bootstrap sweep generator to oscillate at 1 kHz with an amplitude of 8 V. The resulting ramp should be linear within 2%. A positive triggering source of amplitude 3 V and a duration of 0.1 ms is applied. Use a dual power supply of 15 V and transistors with $h_{femin} = 100$.

22. Design a Miller integrator to produce a triangular waveform with $V_{P-P} = 4$ V at 250 Hz. A ±10 V square wave is available. Calculate the lowest possible operating frequencies (assume any missing data).

23. Design a 5 V, 1 kHz ramp generator, using a transistor that has $\beta = 50$, $R_L = 100$ kΩ, and that operates from a 10 V supply. The circuit is to be triggered through a negative pulse having a duty cycle of 10%.

24. Design a linear sweep circuit, using a constant-current source to produce a 4 V ramp (peak) at a frequency of 1 kHz.

3 Analog-Filter Design

3.1 INTRODUCTION

Filters are important—if not indispensable—tools for electronic engineers. Almost every electronic system has at least one filter. A good example is the simple RC filter found in the DC power supply of any electronic system. This simple circuit plays a major role in determining the quality of the system. It reduces the amplitudes of the undesired fluctuations (usually called *ripples*) resulting from the imperfect rectification of the AC (sinusoidal) supply voltage. Such ripples are harmonics of the supply frequency; that is, they are audible. Power supplies of high-fidelity systems should therefore be well filtered to minimize such a disturbing effect, as it greatly affects the output of the system. In some sensitive devices that deal with voltages of very low amplitudes, such as EKGs and EEGs, interferences from the supply frequency may distort the waveform to the extent that the diagnosis is affected. Therefore, a filter is designed and included to highly attenuate the supply frequency component. Filters are used extensively in many fields, including communications, signal processing, biomedical, telemetry, control applications, and others.

A *filter,* in the broad sense, is a device that is employed to shape the frequency and/or the phase spectrum of an input signal. The way and the extent such shaping is carried out depends on the transfer function of the filter, as we shall see later. Circuit implementations of filters have undergone a series of development phases that can be grouped into three epochs [4]:

1. Passive-filter realization or LC realization, where inductors, capacitors, resistors, and transformers are used in the implementation.
2. Inductorless (active) realization, where resistors, capacitors, transistors, and/or operational amplifiers are used.
3. Digital realization, where digital components such as unit delays, multipliers, and adders are employed.

Classical LC filters are considered the best circuit implementations of any filter transfer function, due to their inherent low sensitivity to component tolerance. The technique has been extensively researched over many years, and has matured. LC filters are popular due to the availability of many easy-to-use filter catalogues, where normalized component values are found for any order and type, leading to the development of filters with optimized performance. Through simple transformation formulas, such normalized values could be denormalized to match the actual operating frequency and impedance level. However, due to progress in the field of integrated circuits throughout the last 30 years, LC filters—especially those operating at low frequencies—have become unsuitable. This change is mainly due to their relatively large volume, weight, and disturbing magnetic fields.

On the other hand, the availability of low-cost, miniaturized, and high-performance circuits, such as operational amplifiers, have resulted in the development of new families of inductorless filters, such as *RC* active filters, gyrator filters, switched capacitor filters, and switched resistor filters. Such filters do have several advantages, including:

- The possibility of achieving voltage gain.
- Poles occur in conjugate complex pole pairs, thereby giving more flexibility to the designer to realize all kinds of transfer functions.
- Higher-order filters can be realized through cascading of second-order sections.

However, they have some disadvantages, such as:

- Limited dynamic range
- Limited frequency range
- Power consumption, due to the used active components
- Relatively high sensitivity to component tolerance

Moreover, due to the rapid development of digital techniques and monolithic integrated circuit technology, a new generation of filters has evolved that is based on all-digital circuitry. It is now possible to create a filter with characteristics that were almost impossible with analog filters. This trend is supported by the fact that digital building blocks are becoming cheaper, faster, smaller in size, and less power-hungry, and can be operated at low supply voltages. All these properties have made digital filters good candidates for replacing classical analog filters in the near future.

3.2   DESCRIBING FUNCTIONS

A filter can be represented by a two-port network, as shown in Figure 3.1, having the transfer function $H(S)$, defined as

$$H(S) = \frac{Y(S)}{X(S)} = \frac{N(S)}{D(S)} \qquad (3.1)$$

FIGURE 3.1   A two-port network.

where $N(S)$ and $D(S)$ are polynomials in S, with $S = \sigma + j\omega$. The roots of $N(S)$ and $D(S)$ are the zeros and poles of $H(S)$, respectively. The number of poles determines the order of the filter.

To be realizable, $H(S)$ should be:

1. Causal.
2. Linear and time-invariant (LTI).
3. The order of $N(S)$ should be less than or equal to the order of $D(S)$.

As mentioned previously in Chapter 2, filters can be classified according to their selectivity functions into:

1. Low-pass filters
2. High-pass filters
3. Band-pass filters
4. Band-stop filters

The ideal frequency responses of such basic types of filter functions are depicted in Figure 3.2. Each filter has pass band(s) and stop band(s) defined by the frequency's ω_c's.

Transfer functions that describe any of the earlier ideal (rectangular or brick wall) responses are of infinite orders. To be realizable, however, practical filter functions should have finite orders. Over the last 70 years, mathematicians have developed

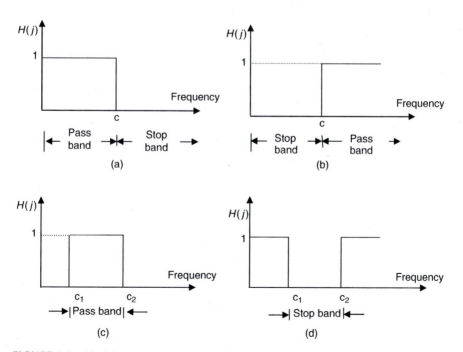

FIGURE 3.2   Ideal frequency responses of basic types of filters. (a) LPF response, (b) HPF response, (c) BPF response, (d) BSF response.

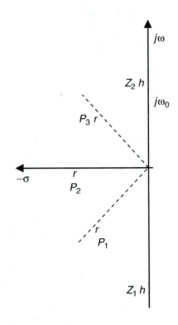

FIGURE 3.3 Poles and zeros of a third-order filter.

several finite-order functions that approximate those ideal responses, as will be shown later. Filters are best described in the S-plane (Laplace domain) by their poles P_i's and zeros Z_i's, as shown in Figure 3.3, where

$$Z_{1,2} = \pm j\omega_{Z_{1,2}}$$

$$P_{1,3} = -\sigma_{P_{1,3}} \pm j\omega_{P_{1,3}}$$

$$P_2 = -\sigma_{P_2}$$

from which a transfer function for the filter can be constructed as follows:

$$H(S) = H_0 \frac{(S + j\omega_{Z_1})(S - j\omega_{Z_2})}{(S + \sigma_{P_1} + j\omega_{P_1})(S + \sigma_{P_3} - j\omega_{P_3})(S + \sigma_{P_2})} \qquad (3.2)$$

where H_0 is a constant that is usually selected to ensure that $H(0) = 1$.

The time response of a filter is described by its impulse response $h(t)$, where

$$h(t) = L^{-1}\{H(S)\} \qquad (3.3)$$

Equation 3.3 indicates that the impulse response of a filter is the inverse Laplace transform of its transfer function $H(s)$. The output $y(t)$ of a filter, in response to an input $x(t)$, can be obtained by convolving the impulse response with the input, or

$$y(t) = h(t) * x(t)$$

$$y(t) = \int_0^\infty h(\tau) \cdot x(t - \tau)dt \tag{3.4}$$

Filters can also be described in the time domain by their differential equations, as follows:

$$\frac{d^N y}{dt^N} + b_{N-1}\frac{d^{N-1}y}{dt^{N-1}} + \cdots + b_1\frac{dy}{dt} + b_0 = a_M\frac{d^M x}{dt^M} + a_{M-1}\frac{d^{M-1}x}{dt^{M-1}} + \cdots + a_0 x \tag{3.5}$$

Laplace-transforming Equation 3.5 gives

$$Y(S)\left\{S^N + b_{N-1}S^{N-1} + \cdots + b_1 S + b_0\right\} = X(S)\left\{a_M S^M + a_{M-1}S^{M-1} + \cdots + a_0\right\}$$

leading to the form

$$\frac{Y(S)}{X(S)} = H(S) = \frac{a_M S^M + a_{M-1}S^{M-1} + \cdots + a_0}{S^N + b_{N-1}S^{N-1} + \cdots + b_0} \tag{3.6}$$

Equation 3.6 represents the general form of a filter transfer function. The exponents M and N are the degrees of the numerator and denominator polynomials, respectively. It is to be noted, as mentioned earlier, that for a filter function to be stable and hence realizable, the condition $M \leq N$ should be satisfied.

EXAMPLE 3.1

Derive the transfer function $H(s)$ of the filter that is described by the following differential equation:

$$\frac{d^2 y}{dt^2} + 5\frac{dy}{dt} + 4y = 5\frac{dx}{dt}$$

Find its poles and zeros, and then sketch its frequency response.

Solution

Taking the Laplace transformation and assuming zero initial conditions, we get

$$(S^2 + 5S + 4)Y(S) = 5S\,X(S)$$

giving

$$H(S) = \frac{Y(S)}{X(S)} = \frac{5S}{(S^2 + 5S + 4)} = \frac{5S}{(S+4)(S+1)}$$

TABLE 3.1

Frequency Response of a BP Filter

ω	$H(j\omega)$
0	0
2	1
∞	0

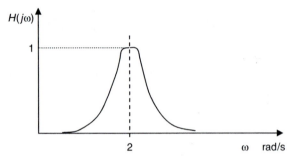

FIGURE 3.4   BPF of Example 3.1.

which indicates that the filter has a zero at the origin and two poles at $S = -1$ and $S = -4$.

To get the frequency response, we put $S = j\omega$, yielding

$$|H(j\omega)| = \left|\frac{j5\omega}{(4-\omega^2)+j5\omega}\right| = \sqrt{\frac{25\omega^2}{(4-\omega^2)^2+25\omega^2}}$$

from which we can estimate the frequency response. Considering only three points, we get the data shown in Table 3.1.

The frequency response is sketched in Figure 3.4, from which it is obvious that the filter attenuates frequencies below and above the frequency $\omega = 2$ rad/s. Thus, it represents the transfer function of a band-pass filter (BPF).

3.2.1   SECOND-ORDER FILTER

As mentioned earlier, the standard basic building block of a filter is a second-order section that can be used to realize higher-order filter functions. The describing function of a generalized second-order section is given by

$$H(s) = \frac{a_2S^2 + a_1S + a_0}{S^2 + b_1S + b_2}$$

$$= K\frac{(S-Z_1)(S-Z_2)}{(S-P_1)(S-P_2)} \tag{3.7}$$

where

$$P_{1,2} = \frac{-b_1 \pm \sqrt{b_1^2 - 4b_2}}{2}$$

and

$$Z_{1,2} = \frac{-a_1 \pm \sqrt{a_1^2 - 4a_0 a_2}}{2a_2} \qquad (3.8)$$

are the poles and zeros of $H(S)$, respectively. It is evident from the earlier equations that the a's and b's coefficients, as they define the poles and zeros locations, will determine the behavior of the filter in the frequency domain. To verify this, let us first define the term "pole quality," Q, as

$$Q = \frac{\text{imaginary part of the pole}}{2 \times \text{real part}}$$

This value is important, because it determines the shape of the filter's frequency response. Rewriting the poles equation from Equation 3.8 in the form

$$P_{1,2} = -\frac{b_1}{2} \pm j\sqrt{\left(b_2 - \frac{b_1^2}{4}\right)} \qquad (3.9)$$

from which the pole quality is

$$Q = \sqrt{\frac{b_2 - (b_1/2)^2}{b_1^2}} = \sqrt{\frac{b_2}{b_1^2} - \frac{1}{4}} \approx \sqrt{\frac{b_2}{b_1^2}}$$

Assume that $b_2 = \omega_c^2$, then $b_1 = \omega_c/Q$; substituting in Equation 3.9 gives the poles' locations at

$$P_{1,2} = -\frac{\omega_c}{2Q} \pm j\sqrt{\omega_c^2 - \left(\frac{\omega_c}{2Q}\right)^2} \qquad (3.10)$$

The absolute value of the poles, $|P_{1,2}|$, can be calculated from

$$|P_{1,2}| = \sqrt{(\text{real part})^2 + (\text{imaginary part})^2} = \sqrt{\frac{\omega_c^2}{4Q^2} + \left(\omega_c^2 - \frac{\omega_c^2}{4Q^2}\right)} \approx \sqrt{\omega_c^2}$$

$$= \omega_c = \text{constant}$$

This equation indicates that the absolute value of any pole is constant and is always equal to the cut-off frequency of the filter ω_c. This indication leads to the conclusion

To get an almost flat response (maximally flat response), no peaks are allowed. To enforce this condition, equate the gain at $S = j\omega_p$ with unity (the gain at $S = 0$). This gives

$$\left|H(j\omega_p)\right| = \frac{Q}{\sqrt{1 - \left(\frac{1}{2Q}\right)^2}} = 1$$

where Q as before is the pole quality leading to

$$4Q^4 - 4Q^2 + 1 = 0$$

giving a value for the pole quality of

$$Q = \frac{1}{\sqrt{2}} \tag{3.13}$$

The magnitude of $H(j\omega)$ at ω_c can be obtained by substituting in Equation 3.11, giving

$$\left|H(j\omega_c)\right| = \frac{\omega_c^2}{\sqrt{\left(\frac{\omega_c^2}{Q}\right)^2}} = \frac{\omega_c^2}{\omega_c^2/Q} = Q = \frac{1}{\sqrt{2}}$$

which on the decibel scale corresponds to

$$A(j\omega_c) = 20\log\frac{1}{\sqrt{2}} = -3\,\mathrm{dB}$$

This equation indicates that the frequency ω_c represents the conventional 3 dB BW or, as it is commonly known in filter terminology, the cut-off frequency, roll-off frequency, or corner frequency of the filter. The response remains almost flat (within 3 dB) from zero frequency up to the cut-off frequency. Above this frequency, the response starts to drop at a rate that is proportional with the order N of the filter (the order of the denominator's polynomial). The drop rate (roll-off rate) of an Nth order filter can be easily solved for as follows: drop rate $= -6N\,\mathrm{dB/octave} = -20N$ dB/decade. Figure 3.7 illustrates this relation.

3.2.3   HPF

The transfer function of a second-order HPF is given by

$$H_{HP}(S) = \frac{a_2 S^2}{S^2 + \left(\frac{\omega_c}{Q}\right)S + \omega_c^2} \tag{3.14}$$

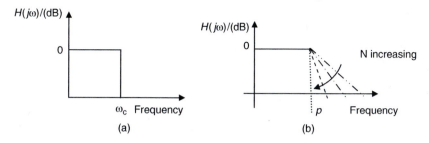

FIGURE 3.7 Frequency response of an LP filter. (a) Response of an ideal filter, (b) finite-order filter.

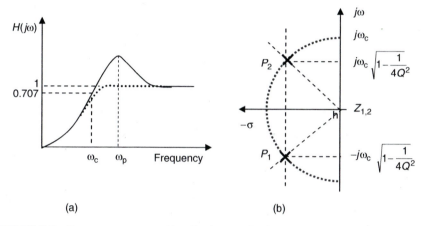

FIGURE 3.8 Frequency response (a) and pole-zero distribution (b) of an HPF.

from which it follows that the filter has two zeros $Z_{1,2}$ at the origin, and one conjugate complex pole-pair $P_{1,2}$, as depicted in Figure 3.8b. The constant a_2 is selected such that the response is unity at $\omega = \infty$. This leads to $a_2 = 1$. The frequency response, plotted in Figure 3.8a, rises at a rate of 20 dB/decade up to the frequency ω_c, where it shows, at a nearby frequency ω_p, a peak having a magnitude that depends on the pole quality. The peak frequency ω_p is given by

$$\omega_p = \omega_c \sqrt{1 - \left(\frac{1}{2Q}\right)^2}$$

in which ω_c, as before, is the 3 dB frequency.

Again, for a maximally flat response (i.e., no peaking), the condition $|H(\omega_p)| = 1|$ should be satisfied, giving a value for the pole quality of $Q = 1/\sqrt{2}$.

3.2.4   BPF

As the name implies, a BPF allows frequencies within a certain band (pass band) to experience zero or minimum attenuation while others outside that range are highly

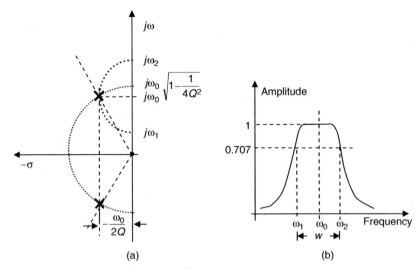

FIGURE 3.9   Pole-zero plot (a) and frequency response (b) of a BPF.

attenuated at a rate that depends on the filter's order. The frequency response of a second-order BPF is described by

$$H_{BP}(S) = \frac{a_1 S}{S^2 + \left(\dfrac{\omega_0}{Q}\right)S + \omega_0^2} \tag{3.15}$$

The pole-zero plot given in Figure 3.8 shows that the filter has one zero at the origin and a conjugate complex pole pair. The magnitude response, plotted in Figure 3.9, illustrates a peak at $\omega = \omega_0$ and two roll-off (corner) frequencies ω_1 and ω_2 located at both sides of ω_0, where the response experiences a drop of 3 dB below its peak. The two frequencies ω_1 and ω_2 are related by the identity $\omega_1 \cdot \omega_2 = \omega_0^2$. The frequency range from ω_1 to ω_2 is the filter's pass band or simply its BW W, where

$$W = \omega_2 - \omega_1 = \frac{\omega_0}{Q} \tag{3.16}$$

The roll-off frequencies ω_1 and ω_2 can be estimated from Figure 3.8 to be

$$\omega_{1,2} = \omega_0 \sqrt{1 - \frac{1}{4Q^2}} \mp \frac{\omega_0}{2Q} \tag{3.17}$$

3.2.5   BSF

A BSF is used to reject (or ideally completely attenuate) a certain band of frequencies of a signal while slightly affecting other frequencies above and below this band.

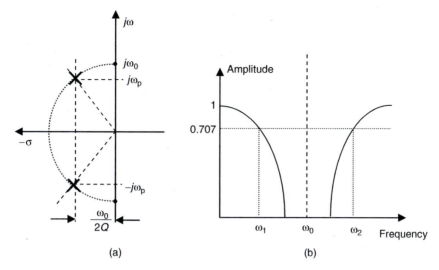

FIGURE 3.10   Pole-zero plot (a) and frequency response (b) of a BSF.

If the band to be rejected is reduced to a single frequency, the filter is then called a *notch filter*. Equation 3.18 describes the transfer function of a second-order BSF and is plotted in Figure 3.10. It is clear from the figure that the filter has two conjugate complex zeros at $\pm j\omega_o$ and two conjugate complex poles at $\pm j\omega_p$. As mentioned before, the slope of the response on both sides depends on the order of the filter and the quality Q.

$$H(s) = \frac{S^2 + \omega_0^2}{S^2 + \left(\dfrac{\omega_0}{Q}\right)S + \omega_0^2} \tag{3.18}$$

The stop band $W = \omega_2 - \omega_1 = \omega_0/Q$.

3.2.6   NOTCH FILTER

In contrast to the previously discussed filters, notch filters represent a class that is mainly intended to filter out (attenuate) a single undesired frequency. Notch filters are widely used in several applications, such as in medical instruments (e.g., an EKG) where the supply frequency can superimpose the "weak" picked-up heart signals, thereby affecting the precision of the diagnosis. Furthermore, such a type of filter is indispensable in other applications, such as high-fidelity audio systems, to insure complete elimination of the hum resulting from insufficiently filtered ripples in the power supply.

The transfer function of notch filters, which are a special case of BSF, is given by

$$H(s) = \frac{a_2 S^2 + a_0}{S^2 + \left(\dfrac{\omega_0}{Q}\right)S + \omega_0^2} \tag{3.19}$$

It is evident that the filter has, beside the common conjugate complex pole pair, two pure imaginary zeros, located at

$$Z_{1,2} = \pm\sqrt{-\frac{a_0}{a_2}} = \pm j\sqrt{\frac{a_0}{a_2}} \qquad (3.20)$$

Based on the location of these two zeros compared to the poles, we can classify three types of notch filters: symmetrical, low-pass, and high-pass.

3.2.6.1 Symmetrical Notch Filters

The pole-zero plot for a second-order symmetrical notch filter is shown in Figure 3.11. It shows that the zeros are located at $\pm j\omega_n$ such that

$$\pm j\omega_n = \pm j\sqrt{\frac{a_0}{a_2}}$$

or

$$\omega_n^2 = \frac{a_0}{a_2}$$

Substitution in Equation 3.18 gives

$$H(s) = \frac{a_2\left(S^2 + \omega_n^2\right)}{S^2 + \left(\dfrac{\omega_0}{Q}\right)S + \omega_0^2} \qquad (3.21)$$

and $\omega_0 = \omega_n$.

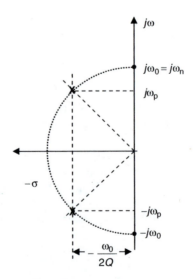

FIGURE 3.11 Pole-zero plot of a symmetrical notch filter.

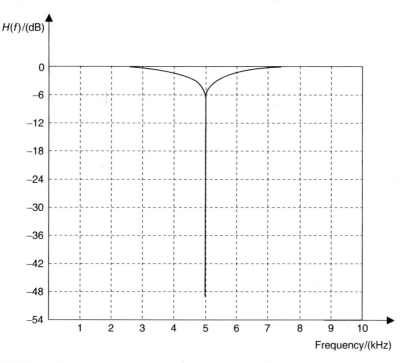

FIGURE 3.12   Frequency response of a symmetrical notch filter.

A plot of such relation would reveal that it reaches a constant value a_2 at $s = 0$ and $S = \infty$ while it vanishes at $S = j\omega_n = j\omega_o$. It follows that

$$H(0) = H(\infty) = a_2$$

and

$$H(j\omega_o) = 0$$

Or in other words, the response shows a dip at ω_o while assuming a constant level at both sides.

The frequency response of a high-quality notch filter is shown in Figure 3.12. The notch frequency (5 kHz) is seen to be highly attenuated. Nearby frequencies experience a maximum attenuation of 6 dB at most.

3.2.6.2   Low-Pass Notch Filters

The transmission zeros of such type occur at $\pm j\omega_n$, such that $\omega_n > \omega_o$, as illustrated in Figure 3.13. Considering again Equation 3.18:

$$H(s) = \frac{a_2 S^2 + a_0}{S^2 + \left(\dfrac{\omega_0}{Q}\right)S + \omega_0^2}$$

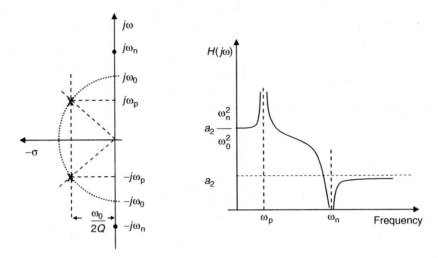

FIGURE 3.13 Pole-zero plot and frequency response of an LP notch.

whose zeros are located, as before at

$$Z_{1,2} = \pm j \sqrt{\frac{a_0}{a_2}} > \omega_0$$

$$j\omega_z = j\omega_n = \pm j \sqrt{\frac{a_0}{a_2}}$$

giving

$$a_0 = \omega_n^2 \cdot a_2$$

substitution in Equation 3.19 yields

$$H(s) = a_2 \frac{S^2 + \omega_n^2}{S^2 + \left(\dfrac{\omega_0}{Q}\right)S + \omega_0^2}$$

from which we can conclude that

1. At $S=0$ $H(0) = a_2 \dfrac{\omega_n^2}{\omega_0^2}$

2. At $S=\infty$ $H(\infty) = a_2$

The pole-zero plot and the transfer function of such types are shown in Figure 3.13.

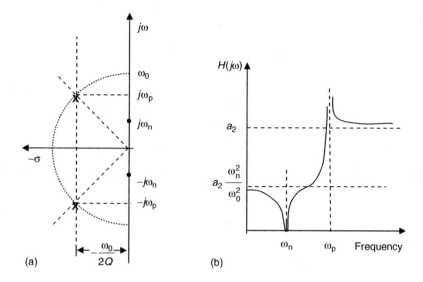

FIGURE 3.14 Pole-zero distribution and frequency response of an HP notch.

3.2.6.3 High-Pass Notch Filter

The pole-zero distribution of a high-pass notch is shown in Figure 3.14a, from which it is evident that the magnitudes of the zero frequencies are in this case less than the pole frequencies; that is, $\omega_n < \omega_o$, giving rise to the frequency response plotted in Figure 3.14b. To find the amplitudes at the extreme frequencies, we again set $a_0 = \omega_n^2 a_2$ leading as before to

$$H(0) = a_2 \frac{\omega_n^2}{\omega_0^2}$$

and

$$H(\infty) = a_2$$

However, because $\omega_n < \omega_0$, the magnitude at $\omega = 0$ is smaller than that at $\omega = \infty$, as it is clear from Figure 3.14b.

3.2.7 ALL-PASS FILTER

The two transmission zeros in an all-pass filter, Figure 3.15, are mirror images of the poles. This implies that they are located on the right-hand side of the S-plane, giving a magnitude response that is flat over all frequencies. The transfer function therefore takes the form

$$H(S) = \frac{S^2 - \left(\dfrac{\omega_0}{Q}\right)S + \omega_0^2}{S^2 + \left(\dfrac{\omega_0}{Q}\right)S + \omega_0^2} \tag{3.22}$$

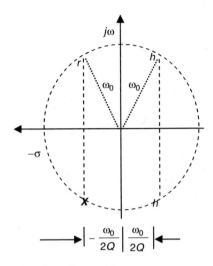

FIGURE 3.15   Pole-zero plot of an all-pass filter.

TABLE 3.3
Numerator Polynomials of Example 3.2

Case	$N(S)$
1	1
2	S
3	S^2
4	$(S^2 + 0.25)$
5	$(S^2 + 0.5)$
6	$(S^2 + 2)$
7	$(S^2 - \sqrt{2}S + 1)$

It is the phase response that is of interest here; hence such a type of filter is usually used for phase error correction—for example, as phase equalizers that are employed to compensate for the phase error distortion resulting from transmission channels.

EXAMPLE 3.2

The following transfer function represents a family of normalized filters ($\omega_0 = 1$):

$$H(S) = \frac{N(S)}{S^2 + \sqrt{2}S + 1}$$

where N(S) is a polynomial that could take the forms given in Table 3.3.

    For each case, plot the pole-zero distribution and the magnitude response, stating its type.

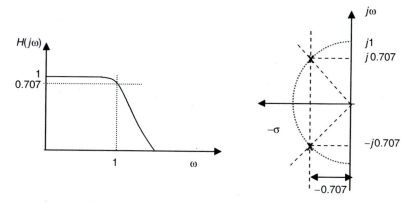

FIGURE 3.16   Frequency response and pole-zero plot of case 1 (an LP filter).

Solution

Because the denominator is the same for all cases, their poles are common. The pole locations can be obtained solving

$$S^2 + \sqrt{2}S + 1 = 0$$

giving

$$P_{1,2} = -\frac{1}{\sqrt{2}} \pm j\frac{1}{\sqrt{2}}$$

Case 1: N(S) = 1
The filter has no zeros; hence it is an all pole case, that is, an LPF (Figure 3.16).

Case 2: N(S) = S
Solving the preceding polynomial reveals that it has a single zero at s = 0; thus, the function represents a BPF (Figure 3.17). The poles, as before, are located at

$$\omega_p = \pm\sqrt{1 - \frac{1}{4Q^2}} \quad \text{with } Q = \frac{1}{\sqrt{2}}$$

which gives

$$\omega_p = \frac{1}{\sqrt{2}} = 0.707$$

Case 3: N(S) = S²
It is evident that the given numerator's polynomial has two zeros at the origin, indicating that we have an HPF (Figure 3.18).

Case 4: N(S) = S² + 0.25
The zeros are located in this case at

$$S = \pm j\,0.5$$

indicating that the filter is a high-pass notch (Figure 3.19).

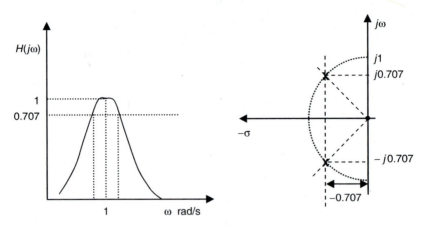

FIGURE 3.17   Frequency response and pole-zero plot of a BP filter.

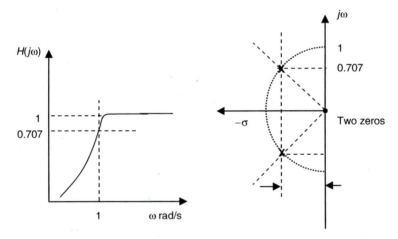

FIGURE 3.18   Frequency response and pole-zero distribution of an HP filter.

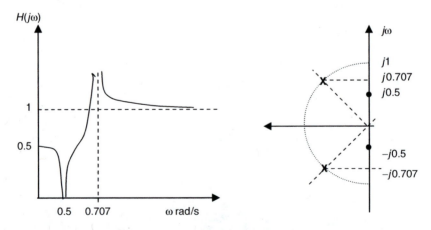

FIGURE 3.19   Frequency response and pole-zero plot of an HP notch filter.

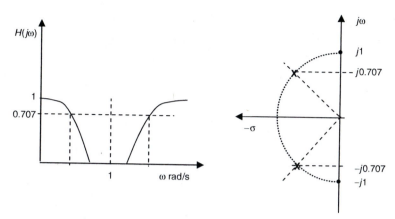

FIGURE 3.20   Frequency response and pole-zero plot of a BSF.

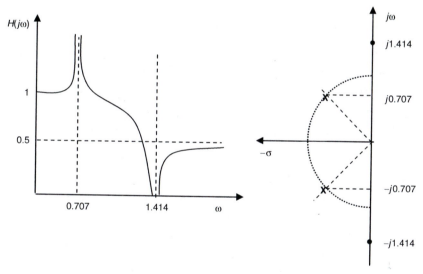

FIGURE 3.21   Frequency response and pole-zero plot of an LP notch filter.

Case 5: $N(S) = S^2 + 1$
The roots of the numerator's polynomial are complex, and are located at

$$S_{1,2} = \pm j1$$

that is, at both ends of the semicircle, giving a response of a BSF (Figure 3.20).

Case 6: $N(S) = S^2 + 2$
Solving the numerator's polynomial, $S^2 + 2 = 0$, gives a conjugate zero pair at

$$S_{1,2} = \pm j\sqrt{2}$$

that is, outside the semicircle (zero frequencies > pole frequencies).

    This provides the response of an LP notch, shown in Figure 3.21. It is evident that to get such a characteristic, the zero frequency should be greater than that of the pole.

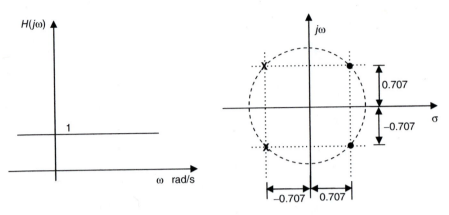

FIGURE 3.22   Frequency response and zero-pole plot of an all-pass filter.

TABLE 3.4
Numerator's Polynomials and Corresponding Filter Functions

Case	Numerator's Polynomial	Filter Function
1	1	LP
2	S	BP
3	S^2	HP
4	$S^2 + 0.25$	HP notch
5	$S^2 + 1$	BS
6	$S^2 + 2$	LP notch
7	$S^2 - \sqrt{2}S + 1$	All pass

Case 7: $N(s) = S^2 - \sqrt{2}S + 1$
The numerator's polynomial has its zeros in this case at

$$S_{1,2} = \frac{1}{\sqrt{2}} \pm j\frac{1}{\sqrt{2}}$$

which are mirror images of the poles of the transfer function, giving rise to an all-pass filter, as depicted in the pole-zero distribution in Figure 3.22.

Table 3.4 summarizes the cases considered earlier for normalized filters. It should be remembered always that it is the numerator's polynomial that determines the function of the filter, while the denominator's polynomial is common to all of them.

3.3   DESIGN PROCEDURES OF AN ANALOG FILTER

The flow chart in Figure 3.23 illustrates the steps usually followed when designing an analog filter; these steps are described in the following sections.

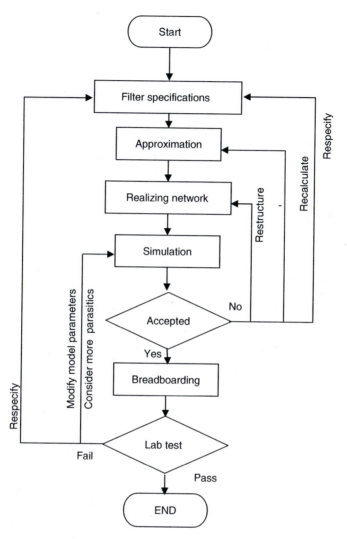

FIGURE 3.23   Design steps for an analog filter.

3.3.1   FILTER SPECIFICATIONS

This set of parameters is used to describe the required filter to the designer. Specifications are mostly concerned with the filter's frequency response, although in some
cases they specify its phase characteristics. As mentioned before, the rectangular
magnitude response of an ideal LPF (the *brick wall* response) is unrealizable practically, as its describing transfer function has an infinite order. Therefore, several
functions that can assume finite orders were introduced to approximate the ideal
response while yielding practically realizable designs. This in turn implies that not
only the filter cut-off frequency is to be specified, but also the decay rate (sharpness
or steepness) of the response, the tolerable fluctuations in the response within the

pass band (usually called *ripples*), as well as the minimum acceptable attenuation in the stop band.

3.3.2 APPROXIMATION

Having relaxed the requirements on the filter to be realized, it became possible to approximate the response through any finite order polynomial to provide a response that meets the specifications. Several approximating functions are present in the literature. However, only popular ones will be discussed later. The outcome of the approximation step are the filter order N, two sets of coefficients describing the numerator's and the denominator's polynomials of the filter's transfer function, and the 3 dB frequency. Tables that give those coefficients for any approximation and order are available in the literature.

3.3.3 REALIZATION

The next step to approximation is circuit implementation of the designed filter, where the obtained filter's transfer function is turned into a network. As mentioned before, there are several generations of circuit implementations for a filter; for example, LC realization, RC-active realization, gyrator realization, and switched capacitor realization. However, the selection of any implementation is primarily based on different factors, such as sensitivity to components' tolerance, size, power consumption, and ease of design.

3.3.4 SIMULATION

The obtained network should first be tested before being constructed to insure that the required specifications are met. Computer simulation is a good means for this step. A usable simulation package should have precise built-in models for the employed components in its library; otherwise, it should be created and verified by the user. There are several powerful network simulation tools such as Pspice, Micro Cap, and Electronics Workbench (EWB) that could be used. A sensitivity analysis is sometimes performed in which all components experience a specified percentage change in their nominal values, one at a time. This analysis may reveal that the values of one or more components are critical and therefore should be selected to possess low tolerance. If unacceptable deviation from the required response is observed, the design should be revised iteratively until a tolerable deviation within the permissible error budget is achieved.

3.3.5 BREADBOARDING

The last station before manufacturing a filter is laboratory testing to verify the simulation result. In this step, which is usually known as *breadboarding*, a prototype filter is constructed on a breadboard and then tested. Such a step is essential, as the measured response might show some deviation from the one obtained through simulation and/or from the desired one. The main cause of such a deviation is the components tolerance. In some other cases, the measured deviation in the response could result either due to some neglected parasitic components or due to improper estimation of their values. Needless to say, small deviations from the required response can be tolerated, as they result from some measuring errors or from ill-defined parameters. However, the degree and extent of this error depends on how critical the specifications are and on the designer's experience.

3.4   SPECIFICATIONS OF A FILTER

As mentioned earlier, practical filters do have frequency responses that deviate from the ideal rectangular brick wall response, so as to get filters having finite orders. Main filter specifications are pass-band attenuation A_p, stop-band attenuation A_r, and pass-band and stop-band edges ω_p and ω_r, respectively. They are best illustrated with the help of a diagram known as the *tolerance structure*, shown in Figure 3.24.

Three distinct properties are to be noticed from the tolerance structure:

1. The response is allowed to fluctuate (ripple) within the pass band, yet within a prescribed limit, giving rise to what is called the *error channel* or the *ripple channel*. The limit of such an error is given by the *maximum allowable attenuation (ripple amplitude)* in the pass band A_p.
2. A finite attenuation A_r is specified that represents the *minimum acceptable attenuation* in the stop band.
3. A third frequency band, the *transition band*, is found between the pass band and the rejection band. This band has the effect of drastically reducing the order of the approximating polynomial. The steepness of the response within this band is proportional with the filter order.

It is evident from Figure 3.24 that the two critical points (ω_p, A_p) and (ω_r, A_r) are enough to describe the filter's transfer function. These values, as we shall see later, are used to determine the order of the filter and hence the order of its approximating polynomial. Usually, the attenuation A_p is less than 3 dB; thus, the cut-off frequency

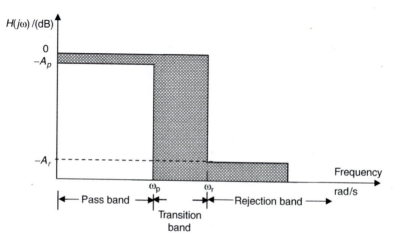

FIGURE 3.24   Tolerance structure of an LP filter.

ω_c lies somewhere between ω_p and ω_r. Based on the amount of permissible tolerance, the filter's transfer function is allowed to ripple within prescribed limits (the hatched area in Figure 3.24) in either the pass band or the stop band—or even both. Despite the resulting amplitude distortion within the band of interest, the roll-off characteristics (mainly the slope) of the response within the transition band are improved. Such improvement is reflected in reduced filter orders. From the engineering point of view, reducing the filter order is recommended, as it is reflected in minimized components count, power consumption, volume, and cost.

3.5   APPROXIMATIONS TO THE IDEAL RESPONSE

As stated before, a filter with a rectangular response is practically unrealizable. Therefore, the requirements on the filter should be relaxed. The allowed deviation is described by the tolerance structure. Actually, any function that can fit within the permissible region in the tolerance structure is an admissible approximating function. The main criterion when comparing possible functions is the resulting filter order. Mathematicians and engineers have provided us over the last 70 years with some useful polynomial functions that could efficiently be used in designing a practical filter. Some important approximating functions will be considered later.

3.5.1   BUTTERWORTH APPROXIMATION

Filters designed using the Butterworth approximation feature a "maximally" flat magnitude response over the pass band. The magnitude response of an Nth order Butterworth LPF is given by

$$|H(j\omega)| = \frac{1}{\sqrt{1 + \varepsilon_p^2 \left(\dfrac{\omega}{\omega_p} \right)^{2N}}} \tag{3.23}$$

in which ε_p is the ripple index in the pass band and ω_p is the pass-band edge.

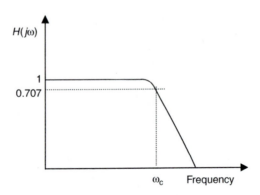

FIGURE 3.25   Magnitude response of a Butterworth LPF.

The ripple index of the pass band, ε_p, can be obtained by substituting the coordinates of the pass-band edge (ω_p, A_p) in Equation 3.23, as follows:

$$|H(j\omega_p)| = \frac{1}{\sqrt{1 + \varepsilon_p^2(\omega_p/\omega_p)^{2N}}} = \frac{1}{\sqrt{1 + \varepsilon_p^2}}$$

The attenuation A_p in dBs is calculated from

$$A_p = 20\log\left(\frac{1}{|H(j\omega_p)|}\right) = 20\log\sqrt{1 + \varepsilon_p^2}$$

$$= 10\log\left(1 + \varepsilon_p^2\right)$$

then

$$(1 + \varepsilon_p^2) = 10^{0.1A_p}$$

giving

$$\varepsilon_p = \sqrt{10^{0.1A_p} - 1} \qquad (3.24)$$

in which A_p, as before, is the maximum permissible drop in the response within the pass band of the filter. Similarly, we can obtain the ripple index in the stop band from

$$\varepsilon_r = \sqrt{10^{0.1A_r} - 1}$$

Substituting in the same equation for the stop band edge coordinates (ω_r, A_r) gives

$$|H(j\omega_r)| = \frac{1}{\sqrt{1 + \varepsilon_p^2(\omega_r/\omega_p)^{2N}}}$$

then

$$20\log|H(j\omega_r)| = -10\log\left(1 + \varepsilon_p^2\left(\frac{\omega_r}{\omega_p}\right)^{2N}\right) \geq -A_r$$

giving

$$\left(\frac{\omega_r}{\omega_p}\right)^N \varepsilon_p \geq \sqrt{10^{0.1A_r} - 1} = \varepsilon_r$$

The $\geq$ sign is used here to indicate that A_r is the minimum acceptable attenuation in the rejection band. Dividing both sides by ε_p, we get

$$\left(\frac{\omega_r}{\omega_p}\right)^N \geq \frac{\varepsilon_r}{\varepsilon_p}$$

Taking the logarithm for both sides gives

$$N \log\left(\frac{\omega_r}{\omega_p}\right) \geq \log\left(\frac{\varepsilon_r}{\varepsilon_p}\right)$$

The order of a Butterworth filter can then be obtained from

$$N_{Butt} \geq \frac{\log\left(\dfrac{\varepsilon_r}{\varepsilon_p}\right)}{\log\left(\dfrac{\omega_r}{\omega_p}\right)} \tag{3.25}$$

The $\geq$ sign means that if the resulting order N is noninteger, the next higher integer should be considered. This step guarantees that the condition of minimum required attenuation A_r at the rejection band is satisfied. Such a choice is feasible, because exceeding the requirement ensures that the response lies inside the error channel. Equation 3.25 can be rewritten as

$$N_{Butt_{LP}} = \left\lceil \frac{\log(\varepsilon_r/\varepsilon_p)}{\log(\omega_r/\omega_p)} \right\rceil \tag{3.26}$$

in which the symbol $\lceil\ \rceil$ denotes the next higher integer. For HPF, to get positive orders, either the numerator or the denominator should be inverted; that is,

$$N_{Butt_{HP}} = \left\lceil \frac{\log(\varepsilon_r/\varepsilon_p)}{\log(\omega_p/\omega_r)} \right\rceil$$

To get the cut-off frequency ω_c, we put $\omega = \omega_c$ and $H(j\omega_c) = 1/\sqrt{2}$ in Equation 3.23 (at ω_c, the transfer function drops to $1/\sqrt{2}$ of its low frequency value), giving

$$\frac{1}{\sqrt{2}} = \frac{1}{\sqrt{1 + \varepsilon_p^2(\omega_c/\omega_p)^{2N}}}$$

which after some manipulation reduces to

$$\left(\frac{1}{\varepsilon_p}\right)^{1/N} = \frac{\omega_c}{\omega_p}$$

We get

$$\omega_c = \omega_p \, \varepsilon_p^{-1/N} \quad \text{for LPF} \tag{3.27a}$$

or

$$= \omega_p \, \varepsilon_p^{1/N} \quad \text{for HPF} \tag{3.27b}$$

Similarly, we can prove that

$$\omega_c = \omega_r \varepsilon_r^{-1/N} \quad \text{for LPF} \tag{3.28a}$$

or

$$= \omega_r \varepsilon_r^{1/N} \quad \text{for HPF} \tag{3.28b}$$

Substituting in either of Equations 3.27a or 3.28a by the integer (rounded) value of N should deliver two different values for the cut-off frequency of an LPF. As a rule of thumb, one can use Equation 3.27a, as it yields the nearest value to the exact one. The same applies to HPF, where Equation 3.27b is recommended, as it yields the nearest value to ω_c. Table 3.5 gives the Butterworth polynomials of different orders, up to order 6.

The magnitude responses of a Butterworth filter having different orders were sketched in Figure 3.7b and repeated here in Figure 3.26 for convenience. As

TABLE 3.5
Butterworth Polynomials

Order	Polynomial
1	$S + 1$
2	$S^2 + 1.414\,S + 1$
3	$S^3 + 2\,S^2 + 2\,S + 1$
4	$S^4 + 2.61\,S^3 + 3.41\,S^2 + 2.61\,S + 1$
5	$S^5 + 3.24\,S^4 + 5.24\,S^3 + 5.24\,S^2 + 3.24\,S + 1$
6	$S^6 + 3.86\,S^5 + 7.46\,S^4 + 9.145\,S^3 + 7.46\,S^2 + 3.86\,S + 1$

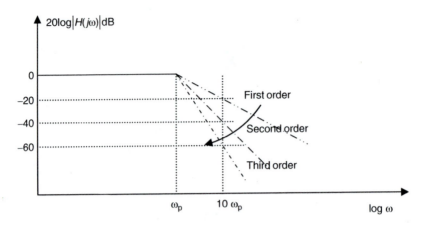

FIGURE 3.26   Log magnitude response of a Butterworth filter.

mentioned before, it can be proven that the decay rate (slope) of the response over the transition band is proportional with the order of the filter, according to the relation

$$\text{Slope} = 6N \text{ dB/octave}$$

$$= 20N \text{ dB/decade}$$

An Nth-order transfer function of an LPF, like the one described by Equation 3.23, is called an *all-pole function*, as it has N poles and no zeros. For Butterworth filters, the N poles were found to be symmetrically distributed along the circumference of a semicircle whose radius is equal to the cut-off frequency ω_c. This explains the unique maximally flat response of Butterworth filters. Consequently, the poles are expected to be separated by an angle θ that is given by

$$\theta = \frac{180°}{N}$$

Figure 3.27 illustrates the poles distribution in the complex S-plane for an even-ordered ($N = 4$) and an odd-ordered ($N = 5$) Butterworth filter.

It is clear from Figure 3.27 that the poles occur in conjugate complex pairs for both cases, except for odd-ordered filters, where a single real pole exists. The location of ith pole can be expressed as

$$P_i = -\sigma_i \pm j\omega_i$$

$$= -\omega_c \left[\sin \varphi_i \pm j \cos \varphi_i \right]$$

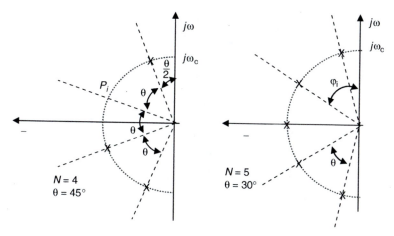

FIGURE 3.27 Pole distribution of a fourth-order (even-ordered) and a fifth-order (odd-ordered) low-pass Butterworth filter.

where φ_i is the angle enclosed between the $j\omega$-axis and the radial line connecting the respective pole with the origin and is given by

$$\varphi_i = \frac{\theta}{2} + (i-1)\theta \quad i = 1, 2, 3, \ldots, \left(\frac{N-1}{2} + 1\right) \quad N = \text{odd}$$

$$= 1, 2, 3, \ldots, \frac{N}{2} \quad N = \text{even}$$

Knowing the poles' coordinates, one can construct the transfer function as follows:

$$H(S) = \frac{H_0}{(S - P_1)(S - P_2)(S - P_3)\cdots(S - P_N)}$$

in which

$$P_N \equiv P_1^* \quad P_{N-1} \equiv P_2^*, \ldots$$

are conjugate poles. The constant H_0 is introduced to ensure that at $H(S)|_{S=0} = 1$
Multiplying each conjugate term together gives, for each conjugate pole pair,

$$D_i(S) = (S + \sigma_i - j\omega_i)(S + \sigma_i + j\omega_i) = [(S + \sigma_i)^2 + \omega_i^2]$$

$$= [S^2 + 2\sigma_i S + (\sigma_i^2 + \omega_i^2)]$$

since $\sigma_i^2 + \omega_i^2 = \omega_c^2$ is always valid along the circumference of the circle, thus we can write for the *i*th-pole pair

$$D_i(S) = S^2 + 2\sigma_i S + \omega_c^2$$

To construct the transfer function of a Butterworth filter, the following rules should be followed:

1. Each conjugate pole pair is represented by a second-order term.
2. The coefficient of S^2 is unity.
3. The coefficient of S is equal to$(-2 \times$ the real part)of the *i*th pole (σ_i).
4. For an LPF (all-pole filter), the numerator is a constant term $H_0 = \omega_c^N$.
5. The absolute term in each second-order term in the denominator is always equal to ω_c^2.

EXAMPLE 3.3

Calculate the constant H_0 for a sixth-order Butterworth LP filter that has a cut-off frequency at 10 kHz.

Solution

The transfer function can be written as

$$H(S) = \frac{H_0}{\left(S^2 + 2\sigma_1 S + \omega_c^2\right)\left(S^2 + 2\sigma_2 S + \omega_c^2\right)\left(S^2 + 2\sigma_3 S + \omega_c^2\right)}$$

To get a unity response at $S = 0$,

$$H(0) = \frac{H_0}{\left[\omega_c^2\right]^3} = \frac{H_0}{\omega_c^6} = 1$$

giving

$$H_0 = \omega_c^N$$

The cut-off frequency is 10 kHz giving

$$\omega_c = 2 \times \pi \times 10^3 = 2\pi \text{ k rad/s}$$

$$H_0 = (2\pi)^6 \times 10^9 = 1555.387 \times 10^9$$

EXAMPLE 3.4

Find the pole locations for a normalized fifth-order Butterworth LP filter, then write its transfer function.

Solution

Given $N = 5$, then $\theta = 180/5 = 36°$. Because the filter is normalized, then $\omega_c = 1$. The angles ϕ_i, as illustrated in Figure 3.28, are 18°, 54°, and 90°.

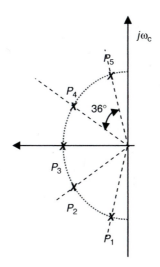

FIGURE 3.28   Pole distribution for a fifth-order Butterworth LPF.

The poles are therefore located at

$$P_{1,5} = \sigma_1 + j\omega_1 = -\sin 18 \pm j\cos 18 = -0.5877 \pm j0.809$$

$$P_{2,4} = -\sin 54 \pm j\cos 54 = -0.809 \pm j0.5877$$

$$P_3 = -\sin 90 = -1$$

The transfer function can then be formulated as follows:

$$H(S) = \frac{H_0}{(S - P_1)(S - P_2)(S - P_3)(S - P_4)(S - P_5)}$$

$$= \frac{H_0}{(S + \sigma_1 - j\omega_1)(S + \sigma_5 + j\omega_5)(S + \sigma_2 - j\omega_2)(S + \sigma_4 + j\omega_4)(S + \sigma_3)}$$

$$= \frac{H_0}{(S^2 + 2\sigma_1 S + 1)(S^2 + 2\sigma_2 S + 1)(S + 1)}$$

For a normalized filter

$$\sigma_1^2 + \omega_1^2 = \sigma_2^2 + \omega_2^2 = \cdots = \omega_c^2 = 1$$

Thus for the condition

$$H(0) = 1$$

to be satisfied, then

$$H_0 = \omega_c^N = 1$$

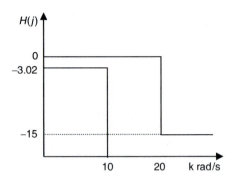

FIGURE 3.29   Tolerance structure for Example 3.5.

EXAMPLE 3.5

Deduce the transfer function $H(S)$ of a Butterworth LPF to meet the specifications described by the tolerance structure shown in Figure 3.29.

Solution

We start by finding the ripple indices ε_p and ε_r from Equation 3.24, giving

$$\varepsilon_p = \sqrt{10^{0.1A_p} - 1} = \sqrt{10^{0.302} - 1} = 1$$

$$\varepsilon_r = \sqrt{10^{0.1A_r} - 1} = \sqrt{10^{1.5} - 1} = 5.533$$

The required filter order, according to Equation 3.26, is

$$N_{\text{Butt}} \geq \frac{\log(\varepsilon_r/\varepsilon_p)}{\log(\omega_r/\omega_p)} \geq \frac{\log(5.533/1)}{\log(20/10)} \geq 2.4568 \cong 3$$

The 3 dB cut-off frequency ω_c can be calculated from Equation 3.27:

$$\omega_c = \omega_p \varepsilon_p^{-1/N} = 10 \text{ k rad/s}$$

The three poles are to be distributed along the circumference of a semicircle having a radius of r = 10 k rad/s. They are separated by an angle $\theta = 180/3 = 60°$. The pole locations are given by

$$P_{1,3} = -\omega_c\left[\sin 30 \pm j\cos 30\right]$$

$$= -10\left[0.5 \pm j0.866\right]$$

$$P_2 = -10$$

The transfer function can then be constructed as follows:

$$H(S) = \frac{H_0}{(S + 10)(S^2 + 10S + 100)} \quad \text{with } H_0 = \omega_c^3 = 10^3$$

3.5.2 CHEBYSHEV (EQUIRIPPLE) APPROXIMATION

In contrast to Butterworth approximation, the response of a Chebyshev filter is allowed to oscillate (ripple) between prescribed limits, either in the pass band or in the stop band. In return, it shows steeper transition response and hence a narrower transition band. Thus for the same slope of the transition band response, the required filter order is lower than that of the Butterworth filter. This is actually an advantage from the engineering point of view, as it means less components and therefore lower cost. However, this savings could be justified only if the resulting amplitude distortion due to the introduced pass band ripples (oscillations) is acceptable; that is, within the tolerance structure. In the Chebyshev approximation I, the response is allowed to ripple in the pass band. On the other hand, in Chebyshev approximation II, the pass band response remains almost flat, while the stop band response is allowed to ripple.

3.5.2.1 Chebyshev I Approximation

The transfer function of a Chebyshev LPF is described by

$$|H(S)| = \frac{1}{\sqrt{1 + \varepsilon_p^2 C_N^2(\Omega)}} \qquad (3.29)$$

in which $C_N(\Omega)$ is the Chebyshev polynomial of order N defined as

$$C_N(\Omega) = \cos(N \cos^{-1}\Omega) \quad \text{for } \Omega \leq 1$$

$$= \cosh(N \cosh^{-1}\Omega) \quad \text{for } \Omega > 1 \text{ where } \Omega = \frac{\omega}{\omega_p}$$

and ε_p is the pass-band ripple index.

In general, the Chebyshev polynomial of any order can be obtained from the following recursive (generating) formula:

$$C_N(\Omega) = 2\Omega C_{N-1}(\Omega) - C_{N-2}(\Omega)$$

Table 3.6 gives the polynomial up to the sixth order. The frequency response of a fifth order Chebyshev-I LP filter is shown in Figure 3.30.

Note that the order of the filter order can be read directly from the number of extrema (maxima and minima) within the pass-band response. Moreover, the response starts differently at $\omega = 0$ for odd- and even-ordered filters; although it begins with zero attenuation for odd orders, that is, $H(0) = 0$, for even orders, it shows an attenuation of

$$\frac{1}{\sqrt{1 + \varepsilon_p^2}}$$

TABLE 3.6

Chebyshev Polynomials

Order	$C_N(\Omega)$
0	1
1	Ω
2	$2\Omega^2 - 1$
3	$4\Omega^3 - 3\Omega$
4	$8\Omega^4 - 8\Omega^2 + 1$
5	$16\Omega^5 - 20\Omega^3 + 5\Omega$
6	$32\Omega^6 - 48\Omega^4 + 18\Omega^2 - 1$

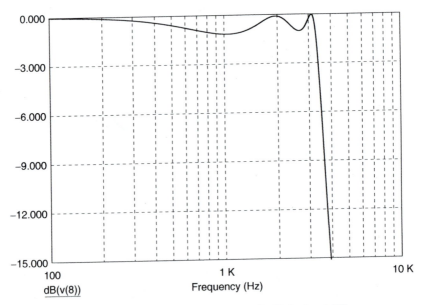

FIGURE 3.30 Frequency response of a fifth-order, 1 dB Chebyshev LPF.

It is also noticeable from the figure that the frequency of oscillations within the pass-band increases and reaches a maximum at the pass-band edge. This results in faster decay of the response over the transition band (better roll-off characteristic), as compared to the roll-off characteristic of a Butterworth filter having the same order. Furthermore, such oscillations (ripples) do have the uniform amplitude within what is usually termed as the "ripple channel" (this explains why such filters are called equiripple filters). This oscillatory behavior of the pass-band response could be referred to the unique distribution of Chebyshev poles along the circumference of an ellipse, rather than a circle, having an approximate principal diagonal length of ω_p. Figure 3.31 illustrates how the pole locations of a third-order Chebyshev filter can be derived from the corresponding Butterworth poles having a cut-off frequency

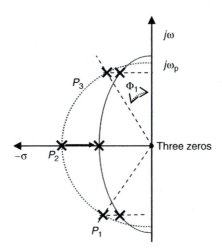

FIGURE 3.31 **(See color insert following page 262.)** Pole-zero distribution of a third-order Chebyshev filter, as derived from Butterworth poles.

that is equivalent to ω_p. From the geometry of the figure, we can find the Butterworth poles as before to be:

$$P_{i\text{Butt}} = -\omega_p \left[\sin \varphi_i \pm j \cos \varphi_i \right]$$

where $\varphi_i = \theta/2, (3/2)\theta, (5/2)\theta, \ldots$ with $\theta = 180/N$.

The corresponding Chebyshev poles can be obtained from the relation

$$P_{i\text{Cheby}} = \sigma_{i\text{Butt}} \cdot \sinh \gamma \pm \omega_{i\text{Butt}} \cdot \cosh \gamma$$

or

$$P_{i\text{Cheby}} = -\omega_p \left[\sin \varphi_i \sinh \gamma \pm j \cos \varphi_i \cosh \gamma \right] \tag{3.30}$$

in which γ is the pole compression factor. It is calculated from

$$\gamma = \frac{1}{N} \sinh^{-1} \frac{1}{\varepsilon_p}$$

To get an expression for the required filter order N for a given set of specifications $(\omega_p, \omega_r, \varepsilon_p, \varepsilon_r)$, consider again Equation 3.29:

$$|H(S)| = \frac{1}{\sqrt{1 + \varepsilon_p^2 C_N^2(\Omega)}}$$

The attenuation A in dB is given by

$$A_{dB} = 20 \log \frac{1}{H(S)} = 10 \log \left[1 + \varepsilon_p^2 C_N^2(\Omega) \right] \qquad (3.31)$$

The following relation is valid within the transition band:

$$C_N(\Omega) = \cosh N \cosh^{-1}(\Omega)$$

From Equation 3.31,

$$A_r \leq 10 \log \left[1 + \varepsilon_p^2 C_N^2(\Omega) \right]$$

with

$$\Omega = \frac{\omega_r}{\omega_p}$$

Then through some mathematical manipulations, we get

$$\sqrt{10^{0.1 A_r} - 1} \leq \varepsilon_p C_N(\Omega)$$

or

$$\varepsilon_r \leq \varepsilon_p C_N(\Omega)$$

Substituting for $C_N(\Omega)$ yields

$$\frac{\varepsilon_r}{\varepsilon_p} \leq \cosh N \cosh^{-1}(\Omega)$$

or

$$\cosh^{-1} \frac{\varepsilon_r}{\varepsilon_p} \leq N \cosh^{-1}(\Omega)$$

giving

$$N_{Cheby} \geq \frac{\cosh^{-1} \varepsilon_r / \varepsilon_p}{\cosh^{-1} \omega_r / \omega_p}$$

or

$$N_{Cheby} = \left[\frac{\cosh^{-1} \varepsilon_r / \varepsilon_p}{\cosh^{-1} \omega_r / \omega_p} \right] \qquad (3.32)$$

As stated before, due to the excellent roll-off characteristics of the Chebyshev filter in the transition band, it was found that they provide greater attenuation in the rejection band than their Butterworth counterparts having the same order. In other words, for the same attenuation level, the required Chebyshev filter's orders are lower than those of the equivalent Butterworth filter. Equation 3.33 gives Chebyshev attenuation as a function of N, the order of the filter, and the Butterworth attenuation [5]:

$$A_{\text{Cheby}} = A_{\text{Butt}} + 6.02(N - 1) \text{ dB} \tag{3.33}$$

Thus for a fourth-order Chebyshev filter, the achievable attenuation is 18 dB more than the attenuation of a Butterworth that has the same order.

EXAMPLE 3.6

Prove that the 3 dB frequency ω_c of a Chebyshev filter is given by

$$\omega_c = \omega_p \cosh\left(\frac{1}{N}\cosh^{-1}\frac{1}{\varepsilon_p}\right)$$

Solution

Considering again the transfer function $H(S)$ at $\Omega = \Omega_c$,

$$|H(j\Omega_c)| = \frac{1}{\sqrt{1 + \varepsilon_p^2 C_N^2(\Omega_c)}} = \frac{1}{\sqrt{2}}$$

yielding

$$\varepsilon_p^2 C_N^2(\Omega_c) = 1$$

thus

$$C_N(\Omega_c) = \frac{1}{\varepsilon_p} = \cosh\frac{1}{N}\cosh^{-1}\Omega_c$$

or

$$\cosh^{-1}\left(\frac{\omega_c}{\omega_p}\right) = N\cosh^{-1}\frac{1}{\varepsilon_p}$$

giving

$$\omega_c = \omega_p \cosh N\cosh^{-1}\frac{1}{\varepsilon_p} \tag{3.34}$$

It is, however, worth mentioning here that the cut-off frequency of a Chebyshev filter is meaningless, and only the pass band and stop band edges ω_p and ω_r are of interest.

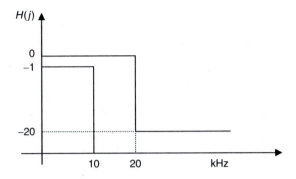

FIGURE 3.32   Tolerance structure for Example 3.7.

EXAMPLE 3.7

Deduce an expression for the transfer function of a Chebyshev filter that satisfies the shown tolerance structure (Figure 3.32).

Solution

We start by calculating the ripple indices in both bands as follows:

$$\varepsilon_p = \sqrt{10^{0.1} - 1} = 0.588$$

$$\varepsilon_r = \sqrt{10^2 - 1} = 9.94987$$

The required filter order is

$$N_{Cheby} = \left\lceil \frac{\cosh^{-1}\varepsilon_r/\varepsilon_p}{\cosh^{-1}\omega_r/\omega_p} \right\rceil = \lceil 2.78 \rceil \cong 3$$

The pole compression factor γ is given by

$$\gamma - \frac{1}{3}\sinh^{-1}\frac{1}{\varepsilon_p} = 0.476$$

The pole separating angle is

$$\theta = \frac{180}{3} = 60°$$

The angles φ_i are 30° and 90°.
Accordingly, the poles are located at

$$P_{1,3} = -10[\sin30\sinh0.476 \pm j\cos30\cosh0.476] = -2.4710 \pm j9.6597 \text{ k rad/s}$$
and

$$P_2 = -10\sinh0.476 = -4.942 \text{ k rad/s}$$

The transfer function can then be constructed as follows:

$$H(S) = \frac{H_o}{(S - P_2)(S - P_1)(S - P_3)}$$

$$= \frac{H_o}{(S + 4.942)(S^2 + 2S(2.471) + ((2.471)^2 + (9.6597)^2)}$$

where, as before,

$$H_0 = 4.942[(9.6597)^2 + (2.471)^2] = 491.312$$

The transfer function becomes

$$H(S) = \frac{491.312}{(S + 4.942)(S^2 + 4.942S + 99.4156)}$$

3.5.2.2 Chebyshev II Approximation

An equivalent filter to that obtained using the Chebyshev I approximation could result if its frequency response is allowed to ripple in an equiripple fashion within the stop band while keeping the pass band response almost flat. The transfer function takes the form

$$|H(S)| = \frac{1}{\sqrt{1 + \dfrac{1}{\lambda^2 C_N^2(\Omega)}}} \tag{3.35}$$

in which $\lambda = 1/\varepsilon_p$. A typical frequency response of the Chebyshev II filter is shown in Figure 3.33.

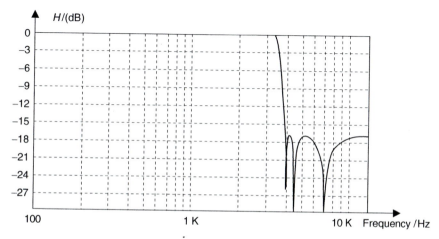

FIGURE 3.33 Frequency response of a Chebyshev II LPF.

with a capacitive load; that is,

$$Z_L = \frac{1}{SC}$$

We get

$$Z_{input} = \frac{SC}{G^2} = S \cdot \frac{C}{G^2}$$

$$= j\omega L_{eq}$$

where

$$L_{eq} = \frac{C}{G^2} \text{ Henery}$$

Figure 3.43 illustrates a gyrator-C realization of a third-order HP section. However, when using such a method in the realization of LP filters, a problem arises: the inductors are floating—that is, they have no terminal connected to ground. To overcome this problem, two gyrators have to be connected, as shown in Figure 3.44, to simulate each floating inductor.

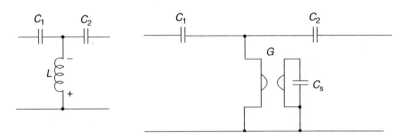

FIGURE 3.43   Gyrator-C realization of a third-order HPF.

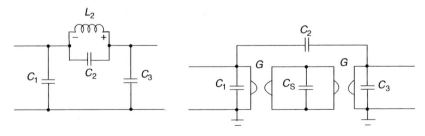

FIGURE 3.44   Simulation of a floating inductor employing two gyrators.

3.6.2.2   Realizations of a Gyrator

When realizing a gyrator, four important issues should be considered:

1. The obtainable quality factor Q of the simulated inductor
2. The frequency range over which the circuit remains stable while providing useable quality factors
3. The possibility of simulating floating inductors (with no terminals connected to ground) as those used in the realization of low-pass and BPF
4. The obtainable dynamic range; that is, the maximum voltage amplitude that can be applied to the filter

Several gyrator realizations have been published[6]. An example of a monolithic-integrated gyrator is the TCA 580 manufactured by Philips. Naturally, the performance of practical gyrators does deviate from theoretical ones. For example, they show finite input and output impedances and have gyration constants G that are frequency-dependent. Inductors simulated using such gyrators show finite Q's that vary with frequency up to a certain limit, where the circuit becomes impractical, due either to undesired oscillations (Q approaches ∞) or to the extremely small values of Q.

3.6.2.2.1   A Single Operational Amplifier Gyrator

Gyrator realization using a single operational amplifier is the most simple and direct realization. As shown in Figure 3.45, two resistors and a capacitor are needed with the operational amplifier.

The phasor diagram, plotted in Figure 3.45, illustrates the current/voltage relationships in the circuit. The phase lag (less than 90°) between the input voltage and current proves that the input impedance consists of an inductor L_{eq} in series with a resistor R_{eq}; that is, the simulated inductor has a finite quality factor Q. The following analysis leads to an expression for the input impedance, from which it will be evident that both the inductor and its series resistor are frequency-dependent.

Assuming an ideal operational amplifier, we can write

$$I^+ = I^- = 0$$

$$V^+ = V^- = V_X$$

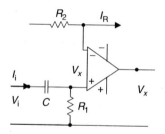
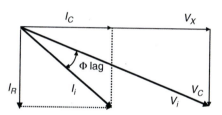

FIGURE 3.45   A single operational amplifier gyrator and its phasor diagram.

$$I_i = I_C + I_R \qquad I_C = \frac{V_X}{R_1} \qquad I_R = \frac{V_i - V_X}{R_2} = \frac{V_C}{R_2}$$

$$V_C = I_C \cdot \frac{1}{SC} = \frac{V_X}{SCR_1}$$

and

$$I_R = \frac{V_X}{SCR_1 R_2} = \frac{V_i - V_X}{R_2}$$

giving

$$V_X \left(1 + \frac{1}{SCR_1} \right) = V_i$$

or

$$V_X = V_i \frac{SCR_1}{1 + SCR_1}$$

$$I_i = I_C + I_R = \frac{V_X}{R_1} + \frac{V_X}{SCR_1 R_2}$$

Substituting for V_X, we get

$$I_i = V_i \frac{SC}{1 + SCR_1} \left(\frac{1 + SCR_2}{SCR_2} \right)$$

giving for the input impedance

$$Z_{i/p} = R_2 \frac{(1 + SCR_1)}{(1 + SCR_2)}$$

Substituting for S by $j\omega$, we get (after some manipulations)

$$= \frac{1 + \omega^2 C^2 R_1 R_2}{1 + \omega^2 C^2 R_2^2} + j\omega C \frac{R_1 - R_2}{1 + \omega^2 C^2 R_2^2} = R_{eq} + j\omega L_{eq}$$

That is, the resulting inductor is frequency-dependant and has a value of

$$L_{eq} = C \frac{R_1 - R_2}{1 + \omega^2 C^2 R_2^2} \qquad (3.36)$$

and a quality factor given by

$$Q = \frac{\omega C(R_1 - R_2)}{1 + \omega^2 C^2 R_1 R_2}$$

with

$$R_1 \neq R_2 \qquad\qquad (3.37)$$

3.6.2.2.2 Differential Amplifier Realization

A rather interesting realization possibility is the use of two specially designed differential amplifiers connected in a back-to-back fashion, as shown in Figure 3.46.

The differential amplifiers should be optimized to satisfy the requirements of a high-quality gyrator; namely, to attain high input and output impedances and wide BWs. This could be achieved, for example, by using high-speed, high-β transistors, as well as active loads (current sources/sinks). The most important characteristic of the resulting gyrator is the obtainable low-frequency quality factor Q_o of a simulated inductor and the frequency range over which the circuit remains stable with a usable value of Q. Figure 3.47 depicts the quality factor of a typical gyrator versus frequency. The frequency f_∞ represents the end of the usable frequency range since above this frequency the quality becomes negative and the circuit oscillates.

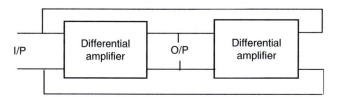

FIGURE 3.46   Differential amplifier realization of a gyrator.

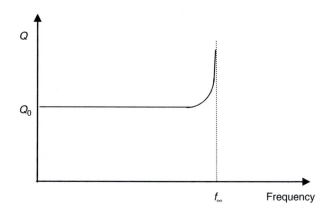

FIGURE 3.47   Quality factor versus frequency of a simulated inductor.

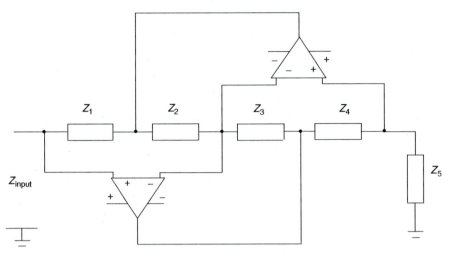

FIGURE 3.48 A generalized impedance converter.

3.6.2.2.3 *Generalized Impedance Converter*

Two operational amplifiers and five impedances, connected as illustrated in Figure 3.48, can be employed to realize a generalized impedance converter, usually known as the Antoniou or Bruton converter[8]. One of its main applications is the simulation of inductors and negative resistances.

A simple analysis of the circuit, assuming ideal operational amplifiers, yields an input impedance of

$$Z_{input} = \frac{V_i}{I_i} = \frac{Z_1 Z_3 Z_5}{Z_2 Z_4} \tag{3.38}$$

The derivation of this formula is left as an exercise for the reader.

Let us consider the case of Z_4 as a capacitive reactance jX_{C4}; while all other elements are resistive, we get by substituting in Equation 3.38 an input impedance of

$$Z_{input} = j\omega C_4 \cdot \frac{R_1 R_3 R_5}{R_2} = j\omega L_{eq}$$

where

$$L_{eq} = C_4 \frac{R_1 R_3 R_5}{R_2}$$

which is the impedance of an inductor whose value is determined by the values of capacitor C_4 and the resistors. The circuit has the excellent feature of being less sensitive to the nonideal features of practical operational amplifiers. However, as it is grounded from one side, its application is limited to the realization of circuits

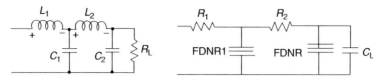

FIGURE 3.49   A fourth-order LP filter and its FDNR equivalent.

containing grounded inductors only, as in the case of HPF. However, floating inductors could be simulated using the same circuit with the following simple trick.

The impedance of every component in the circuit—of an LPF, for example—is
divided by the variable S. In this way, all inductors are turned into resistors, and
resistors into capacitances, whereas capacitors are turned into a new element called
frequency-dependent negative resistance (FDNR). This new component can be
realized by choosing both of Z_2 and Z_4 in Figure 3.48 as pure capacitive reactance,
thereby giving an input impedance of

$$Z_{\text{input}} = (j\omega)^2 C_2 C_4 R_1 R_3 R_5 = -\omega^2 C_2 C_4 R_1 R_3 R_5$$

which has the dimensions of a negative and frequency-dependent resistance. The
idea is illustrated in Figure 3.49 for the case of a fourth-order LP-filter.

3.6.3   *RC*-ACTIVE FILTER REALIZATION

As mentioned before, a filter is a causal, linear, time-invariant system. Thus an
Nth-order filter can be represented by a constant-coefficient Nth-order differential
equation as follows:

$$\frac{d^N y}{dt^N} + b_{N-1}\frac{d^{N-1}y}{dt^{N-1}} + \cdots + b_1\frac{dy}{dt} + b_0 y = a_M \frac{d^M x}{dt^M} + a_{M-1}\frac{d^{M-1}x}{dt^{M-1}} + \cdots + a_0 x$$

$$(3.39)$$

Applying Laplace transforming, we get

$$(S^N + b_{N-1}S^{N-1} + \cdots + b_1 S + b_0) \cdot Y(S) = \left(a_M S^M + a_{M-1}S^{M-1} + \cdots + a_0\right) X(S)$$

which leads to the following transfer function:

$$H(S) = \frac{a_M S^M + a_{M-1}S^{M-1} + \cdots + a_0}{S^N + b_{N-1}S^{N-1} + \cdots + b_1 S + b_0}$$

$$(3.40)$$

where $M \leq N$.

As an example, let us consider the following second-order filter:

$$\frac{d^2 y}{dt^2} + b_1\frac{dy}{dt} + b_2 y = a_0 x$$

For a normalized filter—that is, $\omega_c = 1$—the transfer function becomes

$$H(S) = \frac{1}{S^2 + \left(2/\sqrt{2}\right)S + 1} = \frac{1}{S^2 + \sqrt{2}S + 1}$$

Now comparing the coefficients of the transfer function just derived with those of a Sallen and Key section given in Equation 3.41, we get

$$C_1 C_2 R^2 = 1 \quad \text{and} \quad \frac{2}{RC_1} = \sqrt{2}$$

Thus with $R = 1\ \Omega$, we get

$$C_1 = \sqrt{2} = 1.414\ \text{Farad} \quad \text{and} \quad C_2 = \frac{1}{\sqrt{2}} = 0.707\ \text{Farad}$$

Practical filters do have cut-off frequencies other than 1 rad/s; therefore, some sort of scaling to the desired frequency should be done, as we shall see.

3.7.2.1   Scaling Factors

Normalized filters are hypothetical, as they contain components of impractical values. They should therefore be denormalized by raising their frequencies to the required frequency level. This step usually leads to reasonable component values—consider the capacitances and resistors obtained earlier and imagine the size of a 1 Farad capacitor. Therefore, scaling factors should be introduced to bring the cut-off frequency to the desired level, at the same time giving the components practical values. Two scaling factors are needed; namely, an *impedance scale factor Z* and a frequency scale factor (FSF) $= 2\pi f_c$, where f_c is required cut-off frequency of the filter. The denormalized values of the components, for an LPF, can then be calculated as follows:

$$C_d = \frac{C_n}{FSF \cdot Z}\ \text{Farad} \tag{3.45}$$

and

$$R_d = R_n \cdot Z\ \Omega \tag{3.46}$$

where the subscripts n and d stand for the normalized and the denormalized values, respectively. The following examples illustrate the procedure.

EXAMPLE 3.10

Design a second-order Butterworth LP filter to have a cut-off frequency at 3.4 kHz (note that all telephone calls are band-limited to this frequency).

Solution

The FSF is

$$FSF = 2\pi \times 3.4 \times 10^3\ \text{rad/s} = 21.3628\ \text{k rad/s}$$

considering an arbitrary impedance scale factor Z of 10^4; that is,

$$R_1 = R_2 = 10 \text{ k}\Omega$$

Thus the denormalized capacitors' values are:

$$C_1 = \frac{\sqrt{2}}{10^4 \times 21.3628 \times 10^3} = 6.619 \text{ nF}$$

and

$$C_2 = \frac{1}{\sqrt{2} \times 21.3628 \times 10^3} = 3.3099 \text{ nF}$$

A circuit diagram and frequency response for the filter, obtained through a Micro Cap simulation, are drawn in Figures 3.56 and 3.57, respectively.

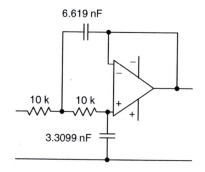

FIGURE 3.56   A second-order Butterworth LP filter.

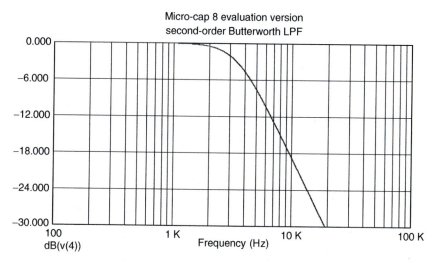

FIGURE 3.57   Frequency response of a second-order Butterworth LPF.

It is clear from the plot that the frequency response remains flat within 3 dB up to 3.4 kHz as required. The decay rate (roll-off) within the transition band occurs at a rate of 12 dB/octave as expected (2 × 6 dB).

EXAMPLE 3.11

Design a third-order LPF to have an equiripple response in the pass band of 1 dB up to 1 kHz.

Solution

Given

$$|A_p| = 1 \, \text{dB}$$

then

$$\varepsilon_p = \sqrt{10^{0.1} - 1} = 0.508$$

According to Equation 3.29, the poles are located at

$$P_{1,3} = -\omega_p [\sinh \gamma \cdot \sin 30 \pm j \cosh \cdot \gamma \cos 30]$$

$$P_2 = -\omega_p \sinh \gamma$$

where

$$\omega_p = 2\pi k \, \text{rad}/s \quad \text{and} \quad \gamma = \frac{1}{3} \sinh^{-1} \left[\frac{1}{0508} \right] = 0.47648$$

leading to

$$P_{1,3} = -2\pi \left[0.2475 \mp j0.96618 \right] k \, \text{rad}/s$$

$$= \left[-1.554 \mp j6.0706 \right] k \, \text{rad}/s$$

and

$$P_2 = -2\pi \times 0.4947 = -3.10829 \, k \, \text{rad}/s$$

The third-order LP filter transfer function has the form

$$H(S) = \frac{H_0}{(S - \sigma_2)(S^2 - 2\sigma_1 S + (\sigma_1^2 + \omega_1^2))}$$

Substituting the preceding values gives

$$H(S) = \frac{3.10829 \times 39.267}{(S + 3.10829)(S^2 + 3.10829S + 39.267)}$$

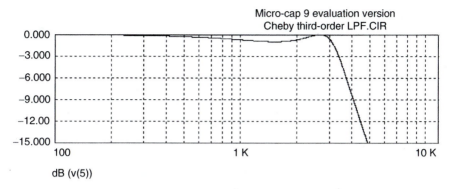

dB (v(5))

FIGURE 3.58 Simulation result of Example 3.8.

Recalling that the transfer function of a third-order LP Sallen and Key section is given by

$$H_{LP}(S) = \frac{1/C_1C_2C_3R^3}{S^3 + S^2\left(\dfrac{2}{C_2R} + \dfrac{2}{C_1R}\right) + \dfrac{S}{C_1R^2}\left(\dfrac{3}{C_3} + \dfrac{1}{C_2}\right) + \dfrac{1}{C_1C_2C_3R^3}}$$

Three simultaneous equations have to be solved to get the values of the three capacitors C_1, C_2, and C_3. This task is left as an exercise for the reader.

Figure 3.58 depicts the response of the filter as obtained from a Micro Cap simulation. It is evident from the plot that the number of maxima and minima is three, which is the order of the filter. It is clear also that the response starts from zero dB at DC (a common property for odd-ordered filter) then drops to a minimum of -1 dB, as required, then rises again to a peak at 0 dB. As specified, the pass-band edge is located exactly at 1 kHz, where the attenuation is 1 dB. The roll-off is seen to be sharper than that of an equivalent third-order Butterworth filter.

3.7.3 REALIZATION USING THE TOLERANCE STRUCTURE

As stated previously, filter specifications are usually summarized in a plot called the tolerance structure, which is described by the two points (ω_p, A_p) and (ω_r, A_r). It was demonstrated previously that the response of the designed filter, irrespective of the used approximation, should be confined within the *ripple channel*, shown hatched. The response can fluctuate "ripple" within this channel without exceeding A_p in the pass band while achieving an attenuation of at least A_r within the stop band. The width of the transition band and the ripple amplitude determine together the order of the filter. The following examples illustrate the procedures. A plot of a tolerance structure of an LPF is shown in Figure 3.59.

EXAMPLE 3.12

Design a Butterworth LPF to meet the specifications described by the tolerance structure shown in Figure 3.60.

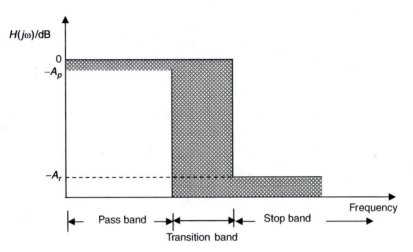

FIGURE 3.59   Tolerance structure of an LP filter.

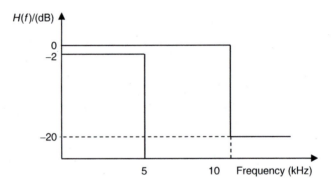

FIGURE 3.60   Tolerance structure of Example 3.12.

Solution

We start by calculating the ripple indices as follows:

$$\varepsilon_p = \sqrt{10^{0.2} - 1} = 0.76478$$

$$\varepsilon_r = \sqrt{10^2 - 1} = 9.9498$$

The required filter order can be obtained from

$$N_{Butt} \geq \frac{\log(\varepsilon_r / \varepsilon_p)}{\log(\omega_r / \omega_p)} \geq \frac{1.1142}{0.301} \geq 3.7 \approx 4$$

The cut-off frequency is located at

$$\omega_c = \omega_p \varepsilon^{-1/N} = 2 \times \pi \times 5 \times 10^3 \times (0.76478)^{-0.25} = 33.5943 \text{ k rad/s}$$

The poles are separated by an angle θ, where

$$\theta = \frac{180}{4} = 45°$$

They are located at

$$P_{1,4} = -\omega_c(\sin 22.5 \pm j\cos 22.5) = -\sigma_1 \pm j\omega_1 = -12.8558 \pm j31.036$$

$$P_{2,3} = -\omega_c(\sin 67.5 \pm j\cos 67.5) = -\sigma_2 \pm j\omega_2 = -31.036 \pm j12.8558$$

The filter's transfer function can be expressed as follows:

$$H(S) = \frac{\omega_c^4}{(S^2 + 2\sigma_1 S + \omega_c^2)(S^2 + 2\sigma_2 S + \omega_c^2)} \tag{3.47}$$

which can be written as

$$H(S) = \frac{\omega_c^2}{\left(S^2 + 2\sigma_1 S + \omega_c^2\right)} \cdot \frac{\omega_c^2}{\left(S^2 + 2\sigma_2 S + \omega_c^2\right)}$$

That is, it could be realized as a cascade of two second-order sections.
Recalling that the transfer function of a second-order Sallen and Key LP section is given by

$$H(S) = \frac{1/R_1 R_2 C_1 C_2}{S^2 + S\dfrac{1}{C_1}\left(\dfrac{1}{R_1} + \dfrac{1}{R_2}\right) + \dfrac{1}{R_1 R_2 C_1 C_2}}$$

which simplifies, taking $R_1 = R_2 = R$, to

$$H(S) = \frac{1/R^2 C_1 C_2}{S^2 + S\dfrac{2}{C_1 R} + \dfrac{1}{R^2 C_1 C_2}} \tag{3.48}$$

Comparing corresponding coefficients of Equations 3.47 and 3.48 gives

$$2\sigma_i = \frac{2}{RC_{1i}} \quad \text{or} \quad C_{1i} = \frac{1}{\sigma_i R}$$

recalling that

$$\omega_c^2 = \frac{1}{C_{1i} C_{2i} R^2} \quad \text{then } C_{2i} = \frac{\sigma_i R}{\omega_c^2 R^2} = \frac{\sigma_i}{\omega_c^2 R}$$

Assuming an impedance scale factor Z of 10 kΩ, we get $C_{11} = 7.7787$ nF, $C_{12} = 3.222$ nF for the first section, and for the second section, $C_{21} = 1.1387$ nF,

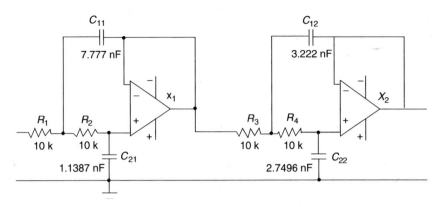

FIGURE 3.61   Filter realization for example 3.9.

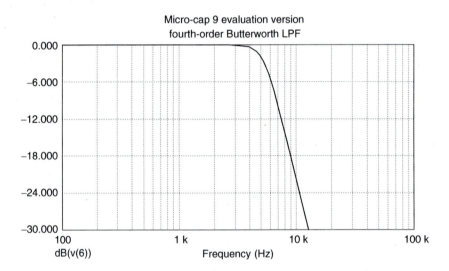

FIGURE 3.62   Frequency response for the filter of Example 3.9.

C_{22} = 2.7496 nF. The circuit diagram and the result of a simulation run on Micro Cap are illustrated in Figures 3.61 and 3.62, respectively.

EXAMPLE 3.13

Estimate the order of a filter equivalent to the one given in Example 3.12 using Chebyshev approximation, then comment on the result.

Solution

The specifications of the preceding example were:

1. Maximum pass band ripples of 2 dB, giving ε_p = 0.76478
2. Minimum stop band attenuation of 20 dB, giving ε_r = 9.9498

3. Pass band edge at $f_p = 5\,\text{kHz}$
4. Stop band edge at $f_r = 10\,\text{kHz}$

Thus, using the same parameters, the required filter order can be calculated using Equation 3.32 as follows:

$$N_{\text{Cheby}} \geq \frac{\cosh^{-1}\varepsilon_r/\varepsilon_p}{\cosh^{-1}\omega_r/\omega_p} \geq \frac{3.25746}{1.31695} \geq 2.47 \approx 3$$

It is clear that the required order for the Chebyshev filter in this example is lower than that required for the Butterworth filter for the same specifications.

3.7.4 REALIZATION USING FILTER CATALOGUES

In the preceding section, we spent a lot of effort to find the required values of the filter components, even for low-order ones. Of course, such effort will become more and more tedious as the order increases, until it becomes impossible without the use of a computer program. Therefore, filters' designers have developed tables in which normalized values of the components (e.g., capacitors for LPFs and resistors for HPFs) are listed for almost all known approximations (given in Appendix A).

To generalize such tables, the listed values are calculated assuming normalized filters; that is, filters that have a cut-off frequency of $\omega_c = 1$ rad/s. These tables list the values of capacitors for LPFs, assuming a common resistance of $R = 1\,\Omega$. For HP filters, these values represent the inverse of the resistors values ($1/R$), assuming a unified capacitance value of $C = 1$ F. The procedure is simple and direct. They could be summarized in the following steps:

1. Calculate the required filter order from the given specifications.
2. Select the proper table for the given approximation.
3. Choose an "impedance scale factor K":

$$K_R = R \text{ Ohms for LPFs}$$

$$K_C = C \text{ Farad for HPFs}$$

4. Calculate the FSF, defined as FSF $= 1/(2\pi f_c)$.
5. Calculate M, the impedance multiplier as follows:

$$M = \frac{1}{K \times \text{FSF}}$$

6. Get the denormalized values by multiplying the corresponding table entries with the multiplier M.

The process is summarized in the Table 3.8—bear in mind that $C_{ni} = 1/R_{ni}$.

TABLE 3.8

Components' Denormalization Table

Type	Z	R_i	C_i
LPF	Z_R	All are equal	$C_{in} \cdot M_c = \dfrac{C_{in}}{FSF \cdot K_R}$
HPF	Z_c	$R_{in} \cdot M_R = \dfrac{R_{in}}{FSF \cdot K_c} = \dfrac{1}{FSF \cdot K_c \cdot C_{in}}$	All are equal

Note: M_C = capacitance multiplier and M_R = resistance multiplier.

To get acquainted with the use of the preceding formulas, assume that we are required to design a second-order Chebyshev LP filter with $f_p = 1$ kHz.

From the corresponding table, we get

$$C_{1n} = 1.638$$

$$C_{2n} = 0.6955$$

Assuming $K_R = 10$ kΩ, the denormalized values C_{id} are calculated from

$$C_{id} = \frac{C_{in}}{2\pi f_p Z_R} = \frac{C_{in}}{6.283 \times 10^7} = 1.5 \times 915 \times 10^{-8} C_{in}$$

Giving

$$C_1 = 26.0695 \text{ nF}$$

and

$$C_2 = 11.069226 \text{ nF}$$

Similarly, for an equivalent HP filter, we get the denormalized resistors R_{id}, assuming $C_R = 10$ nF as follows:

$$R_{id} = \frac{1}{2\pi f_p C_R R_{in}} = \frac{159.1549}{R_{in}} \text{ kΩ}$$

Giving

$$R_1 = 97.164 \text{ kΩ}$$

and

$$R_2 = 228.835 \text{ kΩ}$$

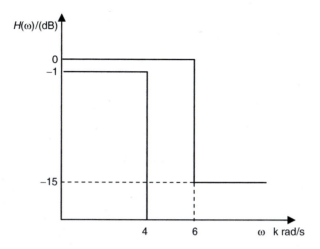

FIGURE 3.63   Tolerance structure of Example 3.14.

EXAMPLE 3.14

Using filter tables, design a Butterworth LPF to satisfy the tolerance structure shown in Figure 3.63, in the form of Sallen-Key sections.

Solution

The ripple factors ε_p and ε_r are first calculated as follows:

$$\varepsilon_p = \sqrt{10^{0.1A_p} - 1} = \sqrt{10^{0.1} - 1} = 0.508$$

$$\varepsilon_r = \sqrt{10^{0.1A_r} - 1} = \sqrt{10^{1.5} - 1} = 5.5337$$

The required filter order is given by

$$N \geq \frac{\log(\varepsilon_r / \varepsilon_p)}{\log(\omega_r / \omega_p)} \geq 5.88 \approx 6$$

The cut-off frequency ω_c can be obtained from

$$\omega_c = \omega_p \varepsilon_p^{-1/N} = 4(0.508)^{-1/6} = 4.478 \text{ k rad/s}$$

The FSF is equal to ω_c; that is, FSF $= 4.478 \times 10^3$.
Considering an impedance scale factor of $K = 10^3$, for the capacitors we can write

$$C_i = \frac{C_{in}}{K \times \text{FSF}} = \frac{C_{in}}{4.478 \times 10^3 \times 10^3} = C_{in} \times 0.2233 \ \mu\text{F}$$

where C_{in} (Table 3.9) are the normalized capacitor values obtained from the entry $N = 6$ of the Butterworth filter tables found in Appendix A1 and Table 3.9.

TABLE 3.9

Normalized Capacitor Values for Example 3.14

I	C_{1i}	C_{2i}
1	1.035	0.966
2	1.414	0.707
3	3.863	0.2588

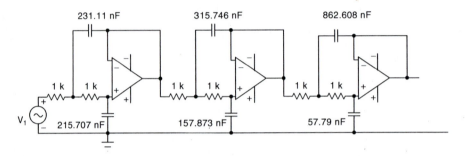

FIGURE 3.64 Circuit diagram for Example 3.14.

The denormalized capacitances are calculated as follows:

$$C_{11} = 1.035 \times 0.2233 = 231.11 \text{ nF}$$
$$C_{21} = 0.966 \times 0.2233 = 215.707 \text{ nF}$$
$$C_{12} = 1.414 \times 0.2233 = 315.746 \text{ nF}$$
$$C_{22} = 0.707 \times 0.2233 = 157.873 \text{ nF}$$
$$C_{13} = 3.863 \times 0.2233 = 862.608 \text{ nF}$$
$$C_{23} = 0.2588 \times 0.2233 = 57.79 \text{ nF}$$

A circuit realization for the filter and its frequency response are shown in Figures 3.64 and 3.65, respectively.

EXAMPLE 3.15

Repeat the previous example for an equivalent HPF.

Solution

The corresponding tolerance structure is shown in Figure 3.66.

Because the rejection band and pass band edges are located at $\omega_r = 4$ k rad/s and $\omega_p = 6$ k rad/s, respectively, the transition BW should the same as before. Also, because the ripple channel has the same depth, we can conclude that the required

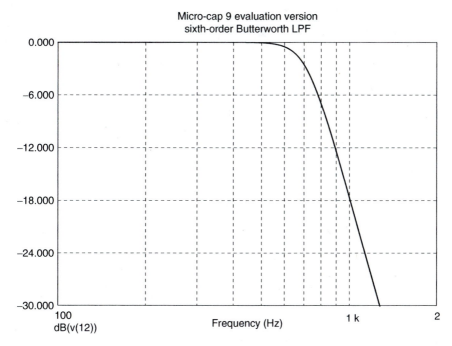

FIGURE 3.65   Simulation result for Example 3.14.

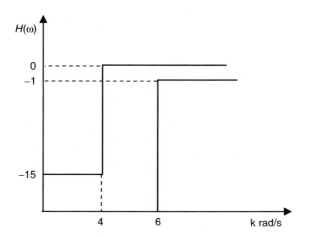

FIGURE 3.66   Tolerance structure for Example 3.15.

filter should have the same order as the preceding case; that is, $N = 6$. However, the cut-off frequency of an HPF is to be calculated from the following equation:

$$\omega_c = \omega_p \varepsilon_p^{1/N} = 6 \times 0.508^{1/6} = 5.3595 \text{ k rad/s}$$

The pole locations are the same as before, with six zeros added at the origin.

Using the same table, considering an impedance scale factor $K_c = 10$ nF $= 10^{-8}$ F, and recalling that the FSF is $\omega_c = 5.3595 \times 10^3$, we can calculate the resistor multiplier from

$$R_m = \frac{1}{K_c \times \omega_c}$$

$$R_{id} = R_{in} \times R_m = \frac{R_{in}}{K_c \times \omega_c} = \frac{1}{K_c \times \omega_c \times C_{in}}$$

where C_{in}, as before, are the normalized capacitor values obtained from the tables at the entry $N = 6$. The denormalized resistors' values are then

$$R_{11} = 18.658 \times \frac{1}{3.863} = 4.83 \text{ k}\Omega$$

$$R_{21} = 18.658 \times \frac{1}{0.2588} = 72.096 \text{ k}\Omega$$

$$R_{12} = 18.658 \times \frac{1}{1.414} = 13.195 \text{ k}\Omega$$

$$R_{22} = 18.658 \times \frac{1}{0.707} = 26.391 \text{ k}\Omega$$

$$R_{13} = 18.658 \times \frac{1}{1.035} = 18.027 \text{ k}\Omega$$

$$R_{32} = 18.658 \times \frac{1}{0.966} = 19.315 \text{ k}\Omega$$

A circuit diagram for the filter is shown in Figure 3.67.

To prove that the filter requirements have been met, the circuit has been simulated using Micro Cap. From the result, illustrated in Figure 3.68, it is clear that the stop band edge and the pass band edge are exactly at 636.6 Hz (4000/2π) and 954.9 Hz, respectively, while proving the required attenuation levels at −15 dB and −1 dB. The slope of the response within the transition band is 36 dB/octave, as expected.

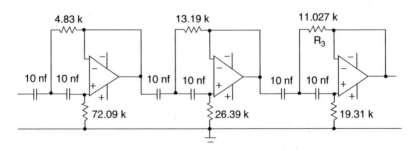

FIGURE 3.67 Circuit diagram for Example 3.12.

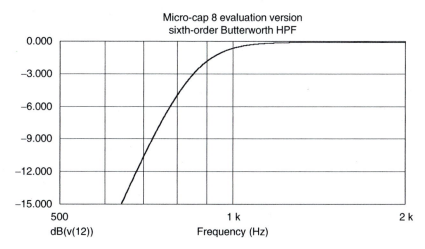

FIGURE 3.68   Frequency response of a sixth-order Butterworth HPF.

EXAMPLE 3.16

Using filter tables, design a Chebyshev LPF to satisfy the shown tolerance structure. Verify the results through computer simulation.

Solution

The ripple factors are

$$\varepsilon_p = \sqrt{10^{0.1} - 1} = 0.508$$

$$\varepsilon_r = \sqrt{10^{2.5} - 1} = 17.75$$

and the filter order is

$$N = \left\lceil \frac{\cosh^{-1}(\varepsilon_r/\varepsilon_p)}{\cosh^{-1}(f_r/f_p)} \right\rceil = \left\lceil \frac{4.2449}{0.8789} \right\rceil = 5$$

from Chebyshev tables given in Appendix A1 for 1 dB ripples and order 5, we get

$$C_{11} = 8.884 \quad C_{21} = 3.935 \quad C_{31} = 0.254$$

$$C_{12} = 11.55 \quad C_{22} = 0.09355$$

If the FSF $F = 2\pi \times 3.4 \times 10^3$ Hz, and considering an impedance scale factor $K_R = 10 \,\text{k}\Omega$, we can find, as before, the denormalized values of the filter capacitors according to the following formula:

$$C_i = M_C \times C_{in} = \frac{C_{in}}{K_R \cdot \text{FSF}} = \frac{C_{in}}{10^4 \times 21.362 \times 10^3} = 4.681 \times C_{in} \text{ nF}$$

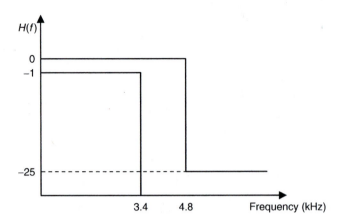

FIGURE 3.69 Tolerance structure for Example 3.16.

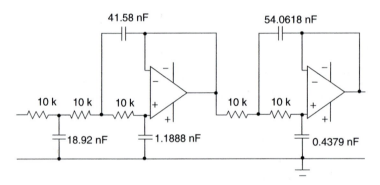

FIGURE 3.70 Circuit diagram of the filter in Example 3.16.

where M_C is the capacitance multiplier and C_{in} is the normalized value of the *i*th capacitor, giving

$$C_{11} = 41.58 \text{ nF} \quad C_{21} = 18.92 \text{ nF} \quad C_{31} = 1.1888 \text{ nF}$$

$$C_{12} = 54.0618 \text{ nF} \quad C_{22} = 0.4379 \text{ nF}$$

It is to be noted here that the third-order section can be located either at the beginning or at the end of the cascade. A circuit diagram and the frequency response of the designed filter are given in Figures 3.70 and 3.71. The shown oscillations along the pass band show increasing frequency as *f* approaches the pass band edge f_p at 3.4 kHz. The number of peaks and bottoms is five, like the number of poles. They all have equal amplitudes—hence the name "equiripple response." The decay rate within the transition band is much faster than that of a Butterworth filter having the same order, as said before.

EXAMPLE 3.17

Design a Chebyshev filter to meet the specifications described by the tolerance structure shown in Figure 3.72. Use simulation to prove that the designed filter satisfies

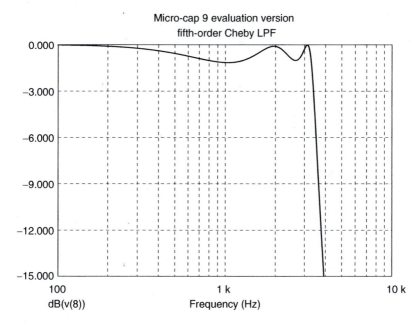

FIGURE 3.71 Frequency response of the filter in Example 3.16.

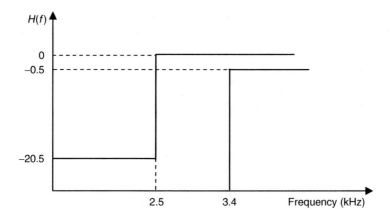

FIGURE 3.72 Tolerance structure for the filter of Example 3.17.

the requirements. Demonstrate the effect of the components' tolerance. It is enough to study the effect of incrementing one component.

Solution

The ripple factors are

$$\varepsilon_p = \sqrt{10^{0.05} - 1} = 0.3493$$

$$\varepsilon_r = \sqrt{10^{2.05} - 1} = 10.545$$

Giving a filter order of

$$N = \left\lceil \frac{\cosh^{-1}(\varepsilon_r / \varepsilon_p)}{\cosh^{-1}(f_p / f_r)} \right\rceil = \left\lceil \frac{4.1}{0.8249} \right\rceil = 5$$

The FSF is $2\pi \times 3.4 \times 10^3 = 21.363 \times 10^3$; thus, considering an impedance scale factor of

$$K_C = 10\ \text{nF} = 10^{-8}\ \text{F}$$

we get a resistor multiplier of

$$M_R = \frac{1}{K_C \times \text{FSF}} = 4.681 \times 10^3$$

The denormalized resistor values can then be calculated, assuming $Z_C = 10$ nF. After denormalizing the values obtained from the filter tables (Appendix A1) for order 5, we get

$$R_{11} = 684.155\ \Omega \quad R_{12} = 1.4112\ \text{k}\Omega \quad R_{13} = 15.4335\ \text{k}\Omega$$

$$R_{21} = 494.68\ \Omega \quad R_{22} = 40.9178\ \text{k}\Omega$$

A circuit diagram for the filter and its frequency response are illustrated in Figures 3.73 and 3.74, respectively.

To illustrate the effect of component tolerance on the response, resistor R_{31} is incremented by 500 Ω. The result is shown in Figure 3.75.

Nonunity Gain Second-Order Sections. The configuration shown in Figure 3.76 can provide a pass-band gain, due to the added resistors R_x and R_y. This could be an advantage if the input signal has a small amplitude.

Band-Pass and Band-Stop Filter Realization. As described earlier, BPFs are realized in the form of cascaded LP and HP sections. BSFs, on the other hand, are realized as a parallel combination of LP and HP sections. Both sections should have half the order of the required filter.

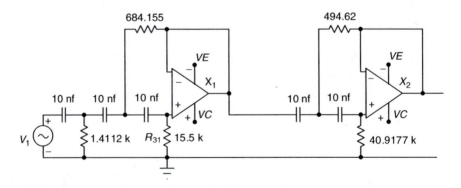

FIGURE 3.73 Circuit diagram of the filter in Example 3.17.

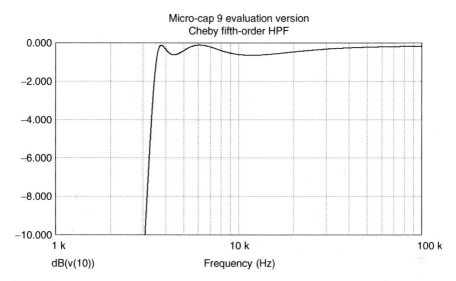

FIGURE 3.74 Frequency response of the filter in Example 3.17.

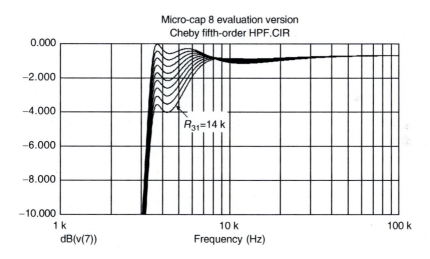

FIGURE 3.75 Effect of component's tolerance on the response of the filter in Example 3.17.

Band-Pass Filters. The frequency response of an ideal BP filter is shown in Figure 3.77. It could be viewed as the product of an HP response having a cut-off frequency of ω_1 and an LP response with a cut-off frequency at ω_2 provided that $\omega_2 > \omega_1$. Thus, a BPF can be constructed from a cascade of an LPF and an HPF whose respective cut-off frequencies are dictated by the required pass band.

As seen in Figure 3.79, the response of the filter reaches its peak at the central frequency ω_0. It then drops by 3 dB at the two frequencies ω_1 and ω_2, the 3 dB frequencies, or the corner frequencies. They define the pass band W of the filter, given by

$$W = \omega_2 - \omega_1 \tag{3.49}$$

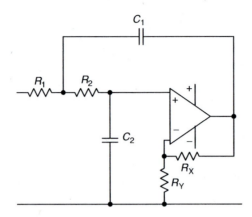

FIGURE 3.76   A nonunity gain second-order LPF.

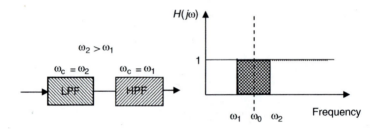

FIGURE 3.77   An ideal BPF and its frequency response.

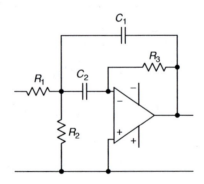

FIGURE 3.78   Circuit diagram for a second-order BPF.

The frequency ω_0 is the geometrical mean of the two corner frequencies; that is,

$$\omega_0 = \sqrt{\omega_1\omega_2} \tag{3.50}$$

The frequency response of a second-order BPF is illustrated in Figure 3.79. A peak is clear at 10 kHz (the central frequency), whereas the roll-off frequencies accrue at 8 and 12.5 kHz, which proves Equation 3.50. The slope of both sides of the response is seen to be 6 dB/octave.

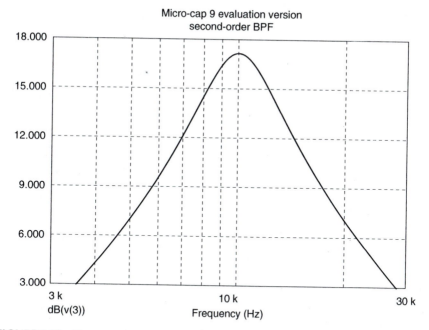

FIGURE 3.79 Frequency response of a second-order BPF.

EXERCISE 3.2

Derive the transfer function H(S) for the BPF shown in Figure 3.79.

Band-Stop Filters. Considering the response of a BSF diagramed in Figure 3.80a, it is easy to conclude that such a response can be obtained by connecting an LPF and an HPF in parallel. The only condition is that the cut-off of the high-pass section should be greater than that of the low-pass section. To do such a connection, we need a summing amplifier, as illustrated in the Figure 3.80b. Also, the same response is obtainable using a BPF, through the simple connection shown in Figure 3.80c. However, due to the expected delay difference between the two signal paths, such a connection should be limited to low-frequency applications. The following example highlights the design process for both types.

EXAMPLE 3.18

Using a filter catalogue, design a Butterworth BPF to satisfy the tolerance structure shown in Figure 3.81.

Solution

The ripple factors and the filter order are obtained from

$$\varepsilon_p = \sqrt{10^{0.301} - 1} = 1$$

$$\varepsilon_r = \sqrt{10^{1.5} - 1} = 5.53378$$

$$N \geq \frac{\log 5.53378}{\log 5/1.25} = \lceil 1.234 \rceil = 2$$

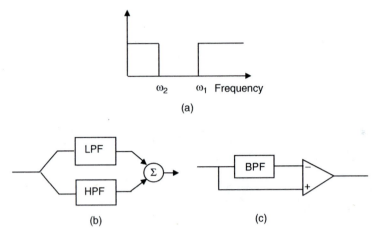

FIGURE 3.80 Frequency response of an ideal BSF and two possible realizations. (a) Ideal frequency response, (b) one possible realization of a BSF, (c) another possible realization for a BSF.

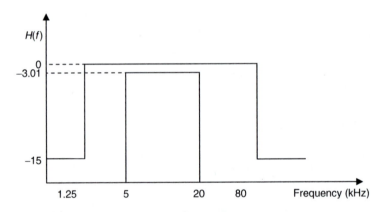

FIGURE 3.81   Tolerance structure for Example 3.18.

The filter will be realized in form of cascaded low-pass and high-pass sections; each has an order of 2.

The Low-Pass Section

The pass-band edge is

$$f_{pl} = 20\ \text{kHz}$$

Taking an impedance scale factor of $10^4\ \text{k}\Omega$, the impedance multiplier is then

$$M_C = \frac{1}{2\pi \times 20 \times 10^3 \times 10^4} = 0.79577 \times 10^{-9}$$

From the Butterworth tables (Appendix A1), after denormalization, we get

$$C_1 = 1.1252 \text{ nF}$$

$$C_2 = 0.5626 \text{ nF}$$

The High-Pass Section

The pass band edge is

$$f_{ph} = 5 \text{ kHz}$$

Assuming an impedance scale factor of 10^{-9}, the corresponding resistance multiplier is

$$M_R = \frac{1}{2\pi \times 510^3 \times 10^{-9}} = 3.1831 \times 10^3$$

Similarly, from Butterworth tables, we denormalized values of the resistors of

$$R_1 = 2.2504 \text{ k}\Omega$$

$$R_2 = 4.5 \text{ k}\Omega$$

The result of a simulation run using Micro Cap is shown in Figure 3.82.

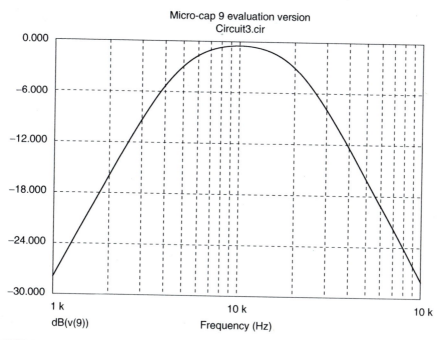

FIGURE 3.82 Frequency response of a fourth-order BPF of Example 3.18.

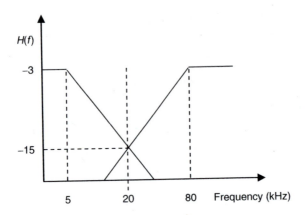

FIGURE 3.83 Specifications of the trap circuit in Example 3.19.

EXAMPLE 3.19

Design a maximally flat trap circuit to satisfy the shown magnitude characteristics (Figure 3.83).

Solution

The ripple factors are

$$\varepsilon_p = \sqrt{10^{A_p} - 1} = \sqrt{10^{0.3} - 1} = 1$$

$$\varepsilon_r = \sqrt{10^{A_r} - 1} = \sqrt{10^{2.4} - 1} = 15.8$$

$$\frac{\omega_n}{\omega_p} = \frac{20}{5} = 4$$

The filter order is

$$N = \left\lceil \frac{\log 15.8}{\log 4} \right\rceil = 2$$

The Low-Pass Section. Assume an impedance scale factor for the resistance of

$$K_R = 10 \text{ k}\Omega$$

and

$$\text{FSF} = 2 \times \pi \times 5 \times 10^3 = 3.14 \times 10^4$$

From the table of Butterworth filters, we get

$$C_{1n} = 1.414, C_{2n} = 0.7071$$

The denormalized capacitances are

$$C_1 = \frac{1.414}{10^4 \times 3.14 \times 10^4} = 4.5 \text{ nF}$$

And similarly,

$$C_2 = 2.25 \text{ nF}$$

The High-Pass Section. Assume $K_C = 1$ nF and

$$SF = 2 \times \pi \times 80 \times 10^3$$

The denormalized resistances are

$$R_1 = \frac{1}{1.414 \times 10^{-9} \times 16 \times \pi \times 10^4} = 1.406 \text{ k}\Omega$$

Similarly,

$$R_2 = 2.813 \text{ k}\Omega$$

To get the required function, the two sections have to be connected in parallel, as shown in Figure 3.84. The obtained frequency response is shown in Figure 3.85.

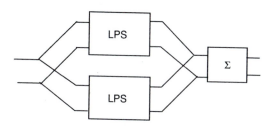

FIGURE 3.84   BSF as a parallel connection of a high-pass section (HPS) and a low-pass section (LPS).

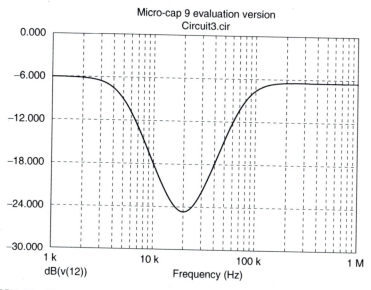

FIGURE 3.85   Frequency response of a fourth-order notch of Example 3.19.

From the figure, it is clear that the notch frequency is 20 kHz, as required, and the slope of any segment is 12 dB/octave, indicating an order of 2.

3.8 SWITCHED CAPACITOR FILTERS

Components used in the construction of active RC filters, especially low-frequency filters, are usually large-valued components. They are therefore uneconomical to produce in monolithic form, as the required silicon area becomes excessive. Instead, they are manufactured as hybrid integrated circuits, where monolithic operational amplifiers—"dies"—are mounted on an insulating substrate (alumina, glass, and so on) together with other elements (resistors and capacitors) that are either deposited under vacuum from the gas phase (*thin film technique*) or screen-printed (*thick film technique*) onto the substrate and then interconnected. The attainable component's precision is usually very high, due to the possibility of trimming each component individually through laser beams. However, the cost and the physical size is of such filters is not to be compared with those of "fully" monolithic circuits. Therefore, technique was introduced in the late 1970s to provide monolithic circuits. The resulting filters are called *switched-capacitor filters*. The technique is based on replacing each resistor in an RC realization by a metal oxide semiconductor (MOS) capacitor and two MOS switches. Because MOS capacitors and MOS switches are easy to manufacture and occupy a small area on the silicon die, the required silicon area would be relatively small and independent of the resistors' values. Figure 3.85 illustrates a resistor and its switched-capacitor equivalent. The two switches shown are controlled by a two-phase nonoverlapping clock operating at a frequency that is several times larger that the cut-off frequency of the filter to avoid errors. During period T_1, capacitor C charges to attain the input voltage. It then delivers this voltage to the output during period T_2. Figure 3.86 illustrates a resistor and its switched-resistor equivalent. To prove the claimed equivalency of the two circuits in this figure, we can conclude that for circuit (b), the charge Q accumulated on the capacitor when switch S_1 is closed is

$$Q = C \cdot V = I \cdot T$$

where $T = 1/f$ is the clock period.

$$\text{Thus } R = \frac{V}{I} = \frac{T}{C} = \frac{1}{f.C}.$$

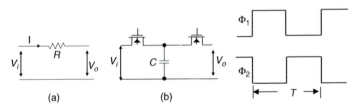

(a)   (b)

FIGURE 3.86   A switched-capacitance equivalent of a resistor.

This equation indicates that a capacitor and two switches that close sequentially can replace a series resistor. Applications of this technique in filter realization are very promising, in view of the rapid development in MOS technology. It is now possible to produce high-quality (low-loss) capacitors and efficient analog switches on monolithic chips at a high packing density and a good yield[1]. Many monolithic programmable switched capacitor filters are now available on the market. An example is the MF10 series from National Semiconductor that is produced in form of a 20-pin DIL package. It includes two independent blocks that could be programmed to provide filters of any type up to fourth-order at frequencies up to 30 kHz. Higher-order filters are obtainable by cascading different modules.

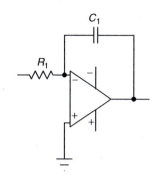

FIGURE 3.87   A simple integrator.

The preceding simple switched-resistor equivalent of a resistor was found to be sensitive to stray capacitances. Because capacitor C is in parallel with any stray capacitance thus they will add up, leading to imprecision. To overcome such deficiency, a different connection is suggested in which four rather than two switches are used and are driven by two-phase nonoverlapping clock. Two modes of operations are possible: inverting and noninverting modes. However, the decision of which one to use is application-dependent. Let us consider the simple integrator depicted in Figure 3.87; the possible modes are illustrated in Figure 3.88.

Operating in the noninverting mode (Figure 3.88a) and during phase I, capacitor C_1 is charged from the input source via the closed switches. During phase II, capacitor C_2 is charged through C_1 via the other pair of switches. Owing to the direction of the current, C_2 is charged such that the output has the same polarity of the input. The opposite takes place in the second connection, shown in Figure 3.88b, where C_1 and C_2 are charged together, leaving an opposite polarity at the output.

3.9   SUMMARY

In this chapter, the different design techniques of analog filters were described in detail. The many design examples allow the reader to master such important design tools. In addition, this chapter might later also facilitate the understanding of the design techniques of their digital counterparts. The chapter started by reviewing all types of filter functions, and giving their describing equations as well as their pole-zero distributions. Design procedures of filters were then given in some detail. Specifications of filters, as summarized by their tolerance structures, were provided. A method was described for deriving a filter's transfer functions by employing different approximations through their associated pole-zero distributions. Analog filter realizations in their chronological order, starting with LC-realization to switched capacitor realizations, were then presented. Numerous design examples were provided, as well as review questions and design problems.

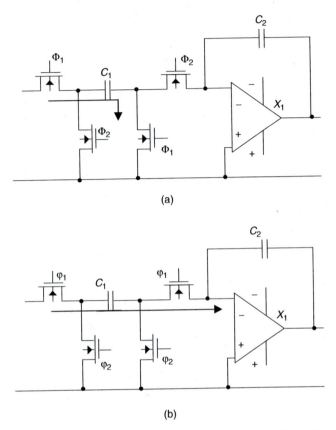

(a)

(b)

FIGURE 3.88 Modes of operation of a switched-capacitor equivalent of an integrator: (a) noninverting mode, (b) inverting mode.

3.10 REVIEW QUESTIONS

1. Explain why *LC* filters, in spite of their superior performance, are not recommended in the realization of modern electronic systems.
2. Poles and zeros are important filter parameters. Which of them is responsible for determining the filter function (LP, HP, BP, or BS)?
3. What are the most important specifications of a gyrator? Give one realization for a high-performance one.
4. Which filter approximation would you select for each of the following cases:
   - Minimum amplitude distortion
   - Minimum filter order
   - Reasonable filter order with allowed ripples in the pass band
   - Linear phase and constant group delay
5. Using an LPF and an HPF, draw a block diagram describing how to realize a fourth-order BP filter having a pass band that extends from ω_1 to ω_2, then repeat for a BS filter that has the same order and the same BW of the BP one.
6. What is meant by (a) pole quality and (b) pole sensitivity?

7. Using sketches, show how you can find out the order of an LPF that uses the following approximations:
   a. Chebyshev I
   b. Cauer

8. Sketch the pole-zero distribution for the following fourth-order Butterworth filter: (a) an LP filter, (b) an HP filter, (c) a BP filter, (d) an all-pass filter, (e) an LP notch, (f) a symmetrical notch and (g) an HP notch.

9. Starting with the pole-zero distribution, deduce the transfer function of a fourth-order Chebyshev filter that has a band-pass ripple index of ε_p.

3.11 PROBLEMS

1. The pole-zero plot of a certain Butterworth filter is shown in the following figure. Design the filter in the form of Sallen and Key sections.

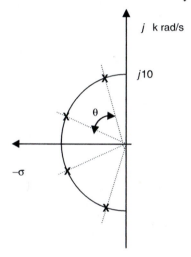

2. Using tables, design a Chebyshev filter to meet the specifications described by the shown tolerance structure. Draw your circuit diagram, indicating the components' values.

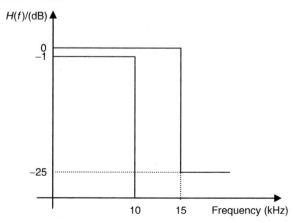

3. Design a Butterworth filter to satisfy the shown tolerance structure. Realize the filter in form of cascaded Sallen and Key sections. Verify the compliance of your design with the requirements for employing computer simulation.

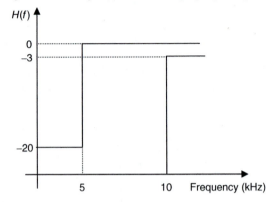

4. Using tables, design a Chebyshev filter to satisfy the requirements set by the shown tolerance structure as a cascade of Sallen and Key sections.

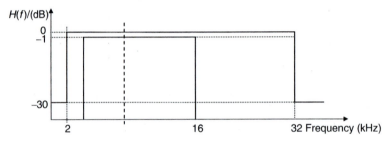

5. Using tables, design a minimum-order *RC* active filter to satisfy the shown tolerance structure. Draw your circuit, giving the values of the components using standard values of the series (1, 1.2, 2.2, 3.3, 4.7, 5.6, 8.2). Use computer simulation to study the effect of rounding the components' values to nearest standard values found on the market. (*Hint:* Use the model of an ideal operational amplifier.)

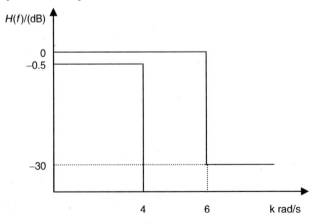

6. Design a filter with a maximally flat magnitude response within the shown tolerance structure. Use computer simulation to verify that the requirements are satisfied.

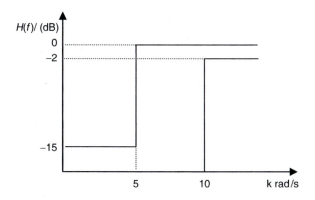

7. Design an RC active filter to satisfy the given tolerance structure as a cascade of Sallen and Key sections using a maximally flat approximation.

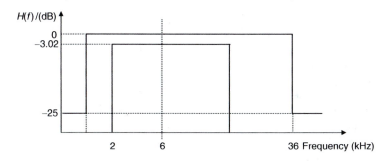

8. Design a Chebyshev RC active filter to satisfy the shown requirements. Then give its switched-capacitor equivalent.

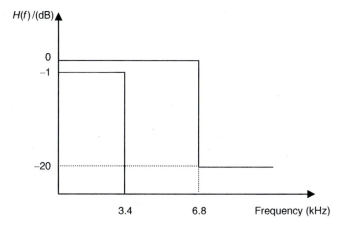

9. Using *RC* active circuits, design a 10 kHz trap to have a maximally flat magnitude response.

10. The following figure shows the tolerance structure of a certain BSF. Design the filter using Butterworth approximation and realize it in the form of *RC* active sections.

Hints:
- The relation $f_o^2 = f_1 \cdot f_2$ holds at any attenuation level.
- Use paralleled sections.

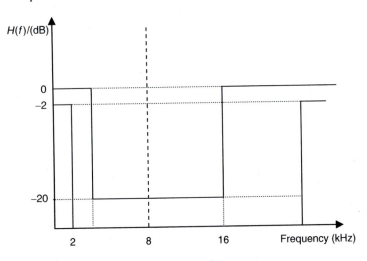

11. To get sound with excellent reproduction quality, low frequencies should be directed to *woofers* (special loudspeakers for low frequencies) and high frequencies to *tweeters* (special loudspeakers for high frequencies). To do this, audio frequency signals should be introduced, in parallel, to audio splitters. A splitter is actually composed of two complimentary parallel-connected filters: an LPF that leads to the woofers and an HPF that is connected to the tweeters. Design such a network to have intersecting responses of the LPF and the HPF at 4 kHz (cross-over frequency) and an attenuation level of −30 dB. Assume a steepness ratio of 2. Draw your circuit diagram.

3.12   MINI PROJECT

Design and construct an LPF to have a cut-off frequency of 3.4 kHz and a transition band that is 1 kHz wide. The minimum acceptable stop band attenuation is 20 dB using LM741 operational amplifiers. Measure the frequency response and verify the results using any circuit simulator (Pspiece, Microcap, EWB, and so on).

4 Data Converters

4.1 INTRODUCTION

The recent tremendous advances in integrated circuit technology have made it possible to produce cheap, high-speed, reliable digital circuits. The drastically reduced feature size (now deep submicron) of very large-scale integrated (VLSI) circuits has resulted in a proportional reduction in the area of silicon dies, which operate at low voltages and consume little power. The degree of integration (packing density) has greatly increased, to the extent that a complete digital system can now be manufactured on a single chip. Processing of signals in digital format has therefore become feasible and even more economical than before. Because the majority of the signals we deal with—for example, audio, video, biomedical, and telemetry signals—are analog by nature, there has been a need for interfacing devices that can transform (convert) such signals from an analog format into a digital format, and vice versa, with minimal loss of information or added noise. The many advantages of digitally processing these signals can justify the resulting hardware, cost, and delay overheads. In this chapter, we are going to study a family of devices called *data converters* that includes ADCs and DACs. Preprocessing operations like band limiting and sampling, which an analog signal should undergo prior to conversion, are first described. The process of quantization and the different quantization techniques are first considered. Due to their simplicity, the most common realizations of DACs are then described. Circuit implementations of the different quantization techniques and ADC techniques are considered in detail, along with some design examples. As the speed and precision requirements of ADCs increase, so does the complexity of their circuit realizations. Therefore, some examples of economical architectures that feature high speed and hardware simplicity are described. Several standard qualitative and quantitative tests that are usually used to judge the performance of data converters are then presented.

4.2 TYPICAL DIGITAL SIGNAL PROCESSING SYSTEM

Digital processing techniques can be applied to analog signals if they are converted into digital format. However, this implies that they should undergo some preprocessing operations to allow the reconstruction of the signal after processing with minimum distortion. As depicted in Figure 4.1, such preprocessing operations include:

1. Band limiting to a frequency f_m in an LPF I
2. Sampling the signal periodically at a specified rate f_s
3. Holding the acquired instantaneous sample amplitudes for a specified period

The digital output (n bits per sample) of the ADC is processed in the block labeled DSP (digital signal processing) before being delivered to the output. The resulting digital output is then DAC-converted and smoothed in an LPF II. The design and construction of such building blocks of the system will be considered in some details in the following section.

4.3   SPECIFICATIONS OF DATA CONVERTERS

Data converters include devices that perform the forward operation of analog-to-digital conversion in which analog signals are converted into strings of digital data. They also include devices that perform the reverse process; that is, digital-to-analog conversion, in which digital data are used to reconstruct the original analog signals. The specifications of data converters can be grouped into:

- Specifications related to input
- Specifications related to internal structure
- Specifications related to output

Because the two operations of ADC and DAC are complementary, the set of specifications describing the input for one of them is equivalent to the set of specifications describing the output of the other and vice versa. Table 4.1 summarizes the most important specifications of ADCs and DACs.

FIGURE 4.1   A typical digital signal processing (DSP) system.

TABLE 4.1
Summary of Important Specifications of Data Converters

Specification	Units	ADC	DAC
FSR	Volts	Input	Output
Bandwidth (f_m)	Hertz	Input	Output
Polarity (unipolar/bipolar)	$\pm V_m$	Input	Output
Resolution (word length)	Bits	Output	Input
Bit format (serial/parallel)		Output	Input
Logic (TTL, ECL, and CMOS)		Output	Input
Speed	Seconds	Conversion time	Settling time
		• Low-speed	
		• Medium-speed	
		• High-speed	
		• Ultra-high-speed	
Hardware complexity		Simple	Simple
		Moderate	Moderate
		Complex	Complex
Power supplies	$\pm E$ Volts	Single/multiple	Single/multiple

4.3.1 SPECIFICATIONS RELATED TO THE ANALOG INPUT/OUTPUT

As it is a measuring device, an ADC has the following input specifications:

- The signal full-scale range (FSR)
- The input signal bandwidth B represented by its maximum significant frequency f_m
- The polarity of the input signal; that is, whether it is bipolar (ranging between $\pm V_m$) or unipolar (ranging from zero to $+V_m$ or from $-V_m$ to zero)

These same specifications are applicable to the output of a DAC.

4.3.2 SPECIFICATIONS RELATED TO THE INTERNAL STRUCTURE

Main specifications that are related to or describe the internal structure are:

- Hardware complexity (number of components)
- Conversion time (conversion speed or throughput rate)
- Number of power supplies needed

The number of required comparators for an ADC structure is an important factor in determining its hardware complexity, and consequently the required silicon area when produced in monolithic form. Directly related to circuit complexity, in addition the cost, is the power it dissipates, which in turn is a function of the power supply voltage(s). The conversion speed is rather an important parameter of ADCs, if not the most important one, as it determines the type of signals the ADC can handle. Modern ADC designs tend to minimize hardware requirements, reducing operating voltages, and hence reducing power dissipation, while operating at the maximum possible speed. This trend has led to the development of what is called "system on chip," where a complete electronic system containing signal-processing components together with all interfacing devices (e.g., the ADC and DAC) are manufactured on a single silicon die.

4.3.3 SPECIFICATIONS RELATED TO THE DIGITAL OUTPUT/INPUT

The following specifications are the most important parameters that describe the digital output of an ADC and the digital input to a DAC:

- Word length (n bits)
- Bit format (serial or parallel)
- Output rate
- Logic level (TTL, CMOS, ECL, and so on)

The *word length*, also called the *resolution* or *precision*, is usually application-dependant. Common resolutions for commercial video signals range from 6 to 8 bits. Audio signals, on the other hand, should be digitized at a resolution that varies between 12 and 24 bits, depending on the minimum acceptable SNR. The format of the output (i.e., whether serial or parallel), as well the voltage levels it assumes (the logic level), are rather important factors in determining a device's compatibility with succeeding devices—that is, devices it can talk with.

4.4   SAMPLING

Representative periodic samples have to be taken from an analog signal before being digitized. The rate at which sampling is carried out is critical, as it determines the possibility of reconstructing a signal from its samples without any loss of information. Sampling is therefore considered a fundamental process prior to digitization. The higher the sampling rate, the better the chances to reconstruct a near replica of the original signal. The minimum acceptable sampling rate (Nyquist rate), as given by the sampling theorem is

$$f_s \geq 2f_m \tag{4.1}$$

where f_m is the maximum significant frequency (bandwidth) of the input signal. The frequency spectrum of a signal before and after sampling is shown in Figure 4.2. It is clear that the spectrum of a sampled signal repeats itself around multiples of f_s; thus, if the previous condition is not satisfied, lower band components of a certain spectrum intercept (overlap) with the higher bands of a preceding one, resulting in a sort of unrecoverable (irreversible) distortion called *aliasing* (shown hatched). Consequently, the reconstructed signal spectrum (base band) will contain some components of the next spectrum. The degree of aliasing depends on how much of the sampling frequency is selected relative to the maximum significant frequency in the base band signal. Therefore, in order to get tolerable amounts of aliasing, it is necessary to sample the signal (with the assumed band limited to f_m) at a frequency f_s such that $f_s \geq 2f_m$. However, the sampling rate should be carefully selected to avoid unnecessarily high data rates.

The frequency f_m of a certain signal is determined by studying its spectrum to determine those components that contribute effectively to the spectrum. Frequency components that are higher than the maximum significant frequency f_m can be attenuated without noticeable effects on the quality of the signal. This would enable us to select the proper sampling rate to either eliminate aliasing completely (by using a sharp cut-off filter) or keep it at minimum, as illustrated in Figure 4.3. Signals that are sampled at a rate lower than $2f_m$ are said to be *undersampled*. They suffer, as said

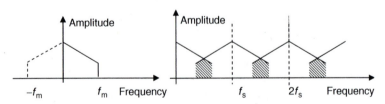

Frequency spectrum of a signal   Aliasing due to insufficient sampling rate

FIGURE 4.2   Frequency spectrum of an analog signal before and after sampling.

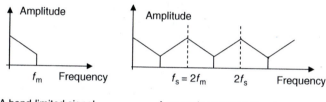

A band-limited signal   A properly sampled signal

FIGURE 4.3   Spectrum of a properly sampled signal.

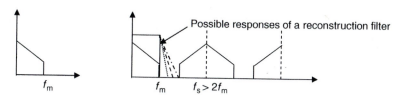

FIGURE 4.4   Effect of oversampling on the reconstruction filter.

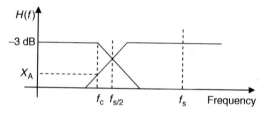

FIGURE 4.5   Aliasing errors in Example 4.1.

before, from irreversible aliasing errors. On the other hand, sampled signals at a rate higher than $2f_m$ are *oversampled*. As demonstrated in Figure 4.4, such signals suffer no aliasing errors; they also relax the conditions imposed on the steepness of the required reconstruction filter response. Oversampled and properly sampled signals can be perfectly reconstructed from their samples through an LPF, as shown in Figure 4.4. In fact, this filter is identical to the front end antialiasing filter LPF I of Figure 4.1.

EXAMPLE 4.1

What are the causes of aliasing errors? Find an expression for the percentage aliasing in a sampled signal as a function of the sampling frequency f_s at the 3 dB frequency (Figure 4.5).

Solution

Aliasing errors are caused by

1. Improper band limiting
2. Insufficient sampling rate

The percentage aliasing $A\%$ is usually measured at the 3 dB frequency f_c, as shown in Figure 4.5. Assuming an Nth order Butterworth filter and a wide-band input signal, the attenuation at any arbitrary frequency f is given by

$$H(f) = \frac{1}{\sqrt{1 + (f/f_c)^{2N}}}$$

The percentage aliasing level A is defined as

$$\%A = \frac{H_A(f_c)}{H(f_c)}$$

where H_A is given by

$$H_A(f_c) = \frac{1}{\sqrt{1 + ((f_s - f_c)/f_c)^{2N}}}$$

Because at $f = f_c$ it is valid that

$$H(f_c) = \frac{1}{\sqrt{2}}$$

then

$$\%A = \frac{\sqrt{2}}{\sqrt{1 + ((f_s - f_c)/f_c)^{2N}}} \times 100 \qquad (4.2)$$

From the previous expression, it is clear that increasing the order N of the band-limiting filter (antialiasing filter) reduces the aliasing level, or that alternatively, for the same aliasing level, it allows the reduction of the sampling frequency.

EXAMPLE 4.2

Write an equation relating the required sampling frequency and the order N of the antialiasing filter, and then plot it for values of $N = 1, 2, 3, 4$, allowing a 1% aliasing error.

Solution

Using the expression obtained in Example 4.1, we can write

$$\sqrt{1 + \left(\frac{f_s - f_c}{f_c}\right)^{2N}} = \frac{\sqrt{2}}{A}$$

Squaring, then taking the Nth root yields

$$\frac{f_s}{f_c} - 1 = \left(\frac{2}{A^2} - 1\right)^{1/2N} \approx \left(\frac{\sqrt{2}}{A}\right)^{1/N}$$

$$f_s \approx f_c \left[\left(\frac{\sqrt{2}}{A}\right)^{1/N} + 1\right]$$

for an aliasing level of $A = 0.01$, we get

$$f_s \approx f_c \left[\left(\frac{\sqrt{2}}{0.01}\right)^{1/N} + 1\right] = f_c \left[(141.4)^{1/N} + 1\right]$$

Thus, trying several values for N, we can construct Table 4.2.

A plot of this equation for a cut-off frequency of 3.4 kHz is given in Figure 4.6. It shows that increasing the order of the antialiasing filter results in a drastic reduction in the required sampling frequency for a specified percentage aliasing level.

TABLE 4.2

Effect of Filter's Order N on f_s

N	1	2	3	4	5	6	7
f_s	$142.4 f_c$	$12.89 f_c$	$6.21 f_c$	$4.448 f_c$	$3.692 f_c$	$3.282 f_c$	$3.028 f_c$

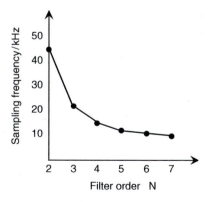

FIGURE 4.6   Effect of increasing N on the sampling frequency.

EXAMPLE 4.3

Calculate the incurred percentage aliasing error in a pulse code modulation (PCM) system that uses a seventh order Butterworth antialiasing filter and a sampling frequency of 12 kHz.

Solution

In a PCM system, speech signals are band limited to 3.4 kHz. Thus, substituting in Equation 4.2 gives

$$\%A = \frac{\sqrt{2}}{\sqrt{1 + ((f_s - f_c)/f_c)^{2N}}} \times 100$$

$$= \sqrt{\frac{2}{1 + (12 - 3.4/3.4)^{14}}} \times 100 = 0.21349\%$$

4.5   SAMPLE AND HOLD CIRCUITS

Sampling a certain analog signal is performed using a switch that closes periodically for a small duration to look at the signal. The switch is controlled by a sampling signal (the sampling command [SC]) at a certain specified repetition frequency f_s (the sampling rate). Because analog signals usually represent phenomena that change continuously, an analog storage device (usually a low-loss capacitor) is needed to hold the acquired amplitude of the signal and keep it constant (hold or freeze it) for the rest of the sampling period T_S. The closure time T_A (acquisition time) of the switch should be selected long enough to allow the hold capacitor to fully charge to the instantaneous amplitude of the sample. The switch is then open during the hold period T_H. The resulting circuit, shown in Figure 4.7, is known as "sample and hold" circuit or simply "S/H."

For this circuit, the following relation is valid:

$$T_S = T_A + T_H$$

where $T_S = 1/f_s$ is the sampling period, T_A is the acquisition period, and T_H is the hold period. As said before, the duration of the period T_A should be long enough to allow C_H to charge the instantaneous amplitude of the sample. However, it should not be excessive, in order to leave enough time for the succeeding ADC to complete its job. Recalling that for a charging capacitor to attain a level that is nearest to the steady state amplitude, we should allow a time of

$$T_{ch} \geq 4\tau_{ch}$$

where τ_{ch} is the time constant of the charging path. However, as a rule of thumb, it is best to let

$$T_A = 7 \cdot \tau_{ch} = 7 \cdot C_H \sum R \qquad (4.3)$$

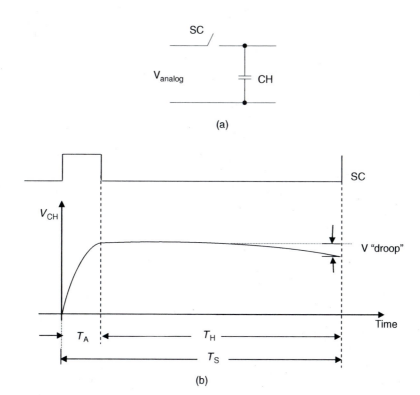

FIGURE 4.7   A basic S/H circuit. (a) A simple circuit and (b) voltage relations.

where ΣR represents the total resistance in the charging path of C_H. This value usually comprises the internal resistance of the analog signal source (the LPF) R_s and the switch-on resistance R_{ON}; that is,

$$\sum R = R_S + R_{ON}$$

Low-loss capacitors are usually employed to ensure minimum leakage of charge during the hold period. Table 4.3 gives examples of some usable capacitor types classified according to the relative permittivities of their dielectric materials.

S/H circuits usually feed finite input impedance ADCs. With the dielectric loss in the capacitor, this would result in a slight drop or "droop" in the voltage stored on the hold capacitor. The droop is measured by the droop rate, defined as follows:

$$\text{Droop rate} = \frac{\Delta V}{T_H} \approx \frac{\Delta V}{T_S} \quad \text{for} \quad T_A \ll T_H \tag{4.4}$$

To reduce the acquisition time, the charging time constant should be kept at minimum. This is possible by choosing switches that have low ON resistance. To get a

TABLE 4.3
Relative Permittivity of Some Dielectric Materials

Dielectric	ε_r
Polystyrene	2
Teflon	2
Silicon dioxide (S_iO_2)	4
Glass	3.7–10
Ceramic	10
Tantalum pentaoxide (Ta_2O_5)	20–25

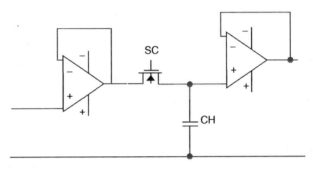

FIGURE 4.8 A practical S/H circuit.

small source resistance, an isolating (buffer) amplifier should precede the switch. Furthermore, to minimize the droop, a second buffer is connected next to the hold capacitor, so as to provide high shunting impedance and hence a large discharge time constant for the capacitor. This arrangement results in the circuit shown in Figure 4.8, with a MOS transistor switch. S/H circuits of such a kind are found on the market in integrated circuit form with the capacitor terminals left open for the user to connect a capacitor of choice. Since on-chip capacitors usually occupy a large silicon area.

The size of the hold capacitor C_H is determined by the amount of tolerable droop rate. It is calculated from the lost charge ΔQ on the capacitor, taking all sources of losses into account. Assuming here that the input bias current of the second Op. Amp. is the most influential one, we can write

$$\Delta Q = C_H \cdot \Delta V = I_{Bias} \cdot T_H \tag{4.5}$$

giving

$$C_H = I_{Bias} \cdot \frac{T_H}{\Delta V} = \frac{I_{Bias}}{\text{droop rate}}$$

EXAMPLE 4.4

Design a S/H circuit to operate at 1 MHz with a duty cycle of 1:10. The permissible droop rate is 0.01 V/s given that the available Op. Amp. draws a bias current of 0.1 nA. Suggest a suitable type for the hold capacitor.

Solution

Given the sampling period $T_S = 1/f_s = 1\,\mu s$, the duty cycle is defined by the ratio

$$\frac{T_A}{T_S} = 0.1$$

then

$$T_A = 100\,\text{ns} \quad \text{and} \quad T_H = 900\,\text{ns}$$

The value of the hold capacitor is obtained, as indicated previously, from

$$C_H = I_{Bias} \cdot \frac{T_H}{\Delta V} = \frac{I_{Bias}}{\text{droop rate}} = \frac{10^{-10}}{0.01} = 10\,\text{nF}$$

A ceramic capacitor is recommended for this case.

4.5.1 SOURCE FOLLOWER S/H CIRCUIT

A rather simple S/H circuit is shown in Figure 4.9. It comprises 3 MOS transistors Q_1, Q_2, and Q_3. While Q_1 is acting as a switch, Q_2 functions as a source follower with Q_3 as its load. The signal SC is used to control the switch. Transistor Q_3 is controlled by the compliment of SC; that is, $\overline{SC}$. The operation of the circuit can be summarized as follows: During the period T_A, switch Q_1 closes, causing C_H to start charging to the signal amplitude while Q_2 is off, thereby providing high impedance at the output. As SC goes low, switch Q_1 turns off, while Q_2 switches on to act as a source follower, thus transferring the frozen voltage on C_H to the output.

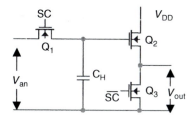

FIGURE 4.9 Source follower S/H circuit.

4.6 QUANTIZATION

After sampling, which could be considered as a time quantization process, analog signals are then quantized in amplitude. In this essential process, the subrange of the FSR to which the input sample belongs is determined. Giving each subrange a code—for example, binary, gray; two's compliment, and so on—one can get the digital equivalent of the sample amplitude. For an n-bit quantizer, the FSR is divided into 2^n equal quantization levels. The size of each step (quantum) Q is then given by

$$Q = \frac{FSR}{2^n} \tag{4.6}$$

4.6.1 QUANTIZER TRANSFER CHARACTERISTIC

The transfer characteristic of a 3-bit quantizer is shown in Figure 4.10. The dotted line represents the transfer characteristic of an ideal quantizer with a resolution $n = \infty$ bits. This means that "each possible analog input amplitude has a corresponding code word." A practical quantizer, however, should have a finite number of levels, and hence the figure shows the solid discontinuous (stepped) characteristic. It is easy to discover from the figure that all input amplitudes (ranging from zero to slightly before Q) are assigned the code 000, which means an increasing amplitude of error as the input increases from zero to Q. At an input-sample amplitude of

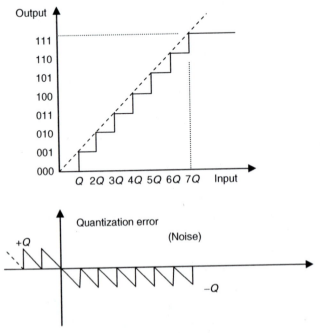

FIGURE 4.10 Transfer characteristic of a 3-bit quantizer.

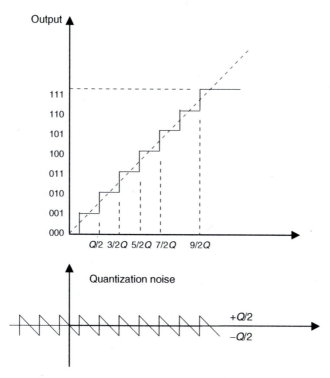

FIGURE 4.11 Midtread transfer characteristics of a quantizer and its associated noise.

exactly Q, the error drops to zero and starts to increase as V_i increases again until it reaches $2Q$. Such an irreversible sawtooth-like error is called the *quantization error* or, more accurately, the *quantization noise*. It is evident from the figure that the peak amplitude of the noise is given by the quantum size and hence by the resolution n. Alternatively, allowing the transitions to occur at points that are midway within each subrange—that is, at $Q/2$, $3Q/2$, $5Q/2$, and so on—would turn the noise into a bipolar one, while keeping the maximum error at $\pm Q/2$ rather than from zero to Q. Doing so would result in a reduction of the noise power to one fourth. The resulting characteristic (usually known as the *midtread* characteristic) is illustrated in Figure 4.11.

EXAMPLE 4.5

Calculate the noise power resulting from a midtread quantizer as compared to an ordinary quantizer.

Solution

The noise power dissipated in a resistor R is given by

$$P_n = \frac{V_{rms}^2}{R}$$

The resulting noise in both cases, as shown before, is a sawtooth wave. Considering each case separately, we get for an ordinary quantizer,

$$V_m = Q$$

$$V_{rms} = \frac{V_m}{\sqrt{3}} = \frac{Q}{\sqrt{3}}$$

giving

$$P_{n1} = \frac{Q^2}{3R}$$

And for a midtread quantizer,

$$V_m = \frac{Q}{2}$$

$$V_{rms} = \frac{Q}{2\sqrt{3}}$$

giving

$$P_{n2} = \left(\frac{Q}{2\sqrt{3}}\right)^2 \cdot \frac{1}{R} = \frac{Q^2}{12R}$$

Comparing the noise powers resulting from each quantizer, one can conclude that the noise power due to the midtread case is one fourth that of the first case.

EXAMPLE 4.6

Calculate the resulting SNR due to an n-bit midtread quantizer for

a. A triangular input signal
b. A sinusoidal input signal

Solution

The noise power as calculated in the previous example was

$$P_n = \frac{Q^2}{12R}$$

The quantum size is

$$Q = \frac{FSR}{2^n} = \frac{FSR}{m}$$

that is,

$$FSR = m \cdot Q$$

Data Converters

a. For a bipolar triangular input that changes from $+mQ/2$ to $-mQ/2$, we have

$$V_{rms} = \frac{mQ/2}{\sqrt{3}}$$

The signal power is

$$P_S = \frac{V_{rms}^2}{R} = \frac{m^2Q^2}{12R}$$

giving an SNR of

$$SNR = 10\log\frac{P_S}{P_n} = 10\log\frac{m^2Q^2}{12R}\cdot\frac{12R}{Q^2} = 10\log m^2$$

$$= 20\log 2^n = 20n\cdot\log 2 = 6n \text{ dB}$$

b. For a sinusoidal input having an amplitude of $\pm mQ/2$,

$$V_{rms} = \frac{mQ/2}{\sqrt{2}} = \frac{mQ}{2\sqrt{2}}$$

$$P_S = \frac{V_{rms}^2}{R} = \frac{m^2Q^2}{8R}$$

The SNR in this case is

$$SNR = 10\log\frac{m^2Q^2}{8R}\cdot\frac{12R}{Q^2} = 10\log\frac{12}{8}m^2 = 6n + 1.76 \text{ dB} \qquad (4.7)$$

4.6.2 NONLINEAR QUANTIZATION

In some applications, as in telephone communications, it is mandatory to keep a constant SNR for all subscribers, irrespective of the level of their sound. This is especially important for telephone subscribers who speak in very low volumes, as their speech could be masked by the quantization noise. To achieve this, a variable rather than a constant quantum size is used, such that low amplitudes are assigned small quantum sizes (i.e., higher resolution), while large signal amplitudes are given larger quantum size (i.e., lower resolution). To do this, the signal should first be "compressed" before being quantized. At the receiving end, however, the signal is then expanded to regain its original relative amplitudes. The whole process is called *companding*, a synthetic term combining "compression" and "expansion." A block diagram illustrating such a process is given in Figure 4.12. This process would result in almost a constant SNR over the whole dynamic range giving what is called "robust quantization." Moreover, as we shall see later, the obtainable overall SNR is higher than that obtainable with linear quantization for the same resolution.

FIGURE 4.12   Block diagram illustrating the process of companding.

The required nonlinear quantizer transfer characteristic is mostly selected to be logarithmic and is realized in a piecewise approximation fashion. Two companding characteristics have been established as industry standards. In the first one, the μ-law, suggested by Bell Labs, the transfer characteristic is described by:

$$v_0 = \frac{\ln(1 + \mu v_i)}{\ln(1 + \mu)} \tag{4.8}$$

where μ is a parameter that determines the slope of the function at $v_i = 0$. Differentiating with respect to v_i, we get

$$\left.\frac{dv_0}{dv_i}\right|_{v_i = 0} = v_0' = \frac{\mu}{\ln(1 + \mu)}$$

Due to companding, the achievable SNRs usually correspond to higher word lengths if linear quantization is used. In other words, for the same word length in both cases, a higher SNR is obtainable. The difference is known as the *companding gain* (CG):

$$CG = SNR|_{comp} - SNR|_{linear} \tag{4.9}$$

Such a value is more or less related to the slope of the quantizer's transfer characteristic. Thus for linear quantization with a slope = 1, the gain is zero, which reaches 6 dB if the slope is doubled. In an early realization of the μ-law, "the T_1 system" μ was chosen to be 100, giving a slope of the quantization characteristic of 21.7. This value leads to a CG of

$$CG = 20\log 21.7 = 26.7 \text{ dB}$$

Recalling that the SNR due to quantization is calculated from

$$SNR = 6n$$

the CG corresponds to an apparent improvement in the resolution of

$$\frac{26.7}{6} \approx 4 \text{ bits}$$

which means that an 8-bit companded quantizer would have the noise performance of a 12-bit one. In a second version, the T_2 system, and the constant μ is increased to 225, giving a slope of 47 and a CG of 33 dB, which corresponds to an increase in word length of 5 bits—that is, the quality of 13-bit.

The European Conference of Postal and Telecommunications Administrations (CEPT) has adopted a different companding characteristic, the "A-law," described by

$$v_0 = \frac{Av_i}{1 + \ln A} \qquad \text{for} \quad v_i < \frac{1}{A}$$

$$= \frac{1 + \ln(Av_i)}{1 + \ln A} \qquad \text{for} \quad v_i > \frac{1}{A} \qquad (4.10)$$

Thus, for $A = 87.6$, the slope at the origin is

$$v_0'\big|_{v_i=0} = 16$$

yielding

$$CG = 20\log 16 = 24 \text{ dB}$$

That is, it represents an improvement in the resolution of 4 bits.

One possible realization for such characteristic is piecewise approximation using 13 segments that are symmetrical around the origin with six segments on both sides. The 13 segments have slopes that are reduced progressively by a factor of 2 in both sides, as shown in Figure 4.13. They are coded by 4 bits with 1 bit for the sign and

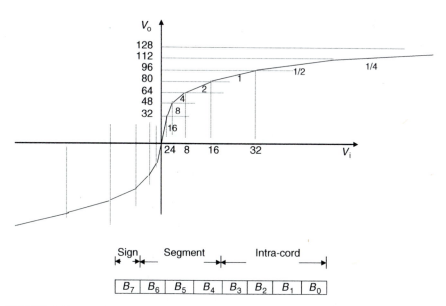

FIGURE 4.13   A 13-segment compressor characteristic.

the other three for eight segments in both directions. However, CEPT has recommended to further subdivide each range to 16 equal intervals, so that an extra 4 bits are required, and a total of 8 bits resolution gives the quality of 12 bits.

4.7 BASIC QUANTIZATION TECHNIQUES

The process of quantization, in essence, is a measurement process. Its main function is to define one level out of m quantization levels within the FSR in which the input sample amplitude is found. Three basic quantization techniques can be used to perform this function. They differ in their speed, reflected in the number of measuring steps, and the complexity of the hardware realizing them, as measured by the required number of references.

4.7.1 COUNTING (LEVEL-AT-A-TIME) METHOD

In this simple technique, a single reference whose amplitude corresponds to the size of a quantum Q is repeatedly compared with the sample amplitude V_s. After the first cycle, if V_s is found to be greater than Q, it is then compared with a reference of double the size—that is, $2Q$. If it is still greater, the size of the reference is increased to $3Q$. The process continues for a number of measuring cycles i until the reference size becomes slightly greater than or equal to V_s. Knowing the cycle time, one can estimate the time required to perform the measurement. As only one reference is required, the hardware needed to implement the technique should be very simple. For n-bit resolution and an input amplitude that is equal to the FSR (worst case), the required number of measuring cycles $i = 2^n - 1$. The technique is illustrated in Figures 4.14 and 4.15.

4.7.2 ITERATION (PUT-AND-TAKE) TECHNIQUE

To get better speed performance for a resolution of n bits, n binary-weighted references of sizes $Q, 2Q, 4Q, 8Q, 16Q, \ldots, 2^{n-1}Q$ are used in this technique. The algorithm implements a weighing-like procedure; it starts by setting the most significant bit (MSB) = 1 and comparing the resulting most significant (MS) weight (1/2 FSR) with

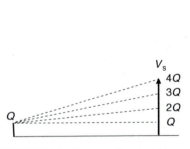

FIGURE 4.14　The counting technique.

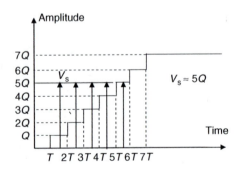

FIGURE 4.15　Measuring steps in the counting technique.

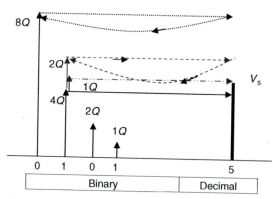

FIGURE 4.16 (See color insert following page 262.) The iteration technique.

the input sample. If the MS weight is found to be larger than the input sample, the MSB is then reset (i.e., changed to 0); otherwise, it is left unchanged. The next lower reference (1/4 FSR) is then tried, where its weight is added to that of the preceding reference. The new reference is then compared with the applied input. Again, if it is found to be greater than the input, the last added reference is removed (reset); otherwise, it is left. The process is repeated until all references are exhausted. The end states of the bits will furnish the digital equivalent of the input amplitude. A unique feature of such technique is that the number of measuring cycles is amplitude-independent and is always equal to the number of references N; that is, $i = N = n$. Thus, it is much faster than the counting technique ($i = 2^n - 1$). Due to its trial-and-error nature, this method is usually called the *iteration* or the *put-and-take technique*. It is easy to conclude that the process illustrated in Figure 4.16 is a 4-bit case, as there are four references; that is, an FSR of $15Q$ with an arbitrary input amplitude of $5Q$.

4.7.3 FLASH (DIRECT OR PARALLEL) TECHNIQUE

In the *flash* (*direct* or *parallel*) technique, the quantization process is performed in a single measuring step—hence the name "flash." However, such high-speed performance is achieved at the expense of increased cost due to increased hardware complexity. The number of references N grows exponentially according to the relation

$$N = 2^n - 1$$

This relation is plotted in Figure 4.17, from which it is easy to conclude that the hardware requirements doubles for each increase in n by 1 bit. Therefore, this technique, although a hardware-demanding one, provides the highest possible speed. In return, the obtainable "ultra-high speeds" are almost amplitude- and resolution-independent. The technique is illustrated in Figure 4.18 for the case of 3-bit resolution.

In Table 4.4, a comparison is made between the mentioned three quantization techniques for a resolution of 10 bits. A figure of merit (FM) is suggested as a basis for comparison. It is expressed as the product of the number of the measuring

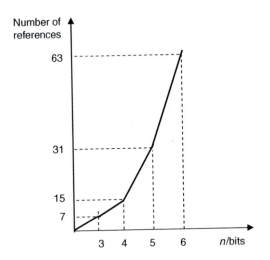

FIGURE 4.17   Hardware requirements versus resolution in FQs.

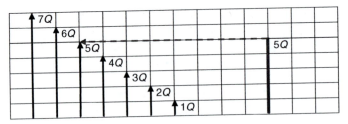

FIGURE 4.18   The direct quantization technique.

TABLE 4.4

Comparing the Basic Quantization Techniques

Technique	Number of References (N)	Number of Steps (i)	FM = $N \cdot i$
Counting	1	$2^n - 1 = 1023$	1023
Iteration	10	10	(100)
Flash	$2^n - 1 = 1023$	1	1023

steps i (representing the speed) times the number of references N (that represents the cost).

It is evident from the last column of this table that the iteration technique furnishes the optimal solution, as it provides the minimum time × cost product. However, the flash method remains the best option whenever ultra-high speeds are required. In general, the decision for any technique is case-dependent.

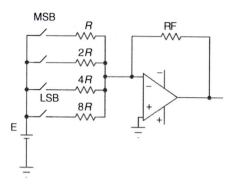

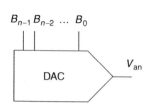

FIGURE 4.19   Circuit symbol
of a DAC.

FIGURE 4.20   A 4-bit weighted-resistor DAC.

4.8   DIGITAL-TO-ANALOG CONVERTERS

The function of a DAC is to convert a string of n-bit digital words into an equivalent analog voltage V_a, where

$$V_a = \left[2^{n-1} B_{n-1} + 2^{n-2} B_{n-2} + \cdots + 2^1 B_1 + 2^0 B_0 \right] \tag{4.11}$$

The term B_i can assume either the values one or zero. Thus, for each string of bits at the input, there will be a unique equivalent output voltage level. Because each change at the input is associated with a new voltage level, it is to be expected that the output voltage V_a will assume a quasi-analog (discontinuous) waveform, rather than a continuous one. Therefore, it should be smoothed to attenuate high frequencies that cause such discontinuity. These are frequencies that are higher than the maximum significant frequency f_m in the original analog signal. Smoothing can be provided by a LPF having a cut-off frequency at f_m. It is to be noted here that the transfer characteristic of a DAC resembles the quantization characteristic, illustrated in Figures 4.10 and 4.11, except that the ordinate and the abscissa are interchanged. In addition, the maximum permissible error is limited as before to $\pm Q/2$ or $\pm 1/2$ LSB. The circuit symbol of an n-bit DAC is shown in Figure 4.19.

4.8.1   WEIGHTED-RESISTORS DAC

A simple and a direct way to implement Equation 4.11 is to sum n binary-weighted currents, then using the sum to develop a proportional voltage in a resistor R_F at the output. The resulting voltage V_{an} can be expressed as

$$V_{an} = -R_F \sum_{i=0}^{n-1} I_i \cdot B_i \tag{4.12}$$

The binary-weighted currents are obtained from a stable voltage reference E and precise binary-weighted resistors. The currents are summed in the virtual ground input of an operational amplifier as diagramed in Figure 4.20. The input bits [B_{n-1}, B_{n-2}, B_{n-3}, ..., B_0] are used to control n analog switches that are connected in series with the resistors. In this way, currents are allowed to flow only in the resistors whose switches are closed.

The selected currents are then added and forced to flow through the feedback resistor R_F. The developed voltage at the output can then be expressed as

$$V_{an} = -R_F \left[\frac{E}{R} B_{n-1} + \frac{E}{2R} B_{n-2} + \frac{E}{4R} B_{n-3} + \cdots + \frac{E}{2^{n-1}R} B_0 \right]$$

$$= -E \frac{R_F}{2^{n-1}R} \left[2^{n-1} B_{n-1} + 2^{n-2} B_{n-2} + \cdots + 2^0 B_0 \right]$$

$$= -R_F \left[\frac{E}{R} B_{n-1} + \frac{E}{2R} B_{n-2} + \cdots + \frac{E}{2^{n-1}R} B_0 \right]$$

$$= -E \frac{R_F}{R} \left[B_{n-1} + \frac{B_{n-2}}{2} + \cdots + \frac{B_0}{2^{n-1}} \right]$$

from which the FS output—that is, the output with all bits high—is given by

$$V_{FS} = -E \frac{R_F}{2^{n-1}R} [2^{n-1} + 2^{n-2} + \cdots + 2^0]$$

$$= -E \frac{R_F}{R} \left[\frac{2^n - 1}{2^{n-1}} \right]$$

The precision of such a circuit is determined mainly by the precision of its weighted resistors, as well as by the stability of the reference source. However, the required precision does increase with the resolution. Therefore, to get a minimum error, the following conditions should be satisfied:

1. The reference voltage source should be highly regulated.
2. The used resistors should have a suitable tolerance and temperature coefficient.
3. The switches should have a low ON resistance, low offset, and reasonable switching speed.
4. The used Op. Amp. should have enough bandwidth, low drifts, low offsets, and low bias currents.
5. A resistor R_b should be connected to the noninverting input of the Op. Amp. to cancel its bias current effects. Its value is selected such that $R_b = R_F//R_{eq}$, where R_{eq} is the equivalent parallel resistance of all weighted resistors.

However, due to the growing ratios between the maximum and the minimum resistances that is given by

$$R_{max}/R_{min} = 2^{n-1}$$

such a type of DACs becomes impractical to realize in monolithic form for resolutions higher than 8 bits. Therefore, hybrid-circuit implementation, where monolithic integrated Op. Amp. and MOS transistor switches are mounted on insulating substrates (e.g., alumina [Al_2O_3]) together with thick or thin film resistors, is considered the most suitable fabrication technology. Film resistors enable trimming using laser beams to adjust their values within tolerances ranging between 0.1 and 0.001% [1].

EXAMPLE 4.7

Design an 8-bit weighted-resistor DAC to have an FS of 2 V, giving proper values to the resistors such that the power delivered by the reference source is limited to 10 mW.

Solution

$$E_{REF} = 2 \text{ V}$$

$$\text{Power delivered} = \frac{E^2}{R_{Eq}} = 10 \times 10^{-3} = \frac{4}{R_{Eq}}$$

giving $R_{Eq} = 400$

$$\frac{1}{R_{Eq}} = \frac{1}{R} + \frac{1}{2R} + \frac{1}{4R} + \cdots + \frac{1}{32R} + \frac{1}{64R} + \frac{1}{128R}$$

$$= \frac{255}{128R}$$

Thus

$$R_{Eq} = 400 = \frac{128R}{255}$$

$$R = 796.87 \ \Omega$$

Take $R = 1 \text{ k}\Omega$:

$$R'_{Eq} = \frac{128}{255} \times 1 \text{ k}\Omega = 500 \ \Omega$$

The permissible resistance tolerance $= \frac{1}{2} \text{LSB} \times 100 = \frac{1}{2 \times 256} \times 100 = 0.1953 \approx 0.1\%$

The full-scale output V_{oFS} is given by

$$V_{oFS} = 2V = \frac{R_F}{R'_{Eq}} E_{REF} = \frac{R_F}{500} \times 2$$

thus

$$R_F = 500 \ \Omega$$

4.8.2 *R-2R* Ladder-Type DAC

To overcome the problems encountered in the previous DAC—namely, the exponentially increasing resistance ratio with resolution—the *R-2R* ladder circuit technique was developed. In such a realization, the required current weighting is achieved through successive division of a constant current in a ladder resistor network having a resistance ratio of 2 (irrespective of the resolution). The concept is illustrated in Figure 4.21, where the current is shared by two equal resistors; that is, divided by two. A circuit implementation of the technique for the case of 3 bits is shown in Figure 4.22. The settling time of the converter is mainly determined by the switching speed of the switches and the slew rate of the Op. Amp. Because the input bits are applied in parallel form, only one switching time—namely, that due to the slowest switch—is taken into consideration.

For an *n*-bit DAC, $2n + 2$ precision resistors, *n*–single-pole, double-throw (SPDT) analog switches, and an Op. Amp. are required. Current division and hence weighting is achieved by applying the MSB to the control terminal of the leftmost switch and the LSB to the control of the rightmost one. It is to be noted at this point that every vertical branch is connected either to a ground (switch is turned to the left contact) or to the virtual ground of the Op. Amp. (switch is turned to the right contact). With all digital inputs low or 0, all currents are returned to ground such that the output

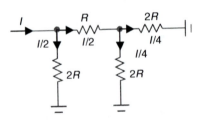

FIGURE 4.21   Current weighting.

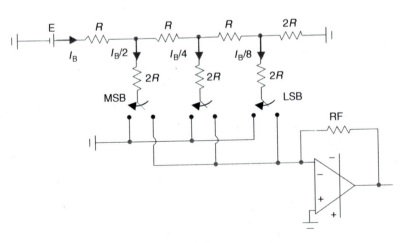

FIGURE 4.22   A 3-bit R-2R ladder-type DAC.

voltage is zero. Any other bit combination yields a proportional output. The equivalent resistance seen by the reference source for any input bit combination is always constant and equal to $2R$. Accordingly, the current drawn from the source is constant (an advantage) and is equal to

$$I_B = \frac{E}{2R}$$

The FS output voltage (with all bits at 1) is given by:

$$V_{FS} = -R_F \sum_{i=1}^{3} I_i = -R_F \left[\frac{I_B}{2} + \frac{I_B}{4} + \frac{I_B}{8} \right]$$

$$= -R_F \cdot I_B \left[\frac{7}{8} \right] = -\frac{7}{8} \cdot R_F \cdot \frac{E}{2R}$$

Choosing $R_F = 2R$, we get

$$V_{FS} = -\frac{7}{8} E$$

The output voltage, for any word Z and resolution n bits, can be expressed as:

$$V_0 = -\frac{Z}{2^n} E$$

where

$$Z = B_{n-1} + B_{n-2} + B_{n-3} + \cdots + B_0$$

and B_i is the weight of the ith bit.

EXAMPLE 4.8

Design a 10-bit R-2R DAC to have an FS output of 2 V such that the maximum dissipated power is 1 mW, then calculate the maximum permissible resistance tolerance. Calculate also the power delivered by the reference source when all even-numbered bits are high.

Solution

$$I_B = \frac{E}{2R}$$

Power delivered:

$$P = E x I_B = \frac{E^2}{2R} = 10^{-3} W$$

$$= \frac{4}{2R}$$

giving

$$R = 2 \text{ k}\Omega$$

and

$$I_B = 1/2 \text{ mA}$$

To find the value of R_F,

$$V_{FS} = 2V = R_F\left[\frac{I_B}{2} + \frac{I_B}{4} + \cdots + \frac{I_B}{2048}\right] = R_F\left[I_B \times \frac{4095}{2048}\right]$$

$$= R_F\left[\frac{1}{2}\text{mA} \times \frac{4095}{2048}\right]$$

yielding

$$R_F = 2 \text{ k}\Omega$$

The permissible resistor tolerance is given by

$$= \frac{\text{LSB}}{2} \times 100 = \frac{1}{2 \times 1024} \times 100 = 0.0488\% \approx 0.04\%$$

The power delivered by the reference source is always constant and is independent from the digital input.

4.8.3 SWITCHED-RESISTOR DAC

A rather simple circuit implementation of a DAC is shown in Figure 4.23 for a 3-bit case. A chain of 2^n equal resistors and a reference voltage source E are employed to provide ascending voltage levels corresponding to all possible quantization levels. A network of $2^n - 1$ SPDT switches is used to select an equivalent voltage level to the digital input. The input bits are applied to control the switches such that the LSB is used to control in-parallel switches S_3 and the MSB to control switch S_1, as depicted in Figure 4.23. A buffer is connected at the output to prevent loading of the resistor chain. One major drawback of such a circuit is the series connection of n switches due to their ON resistances that add together, causing errors. Figure 4.24 illustrates the growth of the number of switches with n.

EXAMPLE 4.9

Design a 4-bit switched-resistor DAC to dissipate a maximum power of 1 mW for an FS of 2 V, then calculate:

1. The number of required switches.
2. The total voltage drop on the series switches if they have an ON resistance of 50 Ω and the used buffer has a bias current of 200 μA, then check suitability of the circuit in view of the resulting error.
3. Find a suitable value for the chain resistance.

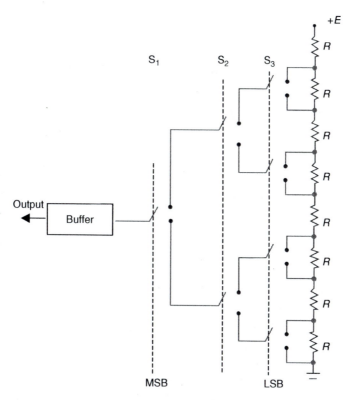

FIGURE 4.23 Switched-resistor DAC.

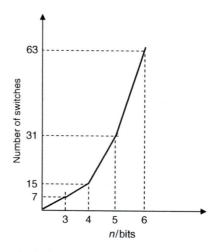

FIGURE 4.24 Number of switches versus resolution in a switched-resistor DAC.

Solution

The number of switches $= [2^n - 1] = 15$.

The voltage drop across each switch $= I_B \times R_{ON} = 200 \times 10^{-6} \times 50 = 10$ mV.

Because there are always four switches connected in series, irrespective of the digital input, the total voltage drop is thus

$$V_d = 4 \times 10 \text{ mV} = 40 \text{ mV}$$

The maximum permissible error $= 1/2 \text{ LSB} = \dfrac{2}{2 \times 16} = 62.5 \text{ mV} > V_d.$

Thus the circuit can be used for a resolution up to 4 bits.

Power dissipated in the chain $= \dfrac{E^2_{REF} = V^2_{FS}}{16R} = \dfrac{4}{16 \times R} = 1 \text{ mW}.$

Power dissipated in the switches $= 4 \times I_B^2 \times R_{ON} = 4 \times 4 \times 10^{-8} \times 50 = 8 \text{ μW}.$

This gives $R = 250 \ \Omega$.

4.9 ANALOG COMPARATORS

Analog comparators are essential building blocks in the construction of ADCs. Their performances greatly affect the overall characteristics of the resulting ADC. The circuit symbol of a comparator, shown in Figure 4.25, indicates that it is a two-input device: the signal input, marked with a plus (+) sign and the reference input, marked with a minus (−) sign. In order to avoid mixing a comparator with an operational amplifier, the comparator will be labeled here with the letter K.

The front end of such a device, like operational amplifiers, is a high-gain, high-input impedance, wide-bandwidth differential amplifier. The output V_o, on the other hand, assumes only the states "1" or "0" based on the sign of the difference $(V_S - V_R)$ such that if

$$V_S \geq V_R \qquad \text{then } V_0 = \text{"1"}$$

$$V_S < V_R \qquad \text{then } V_0 = \text{"0"}$$

This relationship is described by the voltage transfer characteristic shown in Figure 4.26.

The logic swing (LS) expresses the voltage difference between logical 1 and logical 0. It is determined by the logic family (e.g., TTL, ECL, CMOS, and so on) used in the realization of the comparator's output circuit. The width of the transitional region, known as the "gray region" (GR), is determined by the slope of the transfer characteristic over this range, which in turn is dictated by the gain of the front end amplification stage(s). High gain stages provide comparators with narrow GRs. The GR of a comparator is an important parameter, as it determines its resolution; that is, the smallest voltage difference a comparator can detect as well as the maximum

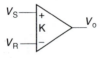

FIGURE 4.25 Circuit symbol of an analog comparator.

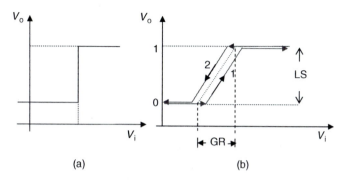

FIGURE 4.26 **(See color insert following page 262.)** Voltage transfer characteristics of an analog comparator. (a) Ideal characteristic and (b) practical characteristic.

resolution of a converter using it. As a rule of thumb, the GR of a comparator should be less than the quantum Q of the ADC; that is,

$$GR < Q \leq LSB$$

Therefore, high-performance comparators include high gain stages to have a small GR. The hysteresis shown in Figure 4.26b (the area bounded between the two inclined parallel lines) results, because a comparator behaves differently depending on the trend of the input voltage. For increasing input, the comparator's output assumes the ascending line 1; for decreasing input, it assumes the descending path 2.

The speed of a comparator is measured by its response time (decision time); that is, the time taken by the comparator to deliver the result of comparison at the output. It is given mainly by the bandwidth of its amplifiers and its logic family. Because the gain $\times$ bandwidth product of any amplifier is a constant quantity, it is difficult for comparators to achieve high speeds and high resolutions simultaneously. Therefore, it is important when designing high-quality comparators to choose circuits having the maximum possible gain $\times$ bandwidth product in order to guarantee satisfactory speed and resolution. A rather important characteristic of an analog comparator is its input bias current I_B. Such current is usually determined by the input impedance of the comparator. Unfortunately, high-speed comparators should have high slew rates, which imply that the input time constant is as small as possible. To guarantee this, the input impedance should be kept as low as possible, causing relatively large input bias currents to flow. Table 4.5 summarizes major performance characteristics of some commercially available analog comparators.

EXAMPLE 4.10

Calculate the voltage gain of a comparator that is required for an 8-bit ADC that has an FSR of 1 V for the following logic levels:

1. TTL
2. ECL

Then comment on the result.

TABLE 4.5

Performance Characteristics of Some Commercial Analog Comparators

Type	GR (mV)	Bias Current	Response Time (ns)	Logic Level
LM339		25 nA	600	Open collector
NE521	7.5		10	TTL
μA710	1.5		40	TTL
μA760	0.5		25	TTL
AD9687			2.7	ECL

Solution

The quantum of the ADC is

$$Q = FSR/2^8 = 1/256 = 3.9 \text{ mV}$$

The required voltage gain $A_V = LS/Q$ equals

1. For TTL logic $A_V = \dfrac{5 - 0.2}{3.9 \text{ mV}} = 1228.8$

2. For ECL logic $A_V = \dfrac{-0.9 + 1.8}{3.9 \text{ mV}} = 230$

This result indicates that for a given gain bandwidth product, because ECL logic requires lower gain than TTL, a larger amplifier bandwidth is made available. This fact explains why ECL comparators are the fastest type.

4.9.1 INTERNAL ARCHITECTURE OF AN ANALOG COMPARATOR

A block diagram illustrating the basic building blocks of an analog comparator is shown in Figure 4.27.

The front end differential amplifier stages (gain stages) determine the magnitude and sign of the difference between the two input signals V_S and V_R multiplied by the voltage gain (which should be high to minimize the GR). The amplifiers should also show sufficiently high input impedance to minimize the input bias current, and a high slew rate to guarantee a wide bandwidth and hence a small response time. The two inputs should also be symmetrical to minimize possible offsets, and above all, these characteristics should be stable; that is, temperature-compensated to minimize drift. The differential amplifier stage is then followed by a single-ended gain stage to provide the required overall amplification. Next to the amplification stage(s), as depicted in Figure 4.27, is a stage that detects the sign of the difference $V_S - V_R$, which in turn determines the output state (either a logical 1 for a positive difference or a logical 0 in

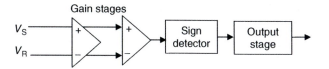

FIGURE 4.27   Internal architecture of an analog comparator.

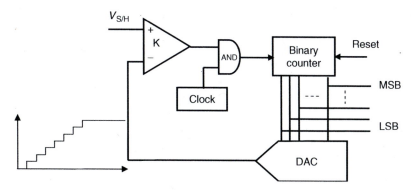

FIGURE 4.28   A counter-type ADC.

the case of negative sign). The speed of the comparator, as said before, is partially determined by the speed of the logic family used in the implementation of such stage. As indicated in Table 4.5, ECL realization provides the highest possible speed.

4.10   BASIC ADC ARCHITECTURES

In the following subsections, we are going to study possible circuit implementations of the previously discussed quantization techniques. In addition, other circuit techniques will be described that aim at providing high resolution at moderate cost and usable speeds for several practical applications.

4.10.1   COUNTER-TYPE ADC

A circuit implementation of the counting quantization technique is shown in Figure 4.28. It consists of a single voltage comparator, an AND gate, an n-bit binary counter, and an n-bit DAC in the feedback loop. A stable clock source is also necessary to adjust the timing. The operation of the circuit starts by receiving a reset signal (the same SC applied to the S/H circuit) to the binary counter. Correspondingly, the DAC output that feeds the comparator's reference voltage will be zero. Comparing $V_{S/H}$ with zero will result in a binary 1 that will enable the AND gate to input the clock pulses to the counter. The DAC output and hence the reference voltage V_R will be incremented by Q with the arrival of each clock pulse, thereby taking the form of a staircase. The reference voltage will continue increasing until it exceeds the input sample amplitude. At this point, the comparator's output goes low, leaving the AND gate disabled while freezing the counter reading at the digital equivalent of the input.

The conversion time (T_{conv}) is given by

$$T_{conv} = i \cdot T_{clock}$$

where i is the number of cycles needed for the reference voltage to reach the applied voltage level and T_{clock} is the clock period that should accommodate the delays of all circuit components in the feedback loop. It is given by

$$T_{clock} \geq T_{comp} + T_{AND} + T_{counter} + T_{DAC} \qquad (4.13)$$

in which

- T_{comp} is the response time of the comparator
- $T_{counter}$ is the propagation delays (pd) of the counter
- T_{AND} is the propagation delay of the AND gate
- T_{DAC} is the settling time of the DAC

It is apparent from this equation that the conversion time is mainly determined by the clock period. Thus, to achieve the best possible speed, the clock period should be kept at its minimum possible value—that is, the sum of the delays of all components. It is noteworthy here that the conversion time is input amplitude–dependent, because the circuit measures actually the time required by the ascending reference voltage to reach the input voltage level.

In order to get a fair basis for comparison with other techniques, the worst case input—that is,

$$V_i = FSR$$

should be considered.

The conversion time is thus given by

$$T_{conv} = (2^n - 1) \cdot T_{clock} \qquad (4.14)$$

EXAMPLE 4.11

Calculate the best possible resolution of a counter-type ADC that is required to digitize a 4 kHz signal using the components given in Table 4.6, then estimate the duty cycle (DC) of the SC.

TABLE 4.6

Example 4.11

Component	Delay (ns)
Analog comparator	$T_{resp} = 50$
DAC	$T_{sett} = 360$
Binary counter	$T_{pd} = 10$
AND gate	$T_{pd} = 5$

Solution

The clock cycle time $T_{clock} = T_{comp} + T_{AND} + T_{counter} + T_{DAC} = 50 + 5 + 10 + 360 = 425$ ns. According to the sampling theorem, the signal should be sampled at

$$f_s = 2f_m = 2 \times 4 = 8 \text{ kHz}$$

The sampling period $T_s = 1/f_s = 1/8 = 125$ μs, which is equal to the maximum possible (worst case) conversion time.

The conversion time is given by

$$T_{conv} = (2^n - 1) \cdot T_{clock}$$

$$(2^n - 1) \, 425 \text{ ns} = 125,000 \text{ ns}$$

$$2^n = \frac{125,000}{425} + 1 = 295.11$$

or

$$n = \left\lfloor \frac{\log 295.11}{\log 2} \right\rfloor = \lfloor 8.25 \rfloor = 8 \text{ bits}$$

giving an actual conversion time of $T_{conv} = 425(2^8 - 1) = 108.375$ μs.

The difference between the available sampling period and the actual conversion time of $125 - 108.375 = 16.625$ μs could be used for acquisition.

The duty cycle of the SC is

$$DC = 16.625/125 \times 100 = 13.3\%$$

The corresponding SC is depicted in Figure 4.29.

EXAMPLE 4.12

Calculate the maximum possible signal frequency that could be digitized employing an 8-bit counter-type ADC using the components given in Table 4.7.

Solution

The clock period $T_{clock} = 40 + 10 + 20 + 130 = 200$ ns.

The worst-case conversion time is given by

$$T_{conv} = T_{clock}[2^n - 1] = 200 [255] = 51000 \text{ ns} = 51 \text{ μs}$$

The maximum sampling frequency is $f_s = 1/T_{conv} = 19.6$ kHz, giving a maximum possible input signal frequency of $f_m = 9.8$ kHz.

TABLE 4.7

Example 4.12

Component	Delay (ns)
Analog comparator	$T_{resp} = 40$
DAC	$T_{sett} = 130$
Binary counter	$T_{pd} = 20$
AND gate	$T_{pd} = 10$

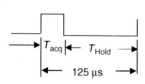

FIGURE 4.29 Clock cycle of Example 4.11.

EXAMPLE 4.13

Write an expression relating the input sample amplitude $V_{S/H}$ and the conversion time T_{conv} for a counter-type ADC that has a resolution of 4 bits and a clock period T_{clock} for inputs ranging from 0 to FSR in steps of 1/4 FSR. Sketch also the relationship for the maximum possible signal frequency. Comment on the results.

Solution

The conversion time is given by

$$T_{conv} = \alpha \cdot T_{clock}$$

with $\alpha = 0, 1, 2, ..., (2^n - 1)$.

In addition, the applied input can be expressed as

$$V_{S/H} = \alpha \cdot Q$$

where

$$Q = FSR/2^n$$

Thus

$$\alpha = V_{S/H}/Q = 2^n V_{S/H}/FSR$$

Substituting in the previous equation, we get

$$T_{conv} = 2^n T_{clock} V_{S/H}/FSR$$

For $n = 4$ bits,

$$T_{conv} = 16 T_{clock} V_{S/H}/FSR$$

Assuming a Nyquist rate,

$$f_s = 2 f_m = 1/T_{conv}$$

from which we can write

$$f_m = 1/2T_{conv} = \frac{FSR}{32 \cdot V_{S/H} T_{clock}} = \frac{FSR}{V_{S/H}} \frac{f_{clock}}{32}$$

Substituting for the given inputs, we can construct Table 4.8.

It is clear from the results that the relation T_{conv} versus $V_{S/H}$ is a linear relationship, and that of f_m versus $V_{S/H}$ is a parabolic one. Both relations are plotted in Figure 4.30.

TABLE 4.8

Example 4.13

$V_{S/H}$	0	1/4 FSR	1/2 FSR	3/4 FSR	FSR
T_{conv}	0	$4 T_{clock}$	$8 T_{clock}$	$12 T_{clock}$	$16 T_{clock}$
f_m	—	$0.125 f_{clock}$	$0.0625 f_{clock}$	$0.0416 f_{clock}$	$0.03125 f_{clock}$

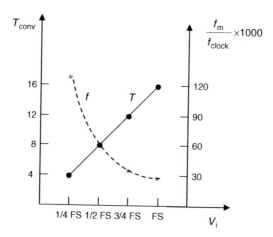

FIGURE 4.30 Results of Example 4.13.

4.10.2 SUCCESSIVE APPROXIMATION ADC

A circuit implementation of the iteration quantization technique, also called successive approximation or the put-and-take technique, is shown in Figure 4.31a. As described before, the method is based on successively comparing the sample amplitude with combinations of N references having binary-weighted amplitudes (i.e., 1/2 FSR, 1/4 FSR, 1/8 FSR, and so on). Such reference voltages are obtained from an n-bit DAC by setting its inputs to 1 successively, starting with the MSB. Based on the comparator's decision, the last modified bit is either left at 1 (comparator output is high) or reset (comparator's output low). The next bit is then set to 1 to provide the next smaller reference that is added to the already applied reference to perform the next comparison. The process continues until all bits are exhausted. The digital output is given by the bits representing the used references.

The put-and-take logic is furnished by a specially designed block called the *successive approximation register* (SAR), illustrated in Figure 4.31b. The internal architecture of a 4-bit SAR is shown in Figure 4.32 below. It consists mainly of four D flip-flops, FF_0, FF_1, FF_2, FF_3, and a shift register. The comparator output K is applied in parallel to the D inputs of all flip-flops, while a logical "1" is applied to the input of the shift register using the clock Φ. The output S_3 is then used to clock the first flip-flop FF_3 causing the output Q_3 to assume the state 1. Bit B_3 is accordingly set to 1. Based on the updated comparator output k, Q_3 and hence B_3 (Q_3 AND S_3) will either remain high (k high) or be reset (k low). With the arrival of the next clock pulse, Q_2 and B_2 will be set to high. The process is repeated until B_0 is reached and tried. At this point, an end-of-conversion (EOC) signal is developed at the output. The output waveform of the DAC and the generated bits are given in Figure 4.32. The algorithm is illustrated and described in Figure 4.33.

The sequence of events can be summarized as follows. With the arrival of an SC, the SAR output is reset, leaving a zero DAC output. The comparator's output will accordingly be high, and the MSB is then set to 1. The resulting output of the DAC (1/2 FSR) will be compared with the input voltage $V_{S/H}$. If the result of the

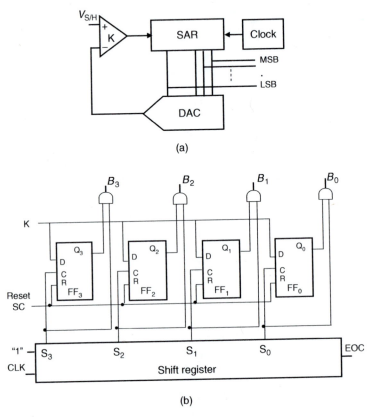

(a)

(b)

FIGURE 4.31 A block diagram for a successive approximation ADC.

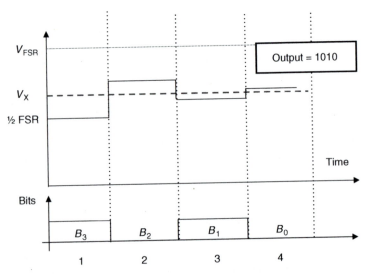

FIGURE 4.32 Generating the output bits in a successive approximation ADC.

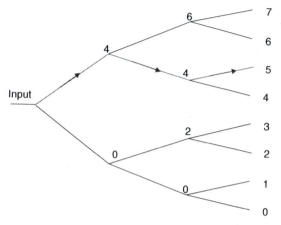

B_2	B_1	B_0
1	0	1

FIGURE 4.33 (See color insert following page 262.) Successive approximation tree for a 3-bit case; input = 5.

TABLE 4.9

Examples of Some Monolithic SARs

Type	Word Length (bits)	Logic
F74LS502	8	TTL
SP74HCT502	8	CMOS
F74LS504	12	TTL
SP74HCT504	12	CMOS

Source: Adapted from U. Tietze and Ch. Schenk, *Electronic Circuits*, Springer-Verlag, Heidelberg, 1991. With permission.

comparison is positive, the next bit is set to 1 and the updated DAC output (3/4 FSR) will be compared again with the input; otherwise, this bit is reset and the next less significant bit is tried. The process continues according to procedures in Figure 4.33 until all bits are tried. The reading of the SAR will represent the digital equivalent of the input. Table 4.9 gives examples of some monolithic SARs.

The main features of successive approximation ADC's are:

1. The conversion time is amplitude-independent and is always given by

$$T_{conv} = n \cdot T_{clock} \qquad (4.15)$$

where n is the word length and T_{clock} is given by

$$T_{clock} = T_{respk} + T_{pd\,SAR} + T_{set\,DAC}$$

2. Compared to counter-type converters, the SARs are much faster and could be classified as either high-speed or very high-speed converters.
3. The circuit is simple and could be produced in monolithic form up to more than 16 bits.

The logical steps involved in the successive approximation algorithm are illustrated in the tree diagram given in Figure 4.33 and in the flow chart of Figure 4.34.

It is to be noted that selecting the upper branch means 1, while following the lower branch means 0. Thus for an analog input of 5, we have the route shown on the tree (marked by arrows), which is equivalent to 101.

Table 4.10 gives examples of some commercially available successive approximation ADCs.

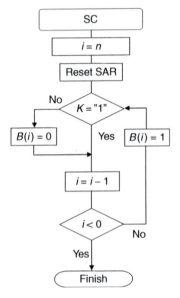

FIGURE 4.34 Flowchart illustrating the successive approximation algorithm.

TABLE 4.10

Examples of Some Successive Approximation ADCs

Type	Resolution (bits)	Throughput (μs)	Power Dissipation (mV)	Technology	Manufacturer
ADC847	8	9	125	CMOS	Datel
ADC910	10	6	400	Bipolar	PMI
AD7672	12	3	110	CMOS	Analog Devices
ADC701	16	2	1500	Hybrid	Burr Brown

Source: Adapted from U. Tietze and Ch. Schenk, *Electronic Circuits*, Springer Science and Business Media, 1991. With permission.

4.10.3   SWITCHED-RESISTOR SUCCESSIVE APPROXIMATION ADC

Based on the successive approximation algorithm and the switched-resistor technique, as described in Section 4.8.3, a rather simple ADC can be realized. As depicted in Figure 4.35, for a 3-bit case, a resistor string of $2^n = 8$ series connected matched (equal) resistors are used to divide the FSR into eight equal quantization levels. Any one of these eight levels can be selected through three series-connected SPDT switches. They are controlled by the output bits of an SAR. A chain of a total of 7 or $(2^n - 1)$ such switches leads to a comparator that feeds the SAR. The conversion process starts with a reset command to the SAR, enforcing its outputs to assume the state "low" and hence turning all switches to position II. This step results in connecting the reference input of the comparator to ground. Any applied analog input V_{in} will cause the comparator output to go to high. The MSB B_2 is thus set to high, while the other bits are left at low. Accordingly, switch S_2 is turned to position I, while other switches remain in position II. A reference voltage that corresponds to 1/2 FSR is then applied to the comparator. Depending on the comparator's decision, bit B_2 either remains high or is reset as before. The second bit B_1 is then set to 1 upon the arrival of the next clock pulse, thereby turning switches S_1 in parallel to position I. The resulting new reference voltage will correspond to either 1/4 or 3/4 FSR. The comparator will then decide whether to leave

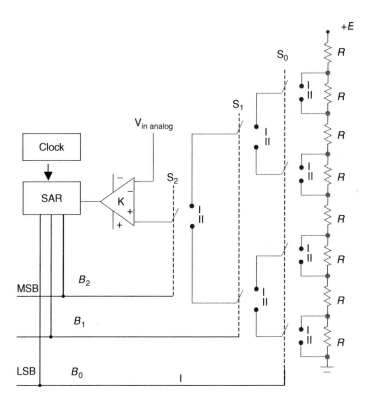

FIGURE 4.35   A 3-bit switched-resistor successive approximation ADC.

B_1 high or to reset it. The process is repeated in order to determine the LSB. As before, the number of cycles is equal to the number of bits and is independent of the analog amplitude. The main source of error is the ON resistance R_{ON} of the series connected switches (n switches each cycle). However, this error could be minimized by selecting a high chain resistance R compared to R_{ON} and by selecting a comparator that has a small bias current. The conversion time is given by

$$T_{conv} = n \cdot (T_{switch} + T_{respk} + T_{SAR})$$

EXAMPLE 4.14

Calculate the conversion time and the maximum possible input-signal frequency to a 1 V FSR, 8-bit switched-resistor ADC, then check for the resulting error if the components listed in Table 4.11 are used.

Solution

Clock frequency = $T_{comp} + T_{SAR} + T_{sw}$ = 10 + 40 + 10 = 60 ns

Conversion time = 8 × 60 = 480 ns

Sampling frequency f_s = 2.0833 MHz

Maximum input signal frequency = 1.04166 ≅ 1 MHz

Voltage drop across the switches = $500 \times 10^{-9} \times 100 \times 8$ = 0.4 mV

1/2 LSB = 1.953 mV > 0.4 mV; therefore OK.

4.10.4 PARALLEL (FLASH) ADC

Parallel or flash ADCs are ultra-high-speed devices. They provide the highest possible conversion speeds among all other techniques. As described before, all possible quantization levels should be available simultaneously for comparison. To do this, $2^n - 1$ references and hence an equal number of comparators are required. The input sample amplitude is applied in parallel to the signal inputs of all comparators, as shown in Figure 4.36. The required voltage references having amplitudes corresponding Q, $2Q$, $3Q$, ..., $(2^n - 1)Q$ are obtained through a chain of 2^n equal (matched) resistors and a stable voltage reference E_{REF}. The comparator's outputs will be a string of ones and zeros. The number of ones is proportional with the analog amplitude, thus resembling the mercury in a thermometer. The resulting code is therefore called the *thermometer code*.

An additional circuit is needed to convert such $2^n - 1$ output code into an n-bit binary coded word. From the truth table given in Table 4.12, it is clear that

---

TABLE 4.11

Data of Example 4.14

Component	Delay	Misc.
Analog comparator	10 ns	Bias current = 500 nA
SAR	40 ns	
Switching time	10 ns	ON-resistance = 100 Ω

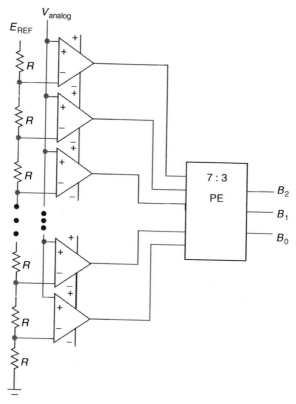

FIGURE 4.36 A 3-bit flash-type ADC.

TABLE 4.12
Truth Table of a 7:3 PE

K_1	K_2	K_3	K_4	K_5	K_6	K_7	B_2	B_1	B_0
0	0	0	0	0	0	0	0	0	0
1	0	0	0	0	0	0	0	0	1
1	1	0	0	0	0	0	0	1	0
1	1	1	0	0	0	0	0	1	1
1	1	1	1	0	0	0	1	0	0
1	1	1	1	1	0	0	1	0	1
1	1	1	1	1	1	0	1	1	0
1	1	1	1	1	1	1	1	1	1

the function of such a circuit, usually called a *priority encoder* (PE), is to count the number of ones resulting from the comparators in the output string and express that number in binary representation. One possible realization of such an encoder is given in Figure 4.37.

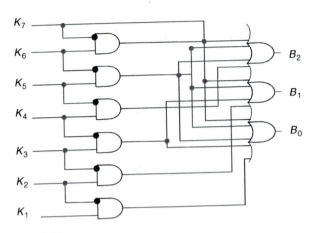

FIGURE 4.37 A 7:3 PE.

The logical expressions that describe the function of the priority encoder are as follows:

$$B_2 = K_4 \cdot \overline{K_5} + K_5 \cdot \overline{K_6} + K_6 \cdot \overline{K_7} + K_7$$

$$B_1 = K_2 \cdot \overline{K_3} + K_3 \cdot \overline{K_4} + K_6 \cdot \overline{K_7} + K_7$$

$$B_0 = K_1 \cdot \overline{K_2} + K_3 \cdot \overline{K_4} + K_5 \cdot \overline{K_6} + K_7$$

An alternative realization for the PE that comprises only two gate levels, irrespective of the number of the inputs, is illustrated in Figure 4.38 together with its logical expressions, again for a 3-bit case.

The outputs are

$$B_2 = X_4 + X_5 + X_6 + X_7$$

$$B_1 = X_2 + X_3 + X_6 + X_7$$

$$B_0 = X_1 + X_3 + X_5 + X_7$$

where

$$X_1 = K_1 \oplus K_2$$

$$X_2 = K_2 \oplus K_3$$

$$X_3 = K_3 \oplus K_4$$

$$X_4 = K_4 \oplus K_5$$

$$X_5 = K_5 \oplus K_6$$

$$X_6 = K_6 \oplus K_7$$

$$X_7 = K_7 \oplus 0$$

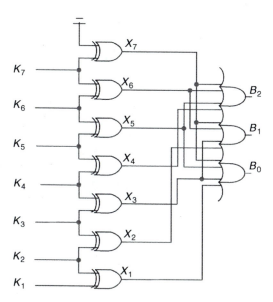

FIGURE 4.38 A two-gate level 7:3 PE.

EXAMPLE 4.15

Write a logical expression for a 15:4 two-gate-level PE.

Solution

$$B_3 = X_8 + X_9 + X_{10} + X_{11} + X_{12} + X_{13} + X_{14} + X_{15}$$

$$B_2 = X_4 + X_5 + X_6 + X_7 + X_{12} + X_{13} + X_{14} + X_{15}$$

$$B_1 = X_2 + X_3 + X_6 + X_7 + X_{10} + X_{11} + X_{14} + X_{15}$$

$$B_0 = X_1 + X_3 + X_5 + X_7 + X_9 + X_{11} + X_{13} + X_{15}$$

where, as before,

$$X_1 = K_1 \oplus K_2$$

$$X_2 = K_2 \oplus K_3$$

$$\vdots$$

$$X_{15} = K_{15} \oplus 0$$

It is evident from Figure 4.38 that only two stages are required to provide the digital output; namely, the bank of comparators and the priority encoder. Being connected in parallel, only one comparator response time T_{respk} is to be considered when calculating the conversion time. Thus we can write

$$T_{\text{conv}} = T_{\text{respk}} + T_{\text{pd PE}} \tag{4.16}$$

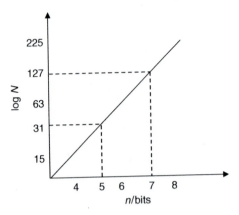

FIGURE 4.39   Number of comparators versus resolution in a flash ADC.

This equation indicates that this family of converters provides the highest possible throughput rates—hence the name "flash-type" converters. On the other hand, a major disadvantage of this type is that the required number of comparators N doubles for an increase in the resolution n by 1 bit, according to the relation

$$N = 2^n - 1$$

Thus, 15 comparators are needed for $n = 4$, which reaches 31 as n is increased to 5 bits. A plot of such an "exponential" relation was given in Figure 4.17 and is given again in Figure 4.39, yet on a semilog scale. The rapid growth in the comparators' count is usually accompanied by the following problems:

1. Increased cost
2. Increased circuit complexity
3. Increased power consumption
4. Increased area of the silicon die
5. Increased errors due to the bias current that flows in the resistor chain
6. Increased input capacitance due to the large number of parallel-connected comparators

Due to these consequences, the resolution of monolithic flash-type ADC converters is usually limited to 8–10 bits. Table 4.13 lists some commercially available monolithic flash converters.

Practical flash-type ADCs suffer from the following major sources of errors:

1. Instability of the reference voltage
2. Mismatch due to tolerance in the resistor chain
3. Input bias current of the comparators
4. Response time mismatch of the comparators
5. Capacitive drive problems

TABLE 4.13

Some Commercially Available Monolithic Flash ADCs

Type	Resolution	Maximum Signal Frequency (MHz)	Logic
AD9048	8	15	TTL
AD9038	8	150	ECL
TDC1025	8	25	ECL
AD9020	10	30	TTL
TDC1020	10	12	TTL

Source: Adapted from U. Tietze and Ch. Schenk, *Electronic Circuits,*
Springer Science and Business Media, 1991. With permission.

EXAMPLE 4.16

For a 3-bit flash ADC, calculate the value of resistance in the resistor chain that can
limit the dissipated power to 5 mW, provided that $E_{REF} = 2$ V.

Solution

The current flowing in the chain is

$$\frac{E_{REF}}{8R}$$

The dissipated power is

$$\frac{E_{REF}^2}{8R} = \frac{4}{8R} = 5 \times 10^{-3} \text{ W}$$

giving a value of

$$R = \frac{4}{8 \times 5 \times 10^{-3}} = 100 \ \Omega$$

4.10.5 INDIRECT ADCs

In addition to the previously discussed conventional conversion techniques, there is
a class of converters that measures the instantaneous amplitude of the samples indirectly; that is, by measuring instead a proportional quantity that is easily or precisely
measurable. Time is a good example of such quantities, as stable time-base circuits could be constructed using, for example, highly stable quartz-controlled clock
sources. The counter-type ADC is an example of this class of converters, as the input
amplitude determines the period over which the AND gate remains enabled in terms
of the counted clock pulses. This means that the process of amplitude measurement

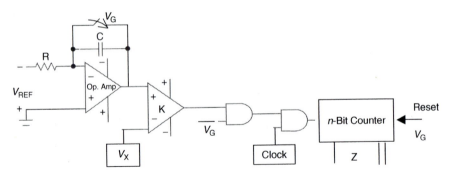

FIGURE 4.43   An improved single-ramp ADC.

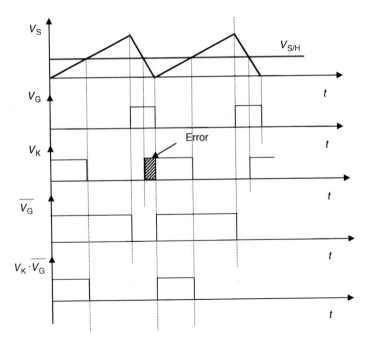

FIGURE 4.44   Waveforms associated with the modified single-ramp ADC.

A simple analysis for Miller integrator yields

$$V_C = \frac{Q}{C} = \frac{\int_0^t i \cdot dt}{C} = -\int_0^t \frac{V_{REF}}{R \cdot C} dt = -\frac{V_{REF}}{R \cdot C} \cdot t$$

$$V_S = -V_C = \frac{V_{REF}}{R \cdot C} \cdot t$$

at $t = t_m$

$$V_S = V_{S/H} = \frac{V_{REF}}{R \cdot C} \cdot t_m \qquad (4.17)$$

As before, the counter reading Z at $t = t_m$ is

$$Z = \frac{t_m}{T_{CLK}}$$

From Equation 4.17, we get

$$Z = \frac{1}{T_{CLK}} \cdot V_{S/H} \cdot \frac{R \cdot C}{V_{REF}}$$

$$= f_{CLK} \cdot \frac{R \cdot C}{V_{REF}} \cdot V_{S/H}$$

now making

$$\left| f_{CLK} \cdot R \cdot C \right| = V_{REF}$$

which gives

$$Z = V_{S/H}$$

From these equations, it is clear that the counter reading and hence the ADC output depends on the stability of the clock source, the precision of the circuit components R and C as well as on the reference voltage source. However, it is possible to alleviate most of them if two ramps are used rather than one ramp, as we shall demonstrate in the following realization.

4.10.5.2 Dual-Ramp ADC

The idea of this circuit is to charge the capacitor in the Miller integrator during a fixed time period t_1 from the applied input sample amplitude $V_{S/H}$ to attain a proportional amplitude. The capacitor is then allowed to discharge to ground potential during a period t_m that is proportional with $V_{S/H}$. As shown in Figure 4.45, the core of the circuit is a Miller ramp generator (integrator) together with an analog comparator, two binary counters, a single AND gate, a clock source, and a control logic. The first counter, counter I, is clocked directly from the clock source to furnish proper timing to the shown SPDT switch. The conversion process starts by turning the switch to position I at $t = 0$, where capacitor C starts to charge linearly to a value V_S that is directly proportional with the applied S/H voltage $V_{S/H}$ over a fixed period t_1. At the end of t_1, the control logic turns the switch to position II where capacitor C starts discharging to ground potential via the reference voltage $-V_{REF}$. At the same time, it enables the three-input AND gate.

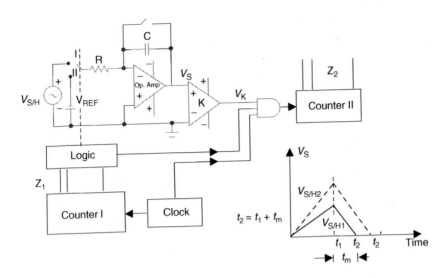

FIGURE 4.45   The dual-slope ADC.

Recall that with $V_S > 0$, the comparator output V_k goes high, thereby allowing counter II to start counting clock pulses received from the clock source. As V_S reaches ground potential, the comparator goes low, causing the counter to stop counting. The reading of counter II Z_2 determines the discharge time t_m and hence the amplitude of $V_{S/H}$. The capacitor voltage is given by

$$V_C = -\frac{Q}{C} = -\int_0^{t_1} \frac{i}{C}\, dt = -\int_0^{t_1} \frac{V_{S/H}}{R \cdot C} \cdot dt = -\frac{V_{S/H}}{R \cdot C} \cdot t \qquad (4.18)$$

For the discharging period t_m, we can write the following relationships:

$$I_d \cdot t_m = Q = C \cdot V_C$$

where

$$I_d = -\frac{V_{REF}}{R}$$

$$t_m = \frac{C \cdot V_C}{I_d} = -\frac{R \cdot C \cdot V_C}{V_{REF}}$$

Then, substituting from Equation 4.18 we get

$$= -\frac{R \cdot C}{V_{REF}}\left(-\frac{V_{S/H}}{R \cdot C}\right) \cdot t_1 = \frac{t_1}{V_{REF}} \cdot V_{S/H}$$

For the counters readings Z_1 and Z_2, we can write

$$Z_1 = t_1 \cdot f_{\text{clock}} = \text{constant}$$

and

$$Z_2 = t_m \cdot f_{\text{clock}}$$

Thus, through substitution we get

$$Z_2 = \frac{t_m}{t_1} \cdot Z_1 = \frac{Z_1}{V_{\text{REF}}} \cdot V_{\text{S/H}}$$

Now, by choosing

$$|Z_1| = V_{\text{REF}}$$

we get

$$Z_2 = V_{\text{S/H}} \tag{4.19}$$

This result indicates that the counter reading Z_2 is made independent of the clock frequency and components values, so better measurement of $V_{\text{S/H}}$ is guaranteed. Moreover, it is also apparent that through integration, any superimposed noise on $V_{\text{S/H}}$ could be partially reduced. Furthermore, power supply hum that might affect the precision could be cancelled or even reduced if period t_1 is a multiple of the power supply period; that is,

$$t_1 = k \cdot \frac{1}{f_{\text{supply}}}$$

Several monolithic integrated dual-slope ADCs are available in the market. They are designed either to directly feed a seven-segment display for applications such as digital multimeters or to be coupled with a μ processor in control systems. Some examples are given in Table 4.14.

TABLE 4.14

Some Commercially Available Monolithic Integrated Dual-Slope ADCs

Type	Resolution (bits)	Conversion Time (ms)	Output
ICL7109	12	40	Binary
TSC80	16	25	Binary
ICL7106	3 1/2	200	Seven-segment
ICL7129	4 1/2	500	Seven-segment

Source: Adapted from U. Tietze and Ch. Schenk, *Electronic Circuits*, Springer Science and Business Media, 1991. With permission.

4.11 PRACTICAL HIGH-SPEED, HIGH-RESOLUTION ADCs

Demands on high-speed, high-resolution ADCs have been recently increased. Of special interest are designs that can be manufactured in monolithic form. Therefore, they should feature circuit simplicity and moderate power consumption. The increased demand on such devices is due to the widespread use of many digital-based modern portable electronic appliances such as camcorders, CD players, digital cameras, MP3 players, cellular phones, and the like. Furthermore, high-performance ADCs play a major role in several industrial and medical applications, such as HDTV, digital broadcasting, satellite communications, and MRI and ultrasound imaging. The goal of converters' architects is to develop designs that can be produced on the smallest possible silicon area or, if possible, that can be integrated together with the digital system where it should serve as an interface (on-chip ADC). Flash-type (pure parallel) ADCs are the best candidates when speed is of prime concern. However, as resolution requirements increase, so does the required number of comparators—exponentially. If manufactured in integrated circuit form, the previously mentioned problems due to the large number of parallel-connected comparators come into play. They are repeated here for convenience:

- Excessive and therefore uneconomical die area
- Errors due to the cumulative effect of the comparators' bias currents flowing through the resistor string
- Excessive parasitic capacitive loading due to the large number of parallel-connected comparators, as their input capacitances will add together
- Excessive power consumption

Therefore, state-of-the art monolithic flash-type ADCs have resolutions that are limited to 8 or 10 bits at most. Such resolutions are satisfactory for most ultra-high-speed applications like image, video, and transient signal processing. On the other hand, there are many other applications that require higher resolutions at lower speeds, as in digital audio, ultrasound imaging, and so on. This requirement makes pure flash converters overqualified for such high-resolution applications. Because economy is an important engineering issue, ADC architects are therefore concerned with the development of designs that satisfy the speed requirements of a wide spectrum of applications at reasonable hardware complexity and reasonable power consumption. Such conditions, if satisfied, make designs amenable for monolithic integration.

One way to reduce hardware complexity of high-resolution ADCs is to generate the output word in two or more cascaded quantization blocks, giving rise to what is called *subranging converters*. Furthermore, stages in the cascade could be either of the same type; that is, all flash or a combination between successive approximation and flash quantizers (FQs). This design should result in a drastic reduction in hardware complexity, although at the expense of speed degradation, which increases as the number of stages in the cascade increases. Because the input and output of each stage in the cascade are incompatible (analog input vs. digital output), interfacing devices are needed. Such devices usually contribute much to the speed loss. Front end stages, usually called *coarse quantizers*, generate the MSBs of the output word. They deliver rough estimates of the applied inputs. Succeeding stages should therefore measure

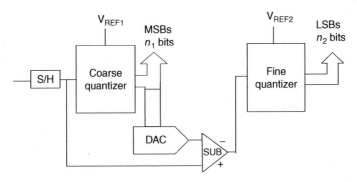

FIGURE 4.46   A basic structure of a subranging ADC.

the true residue (error) of such estimations. This might suggest that the required inter-stage devices should include DACs and subtracting amplifiers (residue amplifiers). These devices are unfortunately time-consuming, such that the delays they produce could be much greater than the time required by the main stages.

4.11.1   CASCADED ADC ARCHITECTURES

This configuration, also known as *subranging architecture*, furnishes a practical design alternative for single-stage high-resolution ADCs, especially where the pure flash solution is economically unfeasible. A block diagram illustrating the concept is given in Figure 4.46. The converter consists of two quantization blocks, labeled "coarse" and "fine" quantizers (not necessarily of the same type or architecture). The analog input is first sampled and held, before being applied to the coarse quantizer to generate n_1 MSBs. These same bits are then applied to an n_1-bit DAC. The resulting DC that represents the nearest analog equivalent of the MSBs is then subtracted, in the block labeled SUB, from the original sample amplitude. The residue will furnish the input to the next "fine" quantizer to generate the LSBs. It is worth mentioning at this point that two different reference sources are used in the two blocks. The first one, V_{REF1}, is equal to the converter's FSR, whereas the reference voltage for the second stage, V_{REF2}, should be equal to the quantum of the first quantizer; that is,

$$V_{REF2} = Q_1 = \frac{V_{FSR}}{2^{n_1}}$$

The schematic diagram given in Figure 4.47 should help describe this equation. It is possible, however, to operate the two quantizers using the same reference sources, if the subtracting amplifier can provide an amplification ratio of 2^{n_1}, as illustrated in Figure 4.48. It should also be stressed here that the precision requirements on the components of the first stage are the same as those of the whole converter, assuming a resolution of $N = n_1 + n_2$.

However, due to this amplification, errors in the coarse quantization and the DAC are also amplified, resulting in serious errors. In such cases, digital error correction (DEC) is usually implemented. One possible scheme is to generate n_1 MSBs

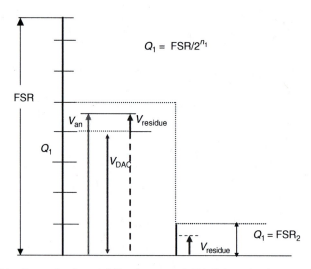

FIGURE 4.47   (See color insert following page 262.) Subranging in cascaded ADCs.

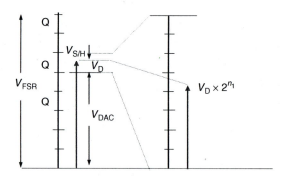

FIGURE 4.48   (See color insert following page 262.) Subranging ADC with interstage amplification.

in the first quantizer and $n_2 + 1$ LSBs in the second block, with an overlap of 1 bit, as shown in Figure 4.49. To introduce digital error correction, the gain of the residue amplifier is to be reduced by the same factor that the quantum of the fine quantizer is reduced—namely, 1/2—while keeping the same FSR for both stages. The redundant bit B_5 in Figure 4.49 can then be used to correct probable errors in the MS part of the output word. Moreover, better error correction can be achieved if more than 1 bit are overlapped.

To compensate for the speed loss due to cascading, a technique called *pipelining* is usually introduced. In this technique, cascaded stages are used to process different samples, yet with a certain degree of overlap. Thus while the first stage is dealing with the ith sample, the second stage should deal with the $(i - 1)$th one. Of course, 100% overlap is impossible, due to the interstage coupling devices. However, in some cases, the conversion time may approach the conversion time of one of them.

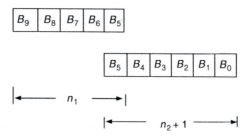

FIGURE 4.49   DEC using bit redundancy.

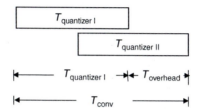

FIGURE 4.50   Pipelining technique.

Figure 4.50 illustrates schematically the technique. The added period, labeled $T_{overhead}$, is called *pipelining overhead*. It depends usually on the degree of overlap.

4.11.2   CASCADED FLASH QUANTIZERS

As said before, cascading of quantization blocks provides an economical alternative for the uneconomical, and sometimes unnecessary super-fast pure flash solution. Although it is not a rule, cascaded flash converters are usually designed and produced in monolithic integration form. Therefore, to ease layout, quantization blocks in the cascade should have a similar structure (same resolution), and they should also consume minimal power and occupy the smallest possible silicon area on the die to economize the production. One of the main design issues is the allocation of the number of bits n_i on the cascaded quantizers and hence the number of stages in the cascade m. This allocation determines the obtainable throughput rate (speed) as well as the resulting hardware complexity and cost. Increasing m does reduce complexity and hence cost, but at the expense of reduced speed. The optimum choice of m and n, besides being application-dependent, is usually given by the cost and delays of the used components. Table 4.15 gives an example of the comparators' count N for six possible configurations of a 12-bit converter. The relationship is illustrated in Figure 4.51. Of special importance here is the overall speed, which should be carefully estimated to check the suitability of the respective design to the application for which it is intended.

The following example highlights how the incurred speed loss increases drastically as m increases.

TABLE 4.15

Number of Comparators in a 12-bit Cascaded-Flash Converter

Number of Blocks "m"	Resolution/Block "n bits"	Number of Comparators "N"
1	12	$(2^{12} - 1) = 4095$
2	6	$2 \cdot (2^6 - 1) = 126$
3	4	$3 \cdot (2^4 - 1) = 45$
4	3	$4 \cdot (2^3 - 1) = 28$
6	2	$6 \cdot (2^2 - 1) = 18$

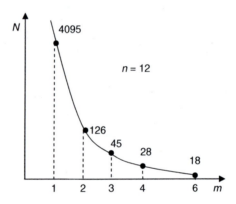

FIGURE 4.51   Number of comparators N versus the number of cascaded stages m.

TABLE 4.16

Data for Example 4.17

Component	Delay
Analog comparators	$T_{resp} = 5$ ns
PEs (any ratio)	$T_{pd} = 5$ ns
DAC (any word length)	$T_{sett} = 100$ ns
Op. Amp.	Slew rate = 1 V/μs

EXAMPLE 4.17

Examine all possible design configurations for a 12-bit, 1 V FSR ADC in the form of similar cascaded flash-type quantizers. The components listed in Table 4.16 are available.

Solution

Because all quantization blocks (m stages) are assumed to be similar and have n-bit resolution, the conversion time of each block T_F is constant and is given by

$$T_F = T_{\text{resp comp}} + T_{\text{pd PE}} = 5 + 5 = 10 \text{ ns}$$

Also, the delay due to the interstage DAC, T_{sett}, is constant irrespective of its resolution. The delay T_{Si} due to the ith subtractor is mainly determined by the slew rate of the used Op. Amp. SR and the quantum Q_i of its preceding stage, where

$$Q_i = \frac{\text{FSR}}{2^{n_i}}$$

Thus it is given by

$$T_{Si} = \frac{Q_i}{\text{SR}} = \frac{\text{FSR}}{2^{n_i} \cdot \text{SR}} = \frac{1\,\text{V}}{2^{n_i} \cdot 1\,\text{V}/\mu\text{s}} = \frac{10^3}{2^{n_i}}\,\text{ns}$$

Now let us consider all design possibilities:

Case I. Pure flash ($m = 1$, $n_1 = 12$)

As there is only one block, the conversion time is that of one FQ: T_F; that is, $T_{\text{conv}} = T_F = 10$ ns.

$$\text{Number of comparators } N = (2^{12} - 1) = 4095$$

Case II. Two quantization blocks; each has a resolution of 6 bits ($m = 2$, $n_1 = 6$). The conversion time is given by

$$T_{\text{conv}} = 2 \cdot T_F + T_{\text{settDAC}} + T_S = 2 \times 10 + 100 + \frac{1000}{2^6} \approx 136\,\text{ns}$$

$$N = 2 \cdot (2^6 - 1) = 126$$

Case III. Three blocks; each has a resolution of 4 bits ($m = 3$, $n_1 = 4$). Two DACs and two subtractors are needed in this case. Thus, we can write

$$T_{\text{conv}} = 3 \cdot T_F + 2 \cdot T_{\text{sett}} + T_{s1} + T_{s2} = 3 \times 10 + 2 \times 100 + \frac{1000}{2^4} + \frac{1000}{2^8} \approx 296.5\,\text{ns}$$

$$N = 3 \cdot (2^4 - 1) = 45$$

Case IV. Four blocks; each has a 3-bit resolution ($m = 4$, $n_1 = 3$). The conversion time in this case will be

$$T_{\text{conv}} = 4 \cdot T_F + 3 \cdot T_{\text{settDAC}} + \frac{T_S}{2^3} + \frac{T_S}{2^6} + \frac{T_S}{2^9} - 40 + 300$$

$$+ 1000 \left[\frac{1}{8} + \frac{1}{64} + \frac{1}{512} \right] = 482.5\,\text{ns}$$

$$N = 4 \cdot (2^3 - 1) = 28$$

Case V. Six blocks; each has a resolution of 2 bits. In this case, we can write

$$T_{conv} = 6 \cdot T_F + 5 \cdot T_{settDAC} + T_s \left[\frac{1}{2^2} + \frac{1}{2^4} + \frac{1}{2^6} + \frac{1}{2^8} + \frac{1}{2^{10}} \right]$$

$$= 60 + 500 + 333 \approx 893 \text{ ns}$$

$$N = 6 \cdot (2^2 - 1) = 18$$

The plot given in Figure 4.52 illustrates the degradation in the conversion speed, as well as the decrease in the number of comparators, as the number of cascaded flash stages increases.

The block diagram given in Figure 4.53 illustrates how FQs are cascaded.

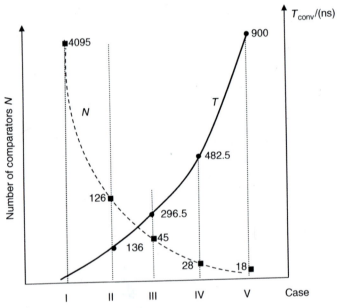

FIGURE 4.52   Comparators' count N and conversion time T for a 12-bit cascaded FQ.

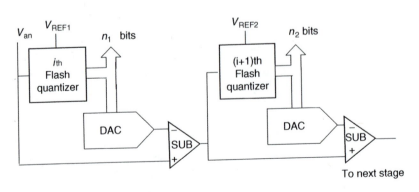

FIGURE 4.53   Cascading of FQs.

TABLE 4.17

Monolithic Half-Flash ADCs

Type	Resolution (bits)	f_{smax} (MHz)	Technology
ADC0820	8	1	CMOS
ADC310	10	20	ECL
AD671	12	2	CMOS

Source: Adapted from Tietze and Schenk, *Electronic Circuits*, Springer Science and Business Media, 1991. With permission.

Comparators are usually the most area-demanding component in flash-type converters, as far as monolithic integration is considered. Therefore, they should be carefully designed to minimize their area, power consumption, and delay. Table 4.17 lists some commercially available ADCs that are based on the half-flash concept.

The following hypothetical example illuminates the impact of the comparators' count on the occupied silicon area and hence the expected cost of an integrated ADC.

EXAMPLE 4.18

Estimate the required silicon area A_{si} for each of the previously considered cases in Example 4.15 as a function of the comparator's area, then plot the obtained results and select the optimum configuration; that is, the one that minimizes the product $T_{conv} \times A_{si}$. The required silicon area for all cases is plotted in Figure 4.54.

Consider the following assumptions:

Area of one comparator	A_k	mm^2
Area of one PE	$(0.1 n_i) A_k$	mm^2
Area of one DAC (resolution n_i bits)	$(1 + 0.2 n_i) A_k$	mm^2
Area of one Op. Amp.	A_k	mm^2

Solution

Case I. 1×12 bits

$$A_{si} = 4095 A_k + (0.1 \times 12) A_k = 4096.2 A_k \text{ mm}^2$$

Case II. 2×6 bits

$$A_{si} = (63 \times 2) A_k + 2(0.1 \times 6) A_k + (1 + 0.2 \times 6) A_k + A_k = 130.4 A_k \text{ mm}^2$$

Case III. 3×4 bits

$$A_{si} = (15 \times 3) A_k + 3(0.1 \times 4) A_k + 2(1 + 0.2 \times 4) A_k + 2A_k = 51.8 A_k \text{ mm}^2$$

Case IV. 4×3 bits

$$A_{si} = (7 \times 4) A_k + 4(0.1 \times 3) A_k + 3(1 + 0.2 \times 3) A_k + 3A_k = 36 A_k \text{ mm}^2$$

Case V. 6×2 bits

$$A_{si} = (3 \times 6) A_k + 6(0.1 \times 2) A_k + 5(1 + 0.2 \times 2) A_k + 5A_k = 31.2 A_k \text{ mm}^2$$

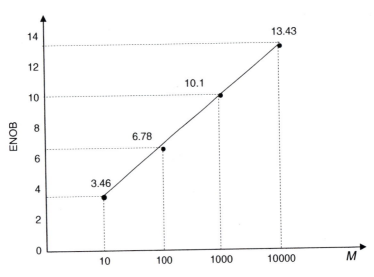

FIGURE 4.61   Effect of OSR (*M*) on the effective number of bits.

4.11.5   Modern High-Speed, High-Resolution ADC Architectures

As mentioned before, high-resolution ADCs are gaining interest in many fields. Applications such as digital audio, medical imaging using ultrasound, and others represent the main customers of this category of converters. For most of these applications, high throughput rate at a reasonable hardware complexity is a must. Conventional ADC architectures become either too slow or too complex as the resolution increases [6]. Modern designs aim at designing low-cost ADCs to have the highest possible resolutions at the maximum possible throughput rates. For best possible speeds, almost all designs generate the output word either by cascading or even recycling low-resolution FQs. However, cascading or recycling such mixed-mode circuits (analog input/digital output) necessitates coupling devices such as DACs and subtracting amplifiers. These interstage coupling devices cause, as mentioned before, a relative deterioration in the speed performance of the resulting converter.

An example of economical high-speed, high-resolution converters that implements the successive approximation algorithm in a flash-like architecture is described in the following section. Through circuit techniques, it minimizes speed loss due to recycling peripherals.

4.11.5.1   An 11-Bit Cyclic-Flash ADC (Example)

As depicted in Figure 4.62 [10], the core of the circuit is a 3-bit FQ that is recycled to generate the 11 bits, and a single 8-bit addressable latch and 8 analog switches. The output word is generated in two phases. In phase I, 8 bits are generated in a successive approximation fashion which are then followed by 3 flash-generated bits during phase II. The voltage references of the comparators are obtained from an *R*-2*R* ladder network. Phase I starts upon applying the sample amplitude, in parallel, to the eight

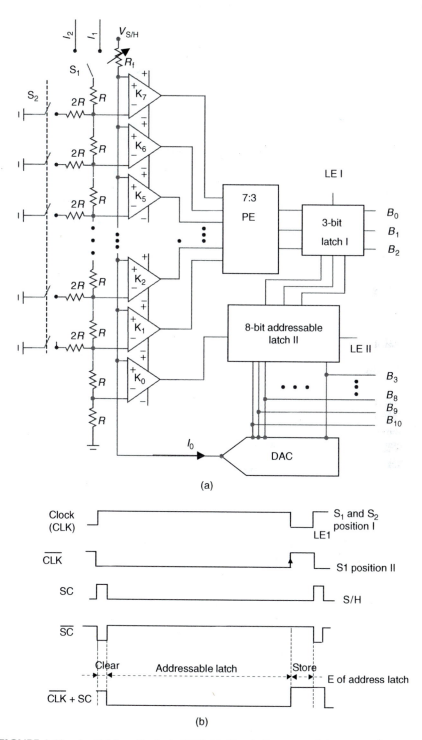

FIGURE 4.62 An 11-bit cyclic-flash ADC. (a) Circuit diagram and (b) timing diagram.

analog comparators K_0–K_7, turning switch S_2 on while switching S_1 to position I. This step lets current I_1 flow through the R-$2R$ ladder network so as to develop binary-weighted reference voltages for the comparators of magnitudes 1/2 FSR, 1/4 FSR, ..., 1/128 FSR. The output of comparator K_0 furnishes the input to the addressable latch L_{II}, while other comparators' outputs are used to generate three addressing bits in the 7:3 PE. The addressed outputs of L_{II} are then used to drive the 8-bit current-output DAC. The generated current I_0 will develop a voltage across resistor R_f (V_{MSB}'s $= I_0 \times R_f$), leaving a difference voltage of $V_D = V_{S/H} - V_{MSB}$'s as an input for the comparators for the next cycle. Resistor R_f is made variable to allow for the compensation of whatever errors there are due to a comparator's bias currents. The resulting difference voltage will produce a new set of comparators' outputs and hence a new 3-bit address to latch II. The process is repeated until comparator K_1 goes low where switches S_2 are turned off and S_1 is turned to position II to drive current I_2. The new selected reference current I_2 will furnish a new set of references for the comparators during the second phase to generate three flash LSBs. It is worth mentioning here that the required number of cycles is amplitude-dependent, in contrast with the conventional successive approximation quantization technique and flash-type converters.

4.12 TESTING OF DATA CONVERTERS

Testing of data converters is essential so that designers can ensure that the requirements of the converter are satisfied. Also it helps to verify the claimed specification by a manufacturer. Testing is divided into qualitative and quantitative testing, and further into static and dynamic testing. The results of such tests are several parameters that define precisely the performance of the converter. Examples of such parameters are offset, gain, differential nonlinearity (DNL) and integral nonlinearity (INL), ENOB, and signal-to-noise and distortion ratio (SNDR). To evaluate or verify the performance of ADCs and DACs, reliable and efficient standard qualitative and quantitative techniques have been devised. However, here we are going to study the performance of ADCs only.

4.12.1 STATIC TESTING

The first and the most important step in static testing of an ADC is finding its transfer characteristic. The test setup shown in Figure 4.63 is usually used. A triangular wave generator is needed to apply a low-frequency signal to the ADC and to the x-input of an oscilloscope operating in the x–y mode as well. On the screen, a transfer characteristic like the one shown in Figure 4.64 (ideal) or Figure 4.65 (practical) will be displayed.

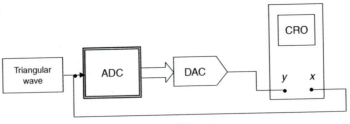

FIGURE 4.63 Test setup for measuring the transfer characteristics of an ADC.

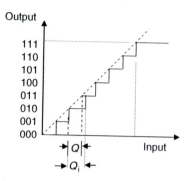

FIGURE 4.64 Transfer characteristics of a 3-bit ADC.

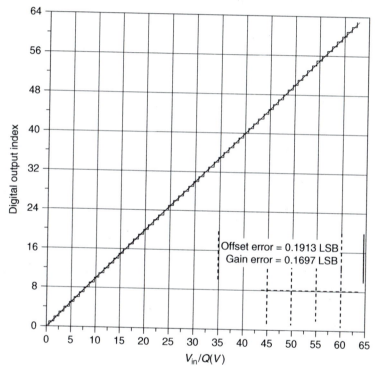

FIGURE 4.65 Input/output characteristic of a 6-bit flash ADC. (Adapted from Hamad, A., Testing of high speed A/D converters, MS Thesis, Mansoura University, 1998.)

From the obtained transfer characteristics, one can calculate the following:

1. Differential nonlinearity.
   The DNL at the ith level is defined as:

$$d_i = \frac{Q_i - Q}{Q}$$

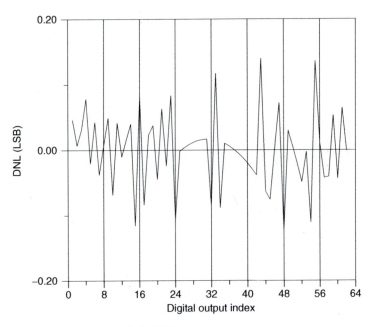

FIGURE 4.66 DNL for a 6-bit flash ADC.

It actually represents the errors due to the unequal step size of the quantizer. Figure 4.66 illustrates the DNL of a 6-bit ADC.

2. Integral nonlinearity.
   The INL of an ADC is calculated from

$$D_i = \sum_{j=1}^{i} d_j$$

The calculated INL again for the same 6-bit converter is plotted in Figure 4.67, where a maximum of 0.183 LSB was detected.

3. The histogram test.
   It is possible to judge the performance of an ADC through the histogram test. Here an input signal is applied that has equally probable amplitudes; for example, a low-frequency triangular signal that is sampled at a much higher frequency. Observing the output codes of the ADC and counting the frequency of occurrence of each code, one can plot a histogram like the one shown in Figure 4.68 for a practical ADC. A nonuniform histogram—that is, one with bars of unequal heights—means errors.

4.12.2 DYNAMIC TESTING

A simple qualitative dynamic test can be performed using the same setup in Figure 4.69. Only the sampling rate is to be gradually increased until the observed transfer characteristics deteriorate. Deterioration is detected when errors like missing codes,

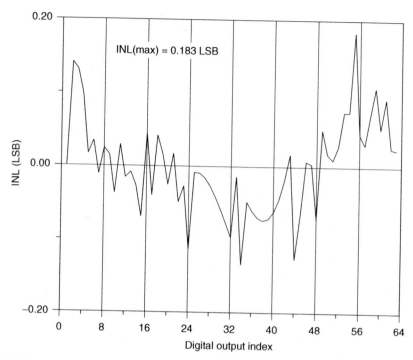

FIGURE 4.67   INL for a 6-bit flash ADC.

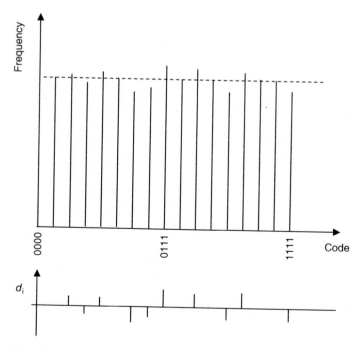

FIGURE 4.68   Histogram and the obtained DNL for a 3-bit ADC.

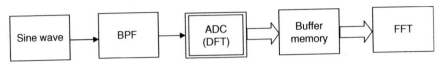

FIGURE 4.69 Test setup for calculating the DFT of an ADC.

nonmonotonicities or large DNL are present. The maximum possible sampling rate (just before errors) is used to estimate the conversion time. Dynamic quantitative tests, on the other hand, study the performance either:

1. In the time domain, by analyzing a histogram, representing the frequency of occurrence of each possible code (code density analysis). This approach is used for evaluating some I/O-related ADC characteristics, such as offset, gain, DNL and INL, SNDR, and ENOB.
2. In the frequency domain, by computing the Fourier transform of the output codes, with a pure sine wave at the input, and analyzing the resulting spectrum [1]. A test setup for measuring the discrete Fourier transform (DFT) of an ADC is shown in Figure 4.69, together with a test result on a 6-bit ADC [12]. Parameters like SNR, ENOB, and total harmonic distortion (THD) could be evaluated at high accuracy, even if few numbers of samples per record are available. Because the SNR of an ADC is roughly estimated from

$$SNR \approx 6 \cdot n$$

where n is the "nominal" resolution (number of bits).

Due to several reasons, the measured SNR is usually less than the one calculated above. Therefore, the ENOB is introduced and can be calculated as follows:

$$ENOB = \frac{measured\ SNR}{6}$$

The test setup shown in Figure 4.69 includes a sine wave generator, a "narrow-band" BPF to eliminate any possible harmonics, the ADC being tested, a buffer memory to store (temporarily) the output, and the fast Fourier transform (FFT) algorithm on any PC. An example of a test result (normalized magnitude DFT) is illustrated in Figure 4.70. Best results should contain the fundamental component and a noise floor.

4.13 REVIEW QUESTIONS

1. Sketch a block diagram for a typical DSP system, giving in a few words the function of each block.
2. What are the main specs of an ADC?
3. Suggest a DAC type for a 12-bit audio system, giving reasons for your selection.
4. Suggest some suitable capacitor types for an S/H circuit.

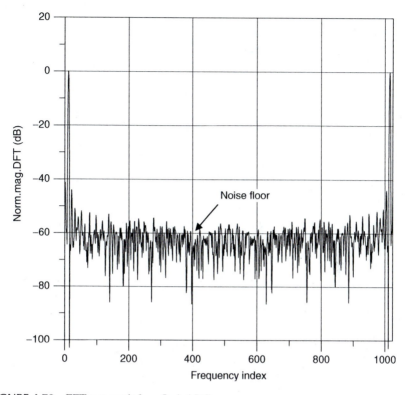

FIGURE 4.70   FFT test result for a flash ADC.

5. What is meant by "aliasing"? How could it be avoided or minimized?
6. You are required to design two ADCs: one for an audio system, and one for a video system. Suggest a suitable converter type for each system, stating the proper sampling rates and the times available for conversion.
7. Explain why it is economical to use the SA quantizer as the front end device in a heterogeneous cascaded converter (SA/flash).
8. Define the following terms: DNL, INL, THD, and ENOB.

4.14 PROBLEMS

1. For a two-buffer S/H circuit, the following data were recorded:

Op. Amp.	Switch	Holding Capacitor	Maximum Signal Frequency
$R_{i/p} = 500\ \text{M}\Omega$	$T_{ON} = 0.5\ \text{ns}$	$C_H = 25\ \text{pF}$	$f_m = 5\ \text{MHz}$
$R_{o/p} = 50\ \Omega$	$R_{ON} = 30\ \Omega$		

Calculate the proper values for the periods T_A and T_H; then find f_s and the duty cycle.

2. Design a two-buffer S/H circuit using the following components:
   - LF521 Op. Amp's with $R_{i/p} = 100$ M Ω, $R_{out} = 50$ Ω, $I_{bias} = 10$ μA
   - MOS switch $R_{ON} = 50$ Ω

   to operate at a sampling frequency of 1 MHz with a duty cycle of 1:10. Calculate the droop rate and draw your circuit.
3. For a 3-bit switched resistor DAC, calculate:
   - The required number of switches.
   - The maximum ON resistance they can have to keep the error within the limit of 1/2 Q, provided that the FSR = 1 V and the used Op. Amp has $I_{bias} = 150$ μA.
   - The chain resistance R such that the dissipated power is limited to 1 mW.
   - Draw a circuit diagram.
4. Calculate the best possible resolution of a counter-type ADC that uses the components given in the following table to operate at sampling frequency of 1 MHz.

Comparators	AND Gate	Binary Counter	DAC
$T_{resp} = 15$ ns	$T_{pd} = 5$ ns	$T_{pd} = 1.66$ ns/bit	$T_{sett} = 40$ ns

5. Calculate the maximum possible sampling rate of the analog input to an 8-bit flash-type ADC that uses the following components:

   NE521 comparators: $T_{Resp} = 10$ ns $\pm$ 10%
   74148 priority encoder: $T_{pd} = 10$ ns

6. Design a 12-bit, 1V FSR, high-speed ADC in form of cascaded quantizers using the components listed in the following table:

Component	Available Quantity	Price ($)	Delay/ns
Analog comparator	126	1	15
63:6 priority encoder	2	2	10
DAC	1	10	50
Op. Amp.	1	1	Slew rate = 1 V/μs

   Suggest a proper design for the cascade, then calculate the maximum possible input signal frequency and the resulting cost.
7. In a 3-bit, R-2R DAC, the Op. Amp. dissipates a power of 0.5 mW and has a slew rate of 10/μs. The used MOS switches have $T_{switches} = 50$ ns. Given that the FSR is 1 V, calculate a proper value for the resistance R such that the power dissipated is limited to 1 mW. Calculate also T_{sett} of the DAC and draw a circuit diagram.
8. Calculate the maximum possible input signal frequency for a 6-bit counter-type

nents:

1 V FS, high-speed ADC in the form of
ligitize a 1 V signal having a maximum
possible configuration, and then evalu-
components are available to you:

Delay	Cost ($)	
ns	2	GR = 0.1 mV
R = 1 V/μs	1	
0 ns	8	
ns	2	

at consists of a successive approximation
ne, find the best possible bit distribution
zes the product time × cost, considering
. The following components are available:

Delay	Cost ($)
0 ns	1
0 ns	2
40 ns	10
100 ns	2/bit
SR = 1 V/μs	1

= 20 ns

= 10 ns

sampling frequency and the corresponding
s of 6, 8, and 12 bits, respectively. Sketch

omical 16-bit, 1 V FSR high-speed ADC
quantizers to operate at a frequency of
mponents:

Component	Quantity	Price ($)	Delay
Analog comparator	Up to 512	1	25 ns
Priority encoder	Up to 4	2	10 ns
DAC	Up to 4	10	100 ns
Op. Amp.	Up to 4	1	Slew rate = 1 V/μs

Suggest a suitable design for the cascade, then calculate the resulting cost.

13. For a 6-bit switched-resistor DAC, determine:
    a. The number of quantization levels.
    b. The percentage resolution.
    c. The maximum permissible resistor tolerance.
    d. The output voltage for the input [1 1 0 1 0 1], provided that $R_{LSB} = R_F = 1$ kΩ and the reference voltage $E = 1$ V.
14. The input sampling rate of a successive approximation ADC is 100 KHz. If the used clock has a maximum guaranteed toggling rate of 0.5 MHz, calculate the best possible resolution of this converter.
15. Calculate the permissible grey region for the comparator in an ADC that has an FSR of 10 V and a resolution of 10 bits. Calculate the gain of the comparator for a TTL and ECL output logic levels.
16. Calculate the best possible resolution of a quantizer if it is required to have an FSR of 6 V at a maximum error of ± 0.05 V. Calculate the resulting S/N ratio for a triangular input.
17. Design an 8-bit high-speed ADC to digitize video signals having $f_{max} = 5$ MHz and FSR of 2V at the minimum possible comparator count. You have the following components available to you:
    • Analog comparators with $t_{resp} = 5$ ns
    • 15:4 priority encoders with $t_{pd} = 5$ ns
    • 4-bit DAC with $t_{sett} = 50$ ns
    • High-speed Op. Amp. with slew rate of 10 V/μs
18. For a 4-bit switched-resistor DAC, calculate:
    a. The required number of switches
    b. The power dissipated in the resistor chain if $R = 10$ Ω and $V_{REF} = 1$ V
    Draw a circuit diagram.
19. In a two–Op. Amp. S/H circuit, the following data were recorded:

Op. Amp.	Switch	Holding Capacitor	Sampling Frequency
$R_{i/p} = 100$ MΩ	$T_{ON} = 0.1$ ns	$C_H = 30$ pF	$f_s = 10$ MHz
$R_{o/p} = 40$ Ω	$R_{ON} = 7.6$ Ω	$\varepsilon_r \gg$	
$I_{bias} = 20$ μA			

    a. Calculate suitable values for the periods T_A and T_H, then calculate the duty cycle.
    b. Calculate the droop rate.
20. Design an economical 16-bit high-speed ADC to digitize a low-quality video signal with $f_{max} = 2$ MHz and an FSR of 2 V. The following components are available to you:
    • Analog comparators with $t_{resp} = 30$ ns
    • 255:8 priority encoders with $t_{pd} = 10$ ns
    • 8-bit DAC with $t_{sett} = 100$ ns
    • High-speed Op. Amp. with slew rate of 2 V/μs

21. For a 10-bit counter-type ADC that uses the following components:
    - Estimate the settling time of a DAC that could be used for the converter to digitize 5 kHz signals.
    - Calculate the conversion time for an input $= 1/2$ FSR.

Comparators	AND Gate	Binary Counter
$T_{resp} = 15$ ns	$T_{pd} = 5$ ns	$T_{pd} = 20$ ns

22. For a 1 V FSR, 12-bit cascaded ADC that consists of a 4-bit successive approximation quantizer followed by an 8-bit flash-type quantizer, calculate:
    a. The maximum possible input signal frequency
    b. The overall cost
       given the components listed in the following table.

Component	Delay	Cost ($)
Comparators	10 ns	2
PE	10 ns	2
SAR	40 ns	10
DAC	90 ns	15
Op. Amp.	3.125 V/µs	1

23. Design an S/H circuit to operate at 10 MHz and a 10% duty cycle, if the available components have the following specifications:

		Component
$R_{o/p} = 40 \ \Omega$	$R_{i/p} = 100 \ M\Omega$	Op. Amp.
$R_{Off} = 500 \ M\Omega$	$R_{ON} = 10 \ \Omega$	MOS switch

Allow eight time constants for the holding capacitor to reach its steady state.

24. You are required to design a 12-bit 1 V FS, high-speed ADC in the form of similar cascaded flash-type quantizers to digitize a 1 V signal having a maximum frequency of 2.5 MHz. Select the best possible configuration and then evaluate the resulting cost. The following components are available to you:

Component	Count	Delay	Cost ($)	
Comparators	Up to 200	20 ns	2	GR = 0.1 mV
Op. Amp.	Any number	SR=1 V/µs	1	
DAC	Up to 3	100 ns	10	
PE	Up to 4	20 ns	2	

25. The input sampling rate of a counter-type ADC is 10 KHz. If the used counter has a maximum guaranteed toggling rate of 0.5 MHz, calculate the best possible resolution of this converter.

26. Given a 12-bit, 1 V FSR ADC that consists of a successive approximation quantizer in cascade with a flash one, find the best possible bit distribution among the two stages that minimizes the product time × cost, considering the cases 4 + 8, 6 + 6, and 8 + 4. The following components can be used:

Component	Delay	Cost ($)
Comparators	10 ns	1
PE	10 ns	2
SAR	40 ns	10
DAC	100 ns	4/bit
Op. Amp.	SR 1 V/μs	1

27. It is required to design a low-cost, high-speed 12-bit ADC to digitize a 1 V, 2.5 MHz signal. The components given in the following table are available to you:

Component	Delay	Count
Analog comparator	20 ns	Up to 126
PE	10 ns	Up to 3
Op. Amp.	SR = 2 V/μs	Up to 3

Select a suitable DAC from the following table:

Type	Settling Time	Resolution
I	50 ns	4 bit
II	100 ns	6 bit
III	120 ns	8 bit

28. Calculate the maximum possible signal frequency that could be applied to a counter-type ADC whose components have the following specs:

Comparator response time = 50 ns, counter prop. delay = 40 ns
DAC settling time = 300 ns, AND gate prop. delay = 10 ns

for resolutions ranging from 4 to 10 bits. Sketch the relation f_m versus n.

29. The following table gives the specs of three types of comparators. Select one type for a flash ADC that has a resolution of 8 bits and an FSR of 2 V. Calculate the maximum possible input frequency if the priority encoder has

a propagation delay of 10 ns.

Type	Response Time	Grey Region
I	90 ns	2 mV
II	40 ns	6 mV
III	10 ns	10 mV

30. In a two–Op. Amp. S/H circuit, the following data were recorded:

Op. Amps.	Hold Capacitor	Sampling Frequency
$R_{i/p} = 200 \ \text{M}\Omega$ $R_{o/p} = 50 \ \Omega$	$C_H = 50 \ \text{pF}$	$f_s = 20 \ \text{MHz}$ Duty cycle = 1:10

Calculate a proper value for the ON resistance of a suitable analog switch. Allow eight time constants for the capacitor to reach its steady state.

31. For an 8-bit, 2 V FSR counter-type ADC that uses the following components:

Comparators	AND Gate	Binary Counter	DAC
$T_{resp} = 20 \ \text{ns}$	$T_{pd} = 5 \ \text{ns}$	$T_{pd} = 15 \ \text{ns}$	$T_{sett} = 85 \ \text{ns}$

a. Calculate the conversion time for an input of 1/2 V.
b. Plot the relation of f_s versus V_{in} for inputs ranging from 0 to FSR in steps of 1/4 FSR.
c. From the plot, find the maximum possible input signal frequency for $V_{in} = 0.6$ FSR.

32. Calculate the maximum permissible resistor tolerance in a 6-bit weighted-resistor DAC. Calculate the required resistor values if $R_{MSB} = 100 \ \Omega$.

33. Given three voltage comparators, find the maximum permissible word length (n) of an ADC that uses each one, assuming an FSR of 4 V.

Type	Grey Region
µA710	1.5 mV
NE521	7.5 mV
µA760	0.5 mV

4.15 MINI PROJECT

Design and construct a 6-bit video-speed ADC using the minimum possible number of the components given in Problem 4.24. Measure the resulting DNL, INL, and offset. Calculate also the resulting cost.

5 Digital Signal Processing

5.1 WHY DIGITAL SIGNAL PROCESSING (DSP)?

As we have seen previously, in Chapter 2, real-life signals such as those obtained from a microphone or a video camera are analog by nature. This means that their magnitude can assume any value within a specified range. Traditional operations such as filtering, amplification, modulation, and so on are usually performed on signals in analog format. However, in the last three decades, great attention has been paid to processing signals in their digital (discrete-time) format. In spite of the added delay and cost due to the necessary interfacing circuits, in several applications, the attainable advantages might justify such overhead. The trend toward all-digital designs is supported by the rapid improvement of the performance characteristics of monolithic integrated digital circuitry, such as the speed-power product, the packing density (how many transistors per chip are introduced), and the cost and yield (ratio of error-free chips to the total number of chips in a batch). Moreover, traditional problems of analog circuits, such as impedance matching, components tolerance, drift, and so on, do not exist with digital circuits. In addition, several tailored hardware products for digital signal processing (DSP) are now on the market. They are becoming more mature, efficient, flexible, reliable, and above all, user-friendly. An example of such a device is the general-purpose TMS320C25 digital signal processor (DSP processor) from Texas Instruments [14]. DSP processors will be described in detail in Chapter 9.

Generally speaking, switching to digital processing of signals brings several advantages and enables designers to achieve characteristics that were impossible to realize with analog circuits. Representative examples are linear-phase filters, time-varying responses, and the time sharing of one processor for several signal sources.

5.2 SOME PRACTICAL APPLICATIONS

DSP techniques are now applied extensively in many fields where they have proved themselves to be efficient, reliable, economical—and sometimes unique and indispensable tools. Examples of DSP applications are given in the following sections, where they are classified according to their respective field. However, in Chapter 10, an overview and a detailed coverage of more practical applications will be discussed.

5.2.1 TELECOMMUNICATIONS

The ever-growing field of telecommunications could be considered one of the main customers of digital processing techniques. The exponential growth in the data that are exchanged between users all over the globe has made applications such as data

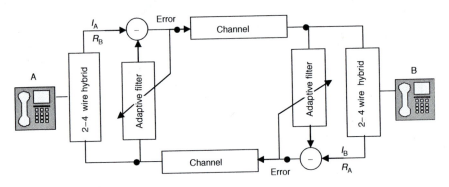

FIGURE 5.1　Echo cancellation.

compression a must. Some compression techniques like JPEG and MPEG are now considered to be industry standards. Practical applications of DSP in communications are many; a few such applications include speech coding in mobile (cellular) communications, data compression and coding in satellite communications, echo cancellation in long-distance telephony, data compression in video conferencing, and ghost cancellation in TV receivers. However, these are only examples of the many current applications.

Telephone calls—especially long distance ones—suffer from an annoying phenomenon called "echo" that results from mismatch at the sending and receiving ends. As illustrated in Figure 5.1, part of the incident signal due to subscriber A, I_A, leaks in the hybrid transformer and is reflected to reach A after a certain delay that depends on the distance it covers. If the reflected signal R_A is not attenuated enough and experiences a delay of around 540 min, subscriber A would receive a delayed version of his voice (echo). The scheme depicted in Figure 5.1 can be employed to minimize echo by subtracting an estimate of the reflected signal that is provided by an adaptive (digital) filter. The resulting error is used in the adaptation of the filter until the minimum error is attained.

5.2.2　Biomedical Engineering

DSP is an indispensable tool in several medical applications, such as diagnosis of heart diseases and brain disorders through EKG and EEG analysis, respectively. It helps also surgeons to determine precisely the coordinates of tumors and to investigate diseased organs, using techniques such as ultrasound imaging, CT scanning, MRI, nuclear magnetic resonance (NMR), and so on. Furthermore, it enables the monitoring of critical cases in ICUs. In radiology, DSP techniques are also successfully applied, where they are used in the enhancement and storage of X-ray images, or detection of disorders in the circulatory systems through Doppler measurements. Moreover, DSP is an important tool in the field of telemedicine, where remote diagnosis, online monitoring, guidance of operations, and drug prescriptions are made possible through live reports received by the physician's cellular phone or on the Internet. Examples of such reports are those expressing heart sounds or an EKG of a remote patient. In fact, the list is growing rapidly, and these areas are just a few examples.

5.2.3 INDUSTRIAL AND E-BANKING APPLICATIONS

Signal-processing applications in the industry are enormous. The following are just examples from a lengthy list. One of the rapidly growing fields of application is *pattern recognition*. It is being applied extensively in quality control tests, robotic vision, man/machine interfacing through speech recognition, and in e-banking (electronic banking). It is also applied in forensic applications, where identity verification is an indispensable tool. Most international airports have recently adopted high-precision identity verification systems that are based on human biometrics such as eye prints (iris textures) to identify blacklisted people. DSP techniques are also necessary in some critical applications such as e-commerce. In such fields, reliable and highly secure schemes should be provided. This is quite important, as credit card numbers, for example, should not be obtainable by Internet hackers to maintain the credibility of the system. Therefore, efficient ciphering algorithms should be developed. DSP techniques are also indispensable in the monitoring and control of critical or sophisticated industrial processes, as in nuclear reactors.

5.2.4 ENTERTAINMENT INDUSTRY

This huge industry is a major customer of DSP products. Many techniques are being applied in a diversity of fields, as in digital audio, to create effects such as theater effects, dome effects, and so on, to provide noise-free recordings (as in CDs and DVDs), and to enable recording of high-quality music at high compression ratios, as is used in MP3 players. DSP techniques are also used in animation, in multimedia applications, and in virtual reality (as in video games), and the like.

5.3 Z-TRANSFORM

Continuous-time systems are described, as mentioned previously, by constant-coefficient differential equations. The Laplace transform was the best tool to deal with this type of equations. On the other hand, discrete-time systems, which we are going to deal with, are represented by constant-coefficients difference equations. To solve such equations, a different transform is used—one that is derived from the Laplace transform; namely, the Z-transform. Through this transform, the whole left-hand side of the Laplace domain (*S*-domain) is mapped into the interior of a unit circle in a new domain, called the Z-domain, with its center at the origin and the $j\omega$ axis representing its circumference. Furthermore, the right-hand side is mapped to the outside, as shown in Figure 5.2. The double-sided Z-transform of a sequence of samples $\{x(nT)\}$ is defined by

$$Z\{x(nT)\} = X(Z) = \sum_{n=-\infty}^{n=\infty} x(nT) \cdot Z^{-n} \tag{5.1}$$

Recall that the Laplace transform $X(S)$ of a time function $x(t)$ is given by

$$X(S) = \int_{-\infty}^{\infty} x(t) \cdot e^{-st} \cdot dt \tag{5.2}$$

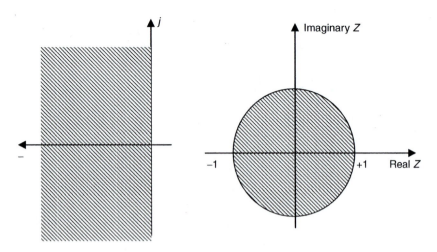

FIGURE 5.2   Mapping the S-domain to the Z-domain.

Thus it is easy to conclude by comparing Equations 5.1 and 5.2 that

$$Z = e^{ST} \tag{5.3}$$

where T is the sampling period. From Equation 5.3 it is evident that Z is a complex variable.

As we are dealing always with causal systems, Equation 5.1 can be rewritten as

$$X(Z) = \sum_{n=0}^{\infty} x(nT) \cdot Z^{-n} = Z\{x(nT)\} \tag{5.4}$$

which represents the one-sided Z-transform.

The inverse Z-transform is defined as

$$x(nT) = \text{INV } Z\{X(Z)\} \tag{5.5}$$

Recalling that

$$X(Z) = Z\{x(nT)\} = \sum_{n=0}^{\infty} x(nT)Z^{-n}$$

and expanding the previous equation, we get

$$X(Z) = x(0) + x(T) \cdot Z^{-1} + x(2T) \cdot Z^{-2} + \cdots \tag{5.6}$$

Thus we can conclude that one way to get the inverse Z-transform is to expand the transfer function as a power series. The coefficients of this power series, Equation 5.6, would represent the inverse Z-transform. Transfer functions in the form of rational functions can be expanded through long division into a power series. It should be mentioned here that the obtained inverse transform of the transfer function of a filter represents in fact its impulse response, as we shall see later.

EXAMPLE 5.1

Find the inverse Z-transform of the following function:

$$H(Z) = \frac{3 + 4Z^{-1} + 5Z^{-2}}{1 - 2Z^{-1} + 3Z^{-2}}$$

Solution

Through long division, we get

$$H(Z) = 3 + 10Z^{-1} + 16Z^{-2} + 2Z^{-3} - 44Z^{-4} + \cdots$$

thus

$$\text{INV } Z\{H(Z)\} = \{3, 10, 16, 2, -44, ...\}$$

The Z-transform of a certain function $G(S)$ can be obtained by applying the substitution

$$Z = e^{ST}$$

or

$$S = \frac{1}{T}\ln Z$$

so that

$$G(Z) = G(S)_{s=\frac{1}{T}\ln Z}$$

As stated earlier, the left-hand side of the S-plane is mapped into the interior of a unit circle in the Z-domain with the $j\omega$ axis of the S-plane representing its circumference. It is clear from Figure 5.2 that the point $+1$ on the real axis is equivalent to the frequencies $\omega = 0, \omega_s, 2\omega_s, ...,$ and the point -1 corresponds to the frequencies $\omega = \omega_s/2, 3\omega_s/2, ...$

It should be noted that the Z-transform of $G(S)$ is an analytic function of Z within its region of convergence (ROC).

EXAMPLE 5.2

The transfer function

$$H(S) = \frac{(S + 1)(S + 3)}{(S + 4)(S + 2 - j)(S + 2 + j)}$$

describes a certain third-order filter. Sketch its pole-zero distribution in the S-plane, then find the corresponding distribution in the Z-plane.

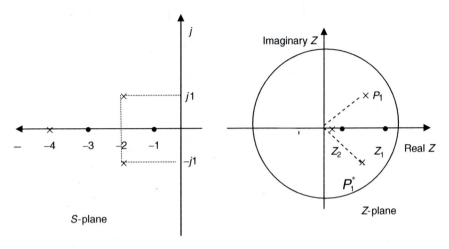

FIGURE 5.3   Pole-zero distribution of Example 5.3.

Solution

The zeros in the S-plane, as depicted in Figure 5.3, are located at $S = -1, -3$, and the poles are located at $S = -4, -2 \pm j1$.

Recalling that

$$Z = e^{ST} = e^{(\sigma + j\omega T)}$$

with $T = 1$, we get

$$Z_{z1} = e^{-1} = 0.3678$$

$$Z_{z2} = e^{-3} = 0.04978$$

and

$$Z_{p1} = e^{-4} = 0.018315$$

$$Z_{p2} = e^{-2 \pm j} = 0.1353\,[\cos 1 \pm j \sin 1]$$

5.3.1 PROPERTIES OF THE Z-TRANSFORM

The followings are some important properties of the Z-transform:

1. Linearity: $Z\{x(nT) + y(nT)\} = Z\{x(nT)\} + Z\{y(nT)\} = X(Z) + Y(Z)$

2. Convolution: $Z\{x(nT)*y(nT)\} = Z\{x(nT)\} \cdot Z\{y(nT)\} = X(Z) \cdot Y(Z)$

3. Time delay: $Z\{x(n - m)T\} = Z^{-m}X(Z)$

TABLE 5.1

Some Useful Z-Transform Formulas

$F(nT)$	$F(Z)$
K	$\dfrac{kZ}{Z-1}$
$\sin(\alpha n)$	$\dfrac{Z\sin\alpha}{Z^2 - 2Z\cos\alpha + 1}$
$\cos(\alpha n)$	$\dfrac{Z(Z-\cos\alpha)}{Z^2 - 2Z\cos\alpha + 1}$
$e^{-\alpha n}\sin(\alpha n)$	$\dfrac{Ze^{-\alpha}\sin\alpha}{Z^2 - 2e^{-\alpha}Z\cos\alpha + e^{-2\alpha}}$
$e^{-\alpha n}\cos(\alpha n)$	$\dfrac{Ze^{-\alpha}(Ze^{\alpha}-\cos\alpha)}{Z^2 - 2Ze^{-\alpha}\cos\alpha + e^{-2\alpha}}$

4. Reversed sequence: $Z\{x(-nT)\} = X(1/Z)$

5. Multiplying by a geometric series: $Z\{\alpha^n x(nT)\} = X(Z/\alpha)$

6. Multiplying by an arithmetic series: $Z\{n \cdot x(nT)\} = -Z(d/dZ)X(Z)$

Table 5.1 lists some important Z-transforms.

EXAMPLE 5.3

Prove that a delay by one sampling period T corresponds to multiplication by Z^{-1} in the Z-domain, then generalize to any number of delays.

Solution

By definition, the Z-transform of a sequence $x(nT)$ is given by

$$Z\{x(nT)\} = \sum_{n=0}^{\infty} x(nT) \cdot Z^{-n}$$

Multiplying both sides by Z^{-1} gives

$$Z^{-1}\{Zx(nT)\} = Z^{-1}\sum_{n=0}^{\infty} x(nT) \cdot Z^{-n}$$

$$= \sum_{n=0}^{\infty} x(nT)Z^{-(n+1)}$$

Changing the lower limit to $n = 1$ gives

$$\text{RHS} = \sum_{n=1}^{\infty} x(nT - T)Z^{-n}$$

Reverting the limit again to $n = 0$, we get

$$\text{RHS} = \sum_{n=0}^{\infty} x(nT - T)Z^{-n} - x(-T)$$

Due to causality, the last term does not exist and should be removed. Thus, we get

$$Z^{-1}X(Z) = \sum_{n=0}^{\infty} x(nT - T) \cdot Z^{-n}$$

Similarly, for a delay of m periods we can write

$$Z^{-m}X(Z) = \sum_{n=0}^{\infty} x(nT - mT)Z^{-n}$$

5.4   SOME BASIC DSP OPERATIONS

5.4.1   CONVOLUTION

Convolution is a widely used operation in many engineering applications, and especially in DSP [4]. It is defined as

$$y(n) = x(n) \otimes y(n) = \sum_{k=0}^{\infty} x(k) \cdot h(n - k) = \sum_{k=0}^{\infty} x(n - k) \cdot h(k)$$

for $n = 0, 1, 2, 3,$ and $(M - 1)$

where $x(n)$ and $h(n)$ are two causal time sequences of lengths N_1 and N_2 with $M = N_1 + N_2$.

Mathematically, it was found that convolution in the time domain is equivalent to multiplication in the frequency domain, and vice versa. Thus it is possible to get the convolution of two time sequences by multiplying their frequency transforms, then inverse transforming the product.

5.4.2 CORRELATION

Correlation, as the name implies, is a measure of similarities and relationships. There are two definitions for correlation [4]:

1. Cross-correlation, which measures the similarities and shared characteristics between two different signals. The cross-correlation function (CCF) of two zero-mean sequences is defined as follows:

$$\rho_{xy} = \frac{r_{xy}(n)}{\sqrt{r_{xy}(0) \cdot r_{yy}(0)}} \qquad n = 0, \pm 1, \pm 2, \ldots$$

where $r_{xy}(n)$ is known as the cross-covariance, defined by

$$r_{xy}(n) = \begin{cases} \dfrac{1}{N} \displaystyle\sum_{k=0}^{N-n-1} x(k) \cdot y(k+n) & n = 0, 1, 2, \ldots \\[2ex] \dfrac{1}{N} \displaystyle\sum_{k=0}^{N+n-1} x(k-n) \cdot y(k) & n = 0, -1, -2, \ldots \end{cases}$$

and

$$r_{xx}(0) = \frac{1}{N} \sum_{k=0}^{N-1} [x(k)]^2$$

$$r_{yy}(0) = \frac{1}{N} \sum_{k=0}^{N-1} [y(k)]^2$$

with N as the length of both sequences.

2. Autocorrelation, which is a special case of cross-correlation. The autocorrelation function (ACF) of a zero-mean sequence is defined as

$$\rho_{xx}(n) = \frac{r_{xx}(n)}{r_{xx}(0)} \qquad n = 0, \pm 1, \pm 2, \ldots$$

in which $r_{xx}(n)$ is known as the *autocovariance*, defined as

$$r_{xx}(n) = \frac{1}{N} \sum_{k=0}^{N-n-1} x(k) \cdot x(k+n) \qquad n = 0, 1, 2, \ldots$$

5.5   DIGITAL FILTERS

A digital filter in the general sense is a linear, discrete-time system that is described by a computational algorithm. It could be implemented either in:

1. Hardware, using special components, namely, unit delays, multipliers, and adders
2. Software that runs on a processor system

Like analog filters, digital filters are used to modify the frequency or phase characteristics of an input discrete-time signal. Due to their unique features, they are extensively used in several DSP applications, as in speech and image processing, amongst others.

Digital filters, as compared to their analog counterparts, provide several advantages—for example, linear-phase response, high accuracy, and time-varying responses. If hardware-implemented, they occupy a small volume and consume less power. Moreover, traditional problems like component tolerance, impedance matching, drift, and so on do not exist if not meaningless. They also posses the unique feature of being able to be multiplexed, where a single filter can be used for several input signals. On the other hand, limitations such as the maximum frequency of an input signal and the relatively high cost are dictated only by the current technology of integrated circuits and should represent no problem in the near future. Table 5.2 presents a comparison between analog and digital filters.

Input signals to digital filters should first be digitized prior to filtering. This implies, however, as mentioned in Chapter 4, that the signals should be band-limited to their maximum significant frequency f_m, employing a simple analog LPF (usually called an anti-aliasing filter). They are then sampled at a rate that is at least twice f_m in an S/H circuit. After being filtered, in the digital filter, the signal is DAC-converted and then reconstructed in an analog LPF (anti-imaging filter) whose cut-off frequency is again located at $f_c = f_m$. A block diagram describing the process is provided in Figure 5.4. The use of LPFs prior to sampling is necessary to prevent aliasing of the repeated signal spectra due to sampling. Also, the reconstruction filter, which is actually an integrator, serves in removing discontinuities in the output of the DAC. In general, the order of such filters gets smaller as the sampling frequency gets higher.

TABLE 5.2
A Comparison between Analog and Digital Filters

Property	Analog Filters	Digital Filters
Frequency domain	S-plane	Z-plane
Time variable	Continuous time t	Discrete time nT
Describing equation	Differential equations	Difference equations

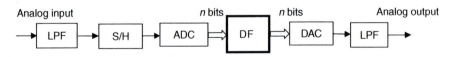

FIGURE 5.4   Block diagram for real-time digital filtering.

5.5.1 Describing Equations

Digital filters are classified according to the duration of their impulse response into:

- Infinite impulse response (IIR) filters
- Finite impulse response (FIR) filters

5.5.1.1 Infinite Impulse Response Filters

As the name implies, the impulse response of this type lasts indefinitely after the excitation has vanished. They are described by the following recursive equation:

$$y(nT) = \sum_{i=0}^{N} a_i x(nT - iT) - \sum_{i=1}^{N} b_i y(nT - iT) \tag{5.7}$$

in which

- N is the filter's order
- $x(nT)$ is the nth input sample
- $y(nT)$ is the nth output sample
- a_i's and b_i's are the filter's coefficients

From Equation 5.7, it is evident that the present nth output is made up of a weighted sum of the present and N past inputs and a weighted sum of N past outputs. IIR filters are mainly used whenever frequency selective responses—that is, a sharp cut-off together with high throughput rates—are required. The goal of the design process is to determine the filter's coefficients; that is, the a's and the b's.

Applying the Z-transform to Equation 5.7, we get

$$Y(Z) = \sum_{i=0}^{N} a_i Z^{-i} X(Z) - \sum_{i=1}^{N} b_i Z^{-i} Y(Z)$$

$$Y(Z)\left\{1 + \sum_{i=1}^{N} b_i Z^{-i}\right\} = X(Z) \cdot \sum_{i=0}^{N} a_i Z^{-i}$$

giving the transfer function $H(Z)$, where

$$H(Z) = \frac{Y(Z)}{X(Z)} = \frac{\sum_{i=0}^{N} a_i Z^{-i}}{1 + \sum_{i=1}^{N} b_i Z^{-i}} \tag{5.8}$$

which is a rational function of two N-th order polynomials in Z. $H(Z)$ can be expressed as

$$H(Z) = H_o \frac{\prod_{i=1}^{N}(Z - Z_i)}{\prod_{i=1}^{N}(Z - P_i)} \tag{5.9}$$

where Z_i's and P_i's are the roots of the numerator's and denominator's polynomials, respectively, or the zeros and poles of the transfer function, respectively.

IIR filters are usually designed recursively; that is, as closed-loop (feedback) systems—hence they are liable to oscillate. Care should therefore be taken to ensure stability over the frequency range of interest.

Through long division, we can rewrite Equation 5.8 in the form

$$H(Z) = h(0) + h(T)Z^{-1} + h(2T)Z^{-2} + \cdots + h(iT)Z^{-i} + \cdots$$

which is an infinite series with $h(iT)$ as the ith sample of the filter's impulse response. This indicates that the response is infinite; hence the name "infinite impulse response."

5.5.1.2 Finite Impulse Response Filters

The impulse response of such type ends after a finite number of samples; that is, it has a limited length (duration). The describing equation contains weighted terms representing the present and $(N - 1)$ past inputs. It can be written as

$$y(nT) = \sum_{i=0}^{N-1} a_i x(nT - iT) \tag{5.10}$$

Due to its finite-duration impulse response, this class of filters is usually called *BIBO filters*; BIBO stands for "bounded-input, bounded-output." Such filters are usually designed nonrecursively; that is, as open loop systems. This means that they are inherently stable. FIR filters enjoy superior phase and delay responses. Therefore, they are mainly used where little or no phase and delay distortion are required.

Z-transforming Equation 5.10, we get

$$Y(Z) = \sum_{i=0}^{N-1} a_i Z^{-i} X(Z)$$

giving

$$H(Z) = \frac{Y(Z)}{X(Z)} = \sum_{i=0}^{N-1} a_i Z^{-i} \tag{5.11}$$

which has $(N - 1)$ zeros.

Again recalling that $H(Z)$ of a filter is the Z-transform of its impulse response; that is,

$$H(Z) = Z\{h(nT)\} = \sum_{n=0}^{N-1} h(nT)Z^{-n} \tag{5.12}$$

Thus by comparing Equations 5.11 and 5.12, we can conclude that

$$h(iT) \equiv a_i$$

That is, the N samples of the filter's impulse response $h(iT)$ are the filter's coefficients a_i's. This conclusion is the basis of FIR filters design, as we shall see later.

EXAMPLE 5.4

For a first-order IIR filter, prove that the coefficients of the impulse response $h(nT)$ are given by

$$y(nT) = (-1)^n \left[-b_1^{n-1} a_1 + b_1^n a_o \right]$$

Solution

The describing equation of a first-order filter is

$$y(nT) = a_o x(nT) + a_1 x(nT - T) - b_1 y(nT - T)$$

For a unit impulse input $x(0) = 1$,

$$y(0) = a_o \cdot 1$$

$$y(T) = 0 + a_1 - b_1 \cdot a_o$$

$$y(2T) = 0 + 0 - b_1(a_1 - b_1 \cdot a_o) = -b_1 a_1 + a_o \cdot b_1^2$$

$$y(3T) = 0 + 0 - b_1(-b_1 a_1 + a_o \cdot b_1^2) = b_1^2 a_1 - a_o b_1^3$$

from which we can conclude that

$$y(nT) = (-1)^n \left[-b_1^{n-1} a_1 + b_1^n a_o \right]$$

5.5.2   TIME DOMAIN ANALYSIS

Consider a filter, Figure 5.5, that is described by its impulse response $h(nT)$. The response $y(nT)$ of the filter due to an excitation $x(nT)$ is given by

$$y(nT) = h(nT) * x(nT)$$

$$= \sum_{i=0}^{N} h(iT) \cdot x(nT - iT)$$

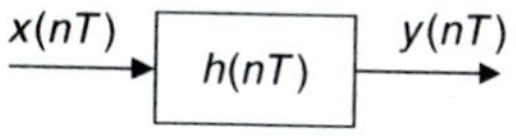

FIGURE 5.5   A digital filter.

where N is the filter's order. Taking the Z-transform, we can write

$$Z\{y(nT)\} = \sum_{i=0}^{\infty} h(iT) \cdot Z^{-i} X(Z)$$

$$= Y(Z)$$

recalling that

$$H(Z) = \frac{Y(Z)}{X(Z)}$$

Thus, by dividing both sides by $X(Z)$, we get

$$H(Z) = \sum_{i=0}^{\infty} h(iT) \cdot Z^{-i} \tag{5.13}$$

which is the Z-transform of $h(iT)$. From these equations, we can draw the following important conclusion:

The transfer function of a certain filter is the Z-transform of its sampled impulse response.

The transfer function of an IIR filter can be expressed as a rational function between two polynomials as shown before; that is, in the form

$$H(Z) = \frac{Y(Z)}{X(Z)} = \frac{\sum_{i=0}^{N} a_i Z^{-i}}{1 + \sum_{i=1}^{N} b_i Z^{-i}}$$

which in turn could be expanded, as before, through long division, into the infinite power series:

$$H(Z) = C_0 + C_1 Z^{-1} + C_2 Z^{-2} + C_3 Z^{-3} + \cdots \tag{5.14}$$

Comparing Equations 5.13 and 5.14, it is easy to detect that the coefficient C_i's correspond in fact to the samples of the filter's impulse response $h(iT)$. Also, the fact that each equation is an infinite series does agree with the name "IIR filters." On the other hand, filters that have transfer functions in the form of Nth-order polynomials do have Nth-term impulse responses—that is, of finite length—hence the name "FIR filters."

EXAMPLE 5.5

A digital filter is described by the following equation:

$$y(nT) = 3 \cdot x(nT) + 2 \cdot x(nT - T) + x(nT - 2T)$$

Determine:

1. The type and order of the filter
2. The output in response to the excitation shown in Figure 5.6.

Solution

1. Because the output is determined by the present and past inputs only, this filter is an FIR one. The maximum delay determines its order. As there are two delays in the transfer function, it is a second-order filter.
2. The output sequence, as described before, is given by

$$y(nT) = \sum_{i=0}^{N-1} h(iT) \cdot x(nT - iT)$$

To ease the calculation, we construct Table 5.3.

A plot of the output $y(nT)$ is given in Figure 5.7.

It is evident from the plot of $y(nT)$ that a bounded input has resulted in a bounded output, as expected. Also, it is evident from Figure 5.7 that the output has experienced a peak at the fourth output. This is a useful result, as this peak is obtained due to the given specific input sequence. Any other input will provide a smaller or no peak. Careful selection of the input sequence for a certain filter usually leads to a distinguished peak among other output samples. Filters that have such property are called "matched filters." This type of filter has several applications in communications.

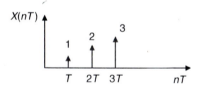

FIGURE 5.6 Input sequence for Example 5.5.

TABLE 5.3
Systematic Calculations of the Response of an FIR Filter

Serial	Input	h(nT)			Output
n	x(nT)	3	2	1	y(nT)
0	0	3 * 0	—	—	0
1	1	3 * 1 +	2 * 0	—	3
2	2	3 * 2 +	2 * 1 +	1 * 0	8
3	3	3 * 3 +	2 * 2 +	1 * 1	14
4	0	3 * 0 +	2 * 3 +	1 * 2	8
5	0	3 * 0 +	2 * 0 +	1 * 3	3
6	0	3 * 0 +	2 * 0 +	1 * 0	0

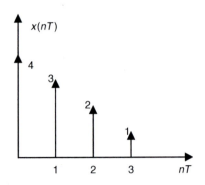

FIGURE 5.9   Input sequence of Example 5.7.

TABLE 5.6

Calculating the Output for Example 5.7

n	x (nT)	4	−7	h(nT) 9	Coeff −10	+10	−10	+10	−10	+10	−10	Y (nT)
0	4	4 × 4	−7 × 0									16
1	3	4 × 3	−7 × 4	9 × 0								−16
2	2	4 × 2	−7 × 3	9 × 4	−10 × 0							23
3	1	4 × 1	−7 × 2	9 × 3	−10 × 4	10 × 0						−23
4	0	4 × 0	−7 × 1	9 × 2	−10 × 3	10 × 4	−10 × 0					21
5	0	4 × 0	−7 × 0	9 × 1	−10 × 2	10 × 3	−10 × 4	10 × 0				−21
6	0	4 × 0	−7 × 0	9 × 0	−10 × 1	10 × 2	−10 × 3	10 × 4	−10 × 0			20
7	0	4 × 0	−7 × 0	9 × 0	−10 × 0	10 × 1	−10 × 2	10 × 3	−10 × 4	10 × 0		−20
8	0	4 × 0	−7 × 0	9 × 0	−10 × 0	10 × 0	−10 × 1	10 × 2	−10 × 3	10 × 4	−10 × 0	20
9	0	4 × 0	−7 × 0	9 × 0	−10 × 0	10 × 0	−10 × 0	10 × 1	−10 × 2	10 × 3	−10 × 4	−20

Solution

Through long division, we get

$$H(Z) = 4 - 7Z^{-1} + 9Z^{-2} - 10Z^{-3} + 10Z^{-4} - 10Z^{-5} + 10Z^{-6} + \cdots (-1)^K Z^{-K}$$

1. The sampled impulse response is thus

$$h(nT) = \{4, -7, 9, -10, 10, -10, 10, -10, 10, \ldots\}$$

   which is, as expected, an infinite series.
2. The output can be obtained through the equation

$$y(nT) = h(nT) * x(nT)$$

That is, by convolving the impulse response with the input series. Alternatively, as a systematic procedure, we can construct a table (Table 5.6).

   The output sequence is shown in Figure 5.10, from which it can be seen that the filter reaches a steady state where it oscillates. Such a result was to be expected from the very beginning, as the order of the numerator is greater than that of the denominator.

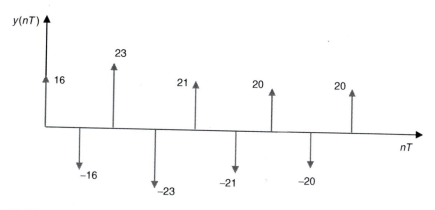

FIGURE 5.10 Output sequence of Example 5.7.

Let $H(Z)$ be the transfer function of an Nth-order filter. The numerator's polynomial $N(Z)$ and the denominator's polynomial $D(Z)$ can each be decomposed into N first-order terms in the form

$$H(Z) = \frac{N(Z)}{D(Z)} = H_0 \frac{\prod_{i=0}^{N}(Z - Z_i)}{\prod_{i=0}^{N}(Z - P_i)} \qquad (5.15)$$

where Z_i's and P_i's are the zeros and poles of $H(Z)$, respectively.

An Nth-order filter is realizable only if it is stable. To prove stability, the following conditions on its poles and zeros should be satisfied:

1. The filter should have a maximum of N zeros.
2. The poles should be distributed inside a unit circle in the Z-plane.

From the above it is evident that IIR filters should be tested for stability prior to their realization. FIR filters, on the other hand, are by nature inherently stable, as their poles are located at the origin. Given in the form of Equation 5.15, one can test the stability of a filter directly by checking that the magnitudes of its poles P_i's are less than unity; that is,

$$|P_i| < 1$$

Higher-order polynomials are usually difficult to decompose into first-order terms so that the earlier simple check for stability becomes infeasible. Therefore, some standard tests have been developed to check the stability of digital filters, as we shall see in the next chapter.

EXERCISE 5.1

Estimate the output of the filter described by

$$Y(nT) = 10 * x(nT) + x(nT - T) + x(nT - 2T)$$

in response to the excitation shown in Figure 5.11.

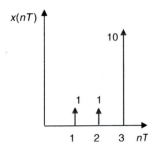

FIGURE 5.11 Input sequence for Exercise 5.1.

5.5.3 Frequency Domain Analysis

5.5.3.1 Graphical Evaluation of the Frequency Response

The frequency response of a continuous-time system can be evaluated on the $j\omega$-axis of the S-plane. Correspondingly, the response of a discrete-time system can likewise be evaluated on the mapped version of such an axis; that is, the circumference of the unit circle, in the Z-domain. Because the circumference is endless, we can also expect that the response will repeat itself every complete revolution on the circle centered on multiples of the sampling frequency ($2f_s$, $3f_s$, $4f_s$, etc.).

To describe how this could be done, let us consider the case of a second-order filter whose pole-zero distribution is shown in Figure 5.12.

The magnitude and phase of the transfer function at any frequency ω are evaluated on a corresponding point on the circumference of the unit circle and are given by

$$H(e^{j\omega T}) = M(\omega)e^{j\theta(\omega)}$$

To get the response at any frequency, we connect the corresponding point on the circumference to all poles and zeros as well as to the origin. The magnitude of the vector connecting this point to pole P_i is M_{Pi} and can be calculated from the triangle of vectors shown in Figure 5.12 as follows:

$$\overline{e^{j\omega T}} = \overline{f_{P1}} + \overline{M(\omega)}$$

Provided that $e^{j\omega T}$ is a unit vector and M_{Pi} is the ith pole frequency,

$$\overline{M_{P1}(\omega)} = \overline{e^{j\omega T}} - \overline{f_{P1}}$$

and the phase is ψ_{P1}.

Doing the same for other poles and zeros, we get the response at that point as follows:

$$\left|H(e^{j\omega t})\right| = H_0 \frac{M_{Z1}M_{Z2}}{M_{P1}M_{P2}}$$

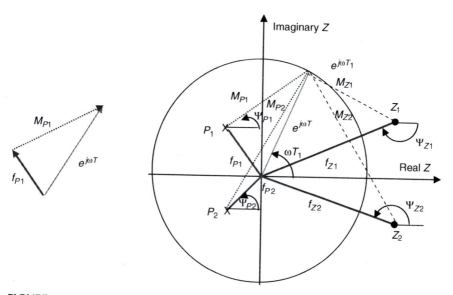

FIGURE 5.12   (See color insert following page 262.) Frequency domain analysis.

and

$$\theta(\omega) = (\psi_{Z1} + \psi_{Z2}) - (\psi_{P1} + \psi_{P2})$$

In general, for an Nth-order transfer function, we can write for the magnitude response

$$\left|H(e^{j\omega t})\right| = H_0 \frac{\prod_{i=1}^{N} M_{Zi}}{\prod_{i=1}^{N} M_{Pi}}$$

and for the phase shift

$$\theta(\omega) = \sum_{i=1}^{N} \psi_{Zi} - \sum_{i=1}^{N} \psi_{Pi}$$

The response could be obtained by repeating the same procedure for different points (frequencies) to cover the whole circumference (one complete revolution). Continuing further the response will repeat itself around multiples of ω_s. Recall that one complete revolution corresponds to

$$\omega T = 2\pi$$

and after n complete revolutions

$$\omega_n T = 2\pi n$$

$$\omega_n = \frac{2\pi n}{T} = 2\pi n f_s = n\omega_s$$

MATLAB® can be used to plot the frequency response, provided that the pole-zero distribution of the filter is given. As an example, the following instructions can be employed to get the frequency response of a fifth-order Butterworth filter that has a normalized cut-off frequency $f_{cn} = 2f_c / f_s = 0.8$.

5.5.3.1.1 MATLAB Assistance

```
[z,p,k]   = butter(5,0.80);
[num,den] = zp2tf(z,p,k);
fvtool(num,den)
```

The transfer function is plotted in Figure 5.13.

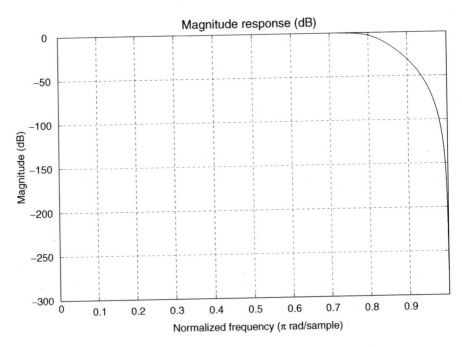

Magnitude response (dB)

FIGURE 5.13 A typical frequency response of a fifth-order Butterworth filter.

EXAMPLE 5.8

Plot the pole-zero distribution and the magnitude response for the filter described by

$$H(Z) = \frac{1 - 0.5Z^{-4}}{1 + 0.5Z^{-4}}$$

Repeat for the case

$$H(Z) = \frac{1 + 0.5Z^{-4}}{1 - 0.5Z^{-4}}$$

and comment on the results.

Solution

MATLAB can be employed to plot the response using the following instructions:

```
num = [1 0 0 0 -0.5];        % numerator's polynomial of H₁
den = [1 0 0 0 0.5];         % denominator's polynomial of H₁
zplane(num,den); grid        % pole-zero plot
fvtool(num,den)              % filter visualization tool
```

The pole-zero distributions and the magnitude response of H_1 are plotted in Figures 5.14 and 5.15, respectively. The corresponding plots for H_2 are depicted in

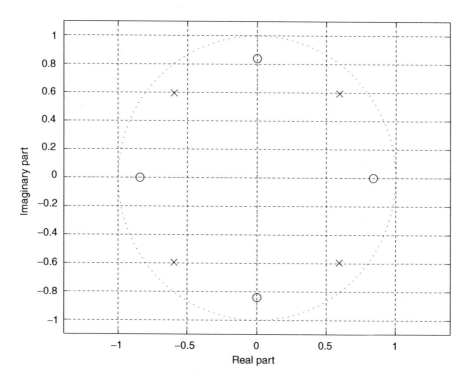

FIGURE 5.14   Pole-zero distribution for H_1, a fourth-order comb filter.

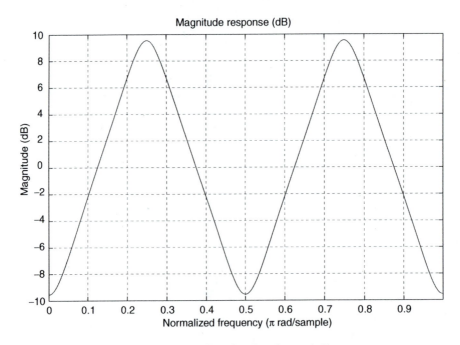

FIGURE 5.15 Magnitude response for H_2, a fourth-order comb filter.

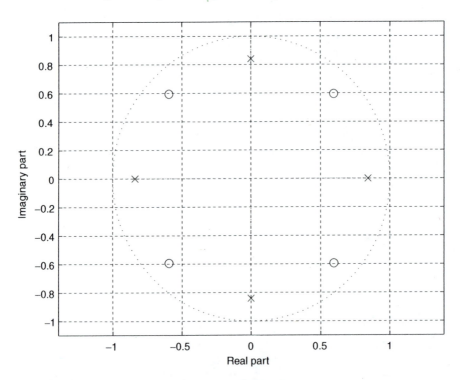

FIGURE 5.16 Pole-zero distribution of filter $H_2(Z)$ in Example 5.8.

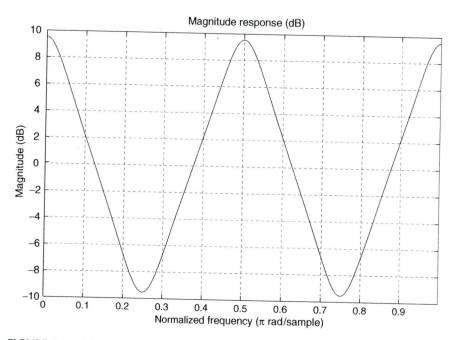

FIGURE 5.17   Magnitude response for filter $H_2(Z)$ in Example 5.8.

Figures 5.16 and 5.17. From the plots, it is evident that although the poles are complex in the first case, two poles are pure real and the other two are pure imaginary in the second case. This means that the poles and zeros of $H_2(Z)$ could be obtained if the poles of H_1 are rotated by $\pi/4$. The magnitude responses of both filters, as depicted in Figures 5.15 and 5.17, have the shape of a sine wave and cosine wave respectively. Thus, the magnitude response of H_2 can be considered similar to that of H_1, yet with a shift of $\pi/2$. Due to the shape of their response, which resembles the tooth of a comb, such filters are called *comb filters*.

5.5.3.2   Numerical Evaluation of the Frequency Response

Transfer functions of orders higher than three are hard to evaluate manually using the graphical method just described. This difficulty is due not only to the associated complicated vector diagram, but also to the need to know "precisely" the pole and zeros frequencies. A rather systematic approach that is purely numerical will be given shortly. Let us consider again an Nth-order transfer function, where

$$H(Z) = \frac{N(Z)}{D(Z)} = \frac{\sum_{i=0}^{M} a_i Z^{-i}}{\sum_{i=0}^{N} b_i Z^{-i}} \quad \text{with} \quad M \le N$$

Substituting for $Z = e^{j\omega T}$ gives

$$H(e^{j\omega T}) = \frac{a_0 + a_1 e^{-j\omega T} + a_2 e^{-j2\omega T} + \cdots + a_M e^{-jM\omega T}}{b_0 + b_1 e^{-j\omega T} + b_2 e^{-j2\omega T} + \cdots + b_N e^{-jN\omega T}}$$

Recalling that $e^{j\omega T} = \cos \omega T + j\sin \omega T$, we get

$$\left|H(e^{j\omega T})\right| = \sqrt{\frac{[a_o + a_1 \cos\omega T + a_2 \cos 2\omega T + \cdots]^2 + [a_1 \sin\omega T + a_2 \sin 2\omega T + \cdots]^2}{[b_o + b_1 \cos\omega T + b_2 \cos 2\omega T + \cdots]^2 + [b_1 \sin\omega T + b_2 \sin 2\omega T + \cdots]^2}}$$

$$\theta(\omega) = \tan^{-1} \frac{a_1 \sin\omega T + a_2 \sin 2\omega T + \cdots}{a_o + a_1 \cos\omega T + a_2 \cos 2\omega T + \cdots}$$

$$- \tan^{-1} \frac{b_1 \sin\omega T + b_2 \sin 2\omega T + \cdots}{b_o + b_1 \cos\omega T + b_2 \cos 2\omega T + \cdots}$$

5.5.4 EVALUATING THE FREQUENCY RESPONSE USING THE FFT

Due to the duality between the frequency and time responses, it is possible to get any one from the other. (The frequency response of a filter is the Z-transform of its impulse response.) Thus, knowing the impulse response $h(nT)$, one can get $H(Z)$ from the equation

$$H(Z) = \sum_{n=0}^{\infty} h(nT) Z^{-n}$$

or

$$H(e^{j\omega T}) = \sum_{n-0}^{\infty} h(nT) e^{-jn\omega T}$$

which is the Fourier transform of $h(nT)$. In Chapter 8, an algorithm (the FFT algorithm) will be described that enables the calculation of the DFT using the minimum number of additions and multiplications. To get a precise estimation of the response, the impulse response should be first sampled at a sufficiently high frequency before it is Fourier-transformed.

5.5.5 DIGITAL FILTER COMPONENTS

Resistors, capacitors, inductors, and operational amplifiers have been used in different ways in the realization of analog filters. Likewise, there are three basic components that are used in the construction of digital filters, as described in the following sections.

5.5.5.1 Unit Delay

Input samples to a unit delay experience a delay corresponding to one sampling period T, where $T = 1/f_s$. This could be expressed as follows:

$$y(nT) = x(nT - T)$$

or

$$Y(Z) = Z^{-1}X(Z)$$

5.5.5.2 Multiplier

This element is used to multiply input samples with a constant m. The output can be expressed as

$$Y(Z) = m \cdot X(Z)$$

5.5.5.3 Adder

The output of such component due to k inputs is given by

$$Y(Z) = \sum_{i=1}^{k} X_i(Z)$$

The symbols used to represent these components are diagrammed in Figure 5.18.

To increase familiarity with the analysis of digital filter networks, the following examples are given.

EXAMPLE 5.9

Deduce the transfer function for the shown filter, stating its order, type, and function (Figure 5.19).

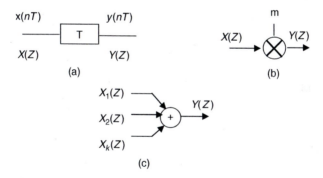

FIGURE 5.18   Digital filter components. (a) Unit delay, (b) multiplier, and (c) adder.

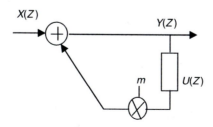

FIGURE 5.19   Filter of Example 5.9.

Solution

Assuming an auxiliary variable $U(Z)$, then from the definition of a unit delay, we can write

$$U(Z) = Z^{-1}Y(Z)$$

such that

$$Y(Z) = X(Z) + mU(Z) = X(Z) + mZ^{-1}Y(Z)$$

$$Y(Z)(1 - mZ^{-1}) = X(Z)$$

giving

$$H(Z) = \frac{Y(Z)}{X(Z)} = \frac{1}{(1 - mZ^{-1})}$$

from which we can conclude that

   i. The filter has only one pole; thus it is a first-order filter.
   ii. The filter has no zeros; thus it is an all-pole filter—that is, an LPF.
   iii. Because $H(Z)$ is a rational function and the network includes a feedback loop, it is an IIR filter.

EXAMPLE 5.10

Prove that the network shown in Figure 5.20 represents an FIR filter, then find its length.

Solution

The transfer function $H(Z)$ is given by

$$Y(Z) = \frac{Y(Z)}{X(Z)}$$

Let us assume the auxiliary variables $X_1(Z)$, $X_2(Z)$, and $X_3(Z)$, where

$$X_1(Z) = Z^{-1} \cdot X(Z)$$

$$X_2(Z) = Z^{-1}X_1(Z) = Z^{-2}X(Z)$$

$$X_3(Z) = Z^{-1}X_2(Z) = Z^{-3}X(Z)$$

Thus the output $Y(Z)$ is given by

$$Y(Z) = a_0X(Z) + a_1Z^{-1}X(Z) + a_2Z^{-2}X(Z) + a_3Z^{-3}X(Z)$$

Dividing both sides by $X(Z)$ gives

$$H(Z) = a_0 + a_1Z^{-1} + a_2Z^{-2} + a_3Z^{-3}$$

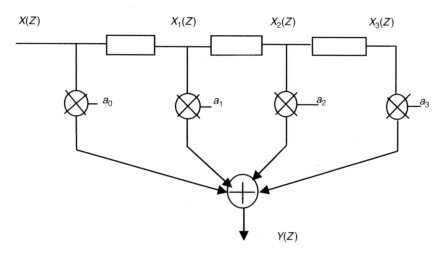

FIGURE 5.20   Filter of Example 5.10.

Because the transfer function is not a rational function, it represents an FIR filter. The three delays indicate that it is a third-order filter and has a length of 4 (four coefficients).

EXAMPLE 5.11

Deduce the transfer function for the filter shown in Figure 5.21, giving its type and order.

Solution

Considering the auxiliary variables $X_i(Z)$ as before, we can write

$$X_2(Z) = Z^{-1}X_1(Z) \quad \text{and} \quad X_3(Z) = Z^{-2}X_1(Z)$$

thus

$$X_1(Z) = X(Z) - b_1Z^{-1}X_1(Z) - b_2Z^{-2}X_1(Z)$$

giving

$$X_1(Z) = \frac{X(Z)}{(1 + b_1Z^{-1} + b_2Z^{-2})}$$

The output can be expressed as

$$Y(Z) = (a_0 + a_1Z^{-1} + a_2Z^{-2}) \cdot X_1(Z)$$

Substituting for $X_1(Z)$, we get

$$H(Z) = \frac{Y(Z)}{X(Z)} = \cdot \frac{a_0 + a_1Z^{-1} + a_2Z^{-2}}{1 + b_1Z^{-1} + b_2Z^{-2}}$$

It is clear from this equation that the filter's transfer function is a second-order rational function; thus it represents a second-order IIR filter. It is noteworthy here that this

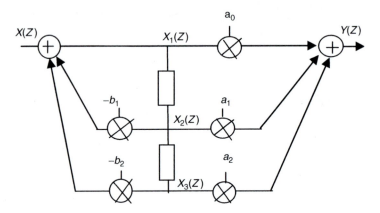

FIGURE 5.21 Filter of Example 5.11.

network structure, usually known as a canonic second-order section, is a basic building block that is used in the construction of higher-order filters. This is possible by connecting second-order sections, either in cascade or in parallel. The reader is therefore asked to keep such a structure in mind, as we shall make use of it frequently.

5.5.6 FIR FILTERS VERSUS IIR FILTERS

Using either FIR filters or IIR filters is questionable. However, as a general rule, IIR filters are usually selected for those applications where selectivity or, more precisely, the magnitude response, is of interest. This means that they are mainly employed to implement conventional frequency selective functions such as low-pass, high-pass, and so on. On the other hand, FIR filters are the best candidates wherever special phase characteristics—for example, linear phase and constant group delay—are required. However, it should be mentioned that such unique characteristics are usually achieved at the expense of a higher filter order, as we shall see in the next chapter. In the design process of IIR filters, the well-established techniques previously employed in designing analog filters are used. They are realized recursively; that is, in the form of closed loops. FIR filters, on the other hand, are usually designed nonrecursively, that is, as open loops.

5.6 SUMMARY

DSP has become an indispensable tool in many fields. To prove this, examples of some practical applications are given, from the fields of telecommunications, biomedical engineering, e-banking and e-commerce, forensic applications, and the entertainment industry. Because all discrete-time systems are better described in the Z-domain, the Z-transform and its important properties are presented. Two fundamental DSP operations—namely, convolution and correlation—are then described. An introductory note on digital filters is provided, where important definitions, types, describing equations, and components, as well as methods of deriving their responses in the time and frequency domains are described.

5.7   REVIEW QUESTIONS

1. Give some practical applications of DSP in communications.
2. What is the relation between the impulse response and the transfer function of a filter?
3. Show how you can get the frequency response of a digital filter from its pole-zero distribution.
4. Sketch the frequency response of the following filter functions:

   An LPF

   A BPF

   A BSF

5. Compare, using equations, between cross-correlation and autocorrelation. Give one application for each.

5.8   PROBLEMS

1. An LPF is described by the transfer function

$$H(Z) = k\frac{1 + Z^{-1}}{1 - CZ^{-1}}$$

   a. Draw a canonic realization for the filter.
   b. Plot, using MATLAB, the magnitude response for $[C = -k = -0.884]$.
   c. Repeat this process for the HPF described by

$$H(Z) = k\frac{1 - Z^{-3}}{1 - CZ^{-3}}$$

2. Derive the transfer function for the filters shown in the following figure.

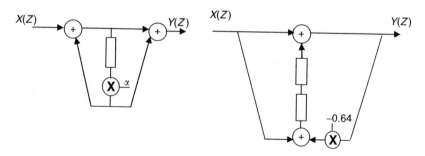

3. Sketch the pole-zero distribution and the magnitude response for the comb filter described by

$$H(Z) = k\frac{1 + Z^{-4}}{1 - \alpha Z^{-4}}$$

   $\alpha = 0.5$ then find the value of k for a unity peak gain. Verify your answer using MATLAB.

4. For the shown all-pass filter:
   a. Derive an expression for the transfer function.
   b. Reconfigure the filter in the form of a standard second-order canonic section.
   c. Compare the two realizations, considering the components count.
   d. Sketch its pole-zero distribution in the Z-domain.

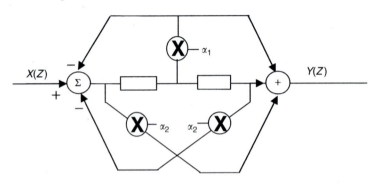

5. Derive an expression for the transfer functions of the filters shown in the following figure, then sketch their magnitude response as well as their pole-zero diagram.

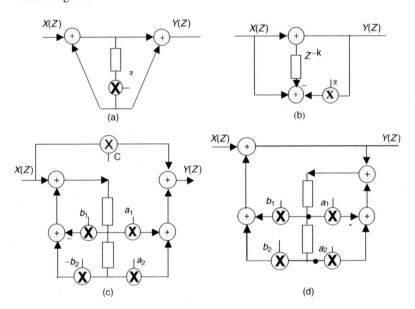

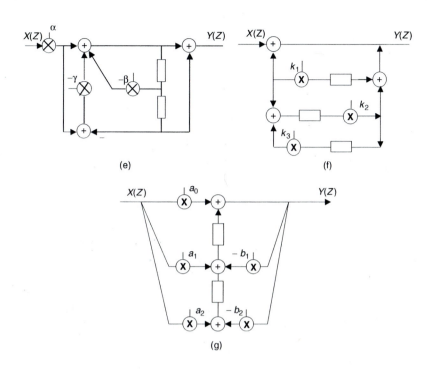

(e)   (f)

(g)

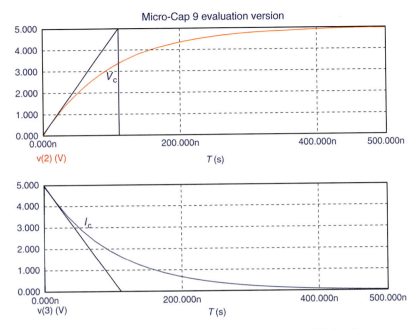

COLOR FIGURE 2.2 Current and voltage waveform in a simple *RC* circuit.

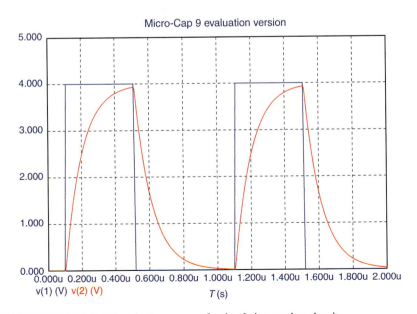

COLOR FIGURE 2.6 Transient response of a simple integrating circuit.

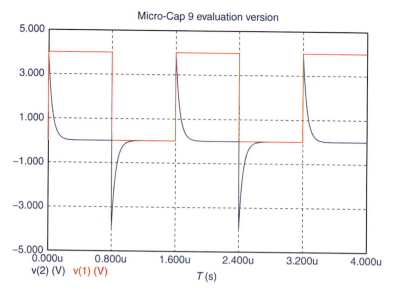

COLOR FIGURE 2.9   Transient response of a differentiating circuit.

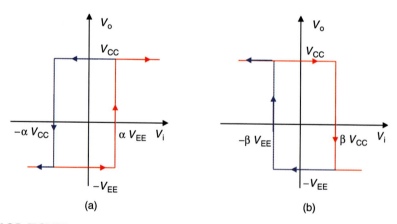

(a)

(b)

COLOR FIGURE 2.40   Transfer functions of Schmitt trigger circuits of Figures 2.39a and 2.39b.

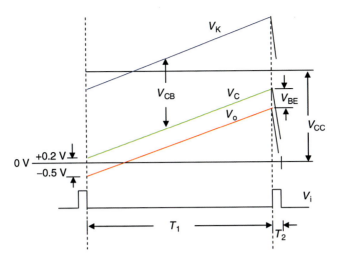

COLOR FIGURE 2.49 Waveforms for a bootstrap sweep generator.

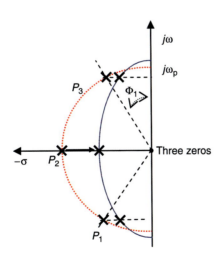

COLOR FIGURE 3.31 Pole-zero distribution of a third-order Chebyshev filter, as derived from Butterworth poles.

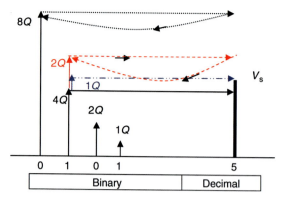

COLOR FIGURE 4.16 The iteration technique.

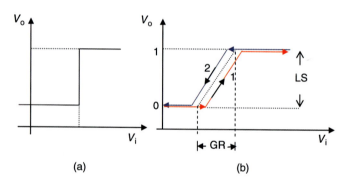

COLOR FIGURE 4.26 Voltage transfer characteristics of an analog comparator. (a) Ideal characteristic and (b) practical characteristic.

B_2	B_1	B_0
1	0	1

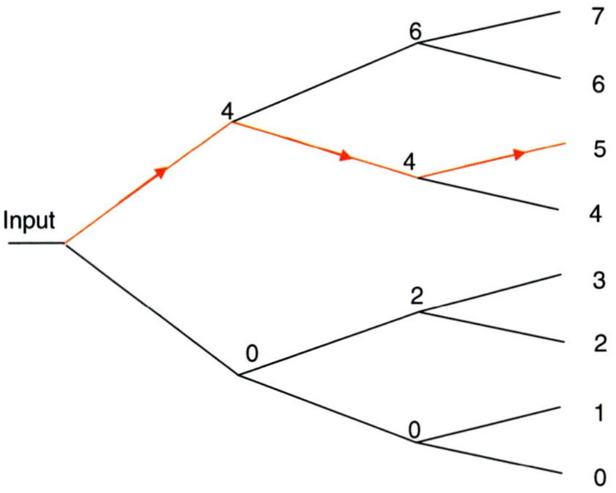

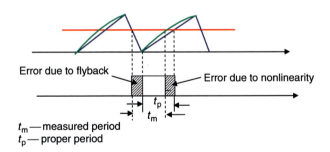

COLOR FIGURE 4.33 Successive approximation tree for a 3-bit case; input = 5.

COLOR FIGURE 4.42 Errors in a single-ramp ADC.

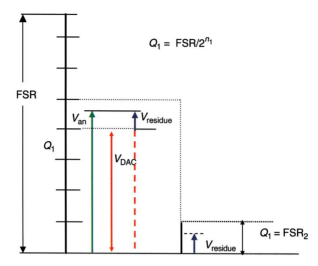

COLOR FIGURE 4.47   Subranging in cascaded ADCs.

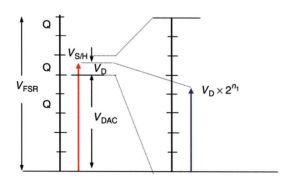

COLOR FIGURE 4.48   Subranging ADC with interstage amplification.

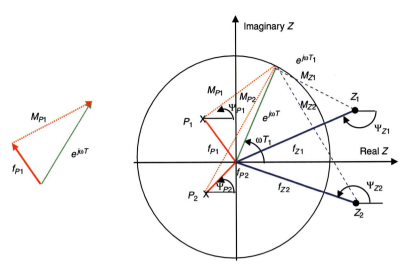

COLOR FIGURE 5.12 Frequency domain analysis.

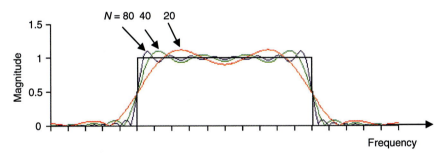

COLOR FIGURE 6.11 Effect of window length on the response.

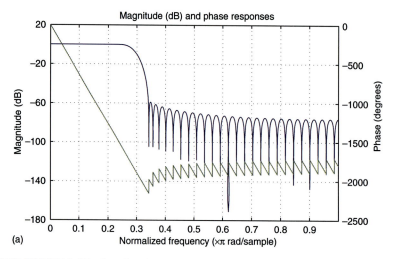

COLOR FIGURE 6.27   Results of Example 6.7. (a) Magnitude and phase response.

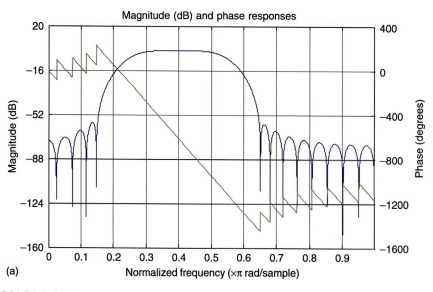

COLOR FIGURE 6.29   Results of Example 6.8. (a) Magnitude and phase response.

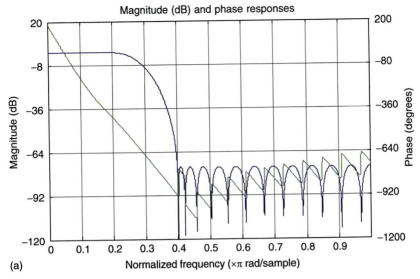

COLOR FIGURE 6.36 Results of Example 6.11. (a) Magnitude and phase response.

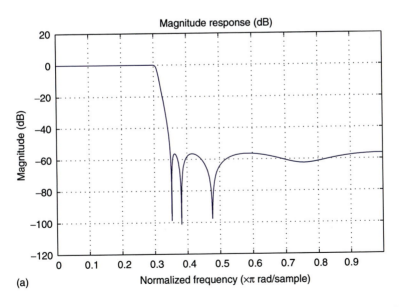

COLOR FIGURE 6.86 MATLAB results for the filter in Example 6.26. (a) Magnitude response.

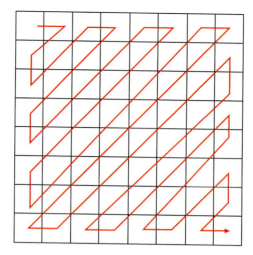

COLOR FIGURE 10.8 Zigzag scanning.

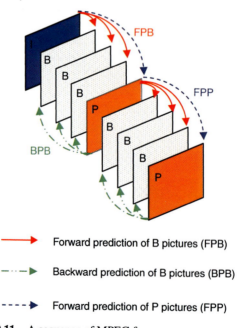

→ Forward prediction of B pictures (FPB)

–·–▶ Backward prediction of B pictures (BPB)

----▶ Forward prediction of P pictures (FPP)

COLOR FIGURE 10.11 A sequence of MPEG frames.

6 Digital Filter Design

6.1 INTRODUCTION

Digital filters represent essential and fundamental components in almost all DSP systems. Like their analog counterpart, they are mostly employed to reshape the magnitude and phase spectrum of an input signal. Besides this major function, they can also be used, for example, to de-noise signals, to disjoin two signals from each other, to limit the bandwidth of a signal, to detect/enhance the contours in a certain image, and so on. In this chapter, we are going to study the design techniques for the two types of digital filters; namely, FIR and IIR filters.

6.2 DESIGN PROCEDURES

Due to their discrete-time nature, the design process of digital filters differs slightly from the classical-filter design process. The flow chart given in Figure 6.1 summarizes the sequence of events encountered in the design process.

6.2.1 FILTER SPECIFICATIONS

The specifications of a filter are usually described using its *tolerance structure*. As depicted in Figure 6.2, this diagram defines the acceptable attenuation levels at certain specific frequencies. It is exactly the same as the one used in Chapter 3 for the design of analog filters, except in the introduction of the sampling frequency ω_s. It is well defined by two points, (ω_p, A_p) and (ω_r, A_r).

A tolerance structure thus defines the tolerable deviation(s) of a filter's response from the ideal "brick wall" response. Along the pass band $(0 - \omega_p)$, the filter response is allowed to fluctuate (ripple) within the so-called *ripple channel* to a maximum of A_p dB (δP in magnitude). The range of frequency between the pass band edge ω_p and the rejection band edge ω_r defines the allowable transition band between the pass band and the rejection band (stop or attenuation band). A_r gives the minimum acceptable attenuation at the rejection band edge ω_r. As digital filters, like any other discrete-time systems, have their frequency responses repeated around the sampling frequency ω_s, to avoid aliasing, ω_s should be much higher than all other frequencies of interest $(\omega_p, \omega_c, \omega_r)$.

6.2.2 APPROXIMATION

In this fundamental design step, a causal discrete-time function is determined, such that it can fit within the error channel, while satisfying the desired specifications concerning amplitude, phase, and delay characteristics. The outcome of this process of components. Hardware complexity determines not only the cost but also the processing speed and hence the delay. This last property is of special importance whenever online processing is concerned.

6.2.5 ERROR BUDGET

Arithmetic devices/components used in the approximation and realization of digital filters by nature have limited accuracy due to their finite word length. Repeated operations, especially multiplications, are accompanied by rounding (truncation) where the resulting output word is continuously subjected to repeated truncation (limiting the results to the nominal word length of the used components). To illustrate this fact, let us consider the multiplication of two n-bit numbers. Because the word length of the used hardware is fixed at n bits, the resulting $2n$-bit word should be truncated by eliminating n least significant bits. Because the processing of each input sample undergoes several consecutive multiplications and additions, the result is subjected to a series of truncations causing deviation from the exact value.

Excessive errors may lead to a filter response that does not fulfill the specifications set by the tolerance structure. Therefore, it is mandatory to check that the incurred overall errors from all sources together are kept within the permissible error budget; otherwise, the design process should be revised. The flow chart given in Figure 6.1 depicts where the design process should be interrupted to repeat a design step. Corrective actions include:

- Restructuring the network by selecting another topology that is less sensitive to most error sources
- Using a larger word length
- Revising the approximation step
- Relaxing the specifications (e.g., allowing wider transition band, larger ripple amplitude) to reduce the filter order and hence reduce the number of operations involved

In addition to truncation errors, there are other error sources, namely:

- Quantization errors as a result of the limited resolution of the used ADC. Such errors could be reduced by increasing the resolution of the converter and hence the word length of all other components.
- Sampling frequency instability due to drifts in the frequency of the sampling source. This can be rectified by using a highly stable clock source; for example, a quartz-controlled one.

6.2.6 IMPLEMENTATION

As mentioned before, digital filters can be implemented either in:

- *Hardware.* The filter is realized using standard digital components. The resulting hardware (dedicated hardware) usually implements a filter function using a specific approximation of proper order.

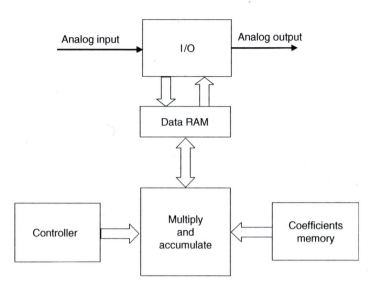

FIGURE 6.3 A typical processor system for digital filter implementation. (From Ludeman L., *Fundamentals of Digital Signal Processing*, Harper & Row Pub., New York, 1986.)

- *Software*. The filter function is simulated on a general-purpose processor system. This approach is flexible, as filter parameters such as order, cut-off frequency, approximation, and ripple amplitude could be changed by simply modifying the program (code). Furthermore, it allows special modes of operation, such as a time-dependent filter response.

A block diagram for a typical processor system that is designed to implement a digital filter is given in Figure 6.3. The core of the system is the multiply and accumulate (MAC) where the basic filter calculations are performed. The input/output unit contains an ADC and a DAC as interfacing devices. The filter's coefficients are stored in ROM or even RAM. The whole process is operated by the controller where a sequencer is included to insure the execution of all operations in the proper order.

The sequence of steps involved in calculating a certain output are illustrated in Example 6.1.

EXAMPLE 6.1

A second-order IIR filter is described by the following recursive equation:

$$y(nT) = a_0 \cdot x(nT) + a_1 \cdot x(nT - T) + a_2 \cdot x(nT - 2T) - b_1 \cdot y(nT - T) - b_2 \cdot y(nT - 2T)$$

Implement the filter in hardware, showing main blocks, then describe the sequence of executing the different instructions.

Solution

A simple hardware realization for the equation is depicted in Figure 6.4. The filter coefficients a_i's and b_i's are stored in the coefficient's RAM. The input samples $x(nT)$ and the two final outputs $y(nT - T)$ and $y(nT - 2T)$ are buffered in the X and Y registers,

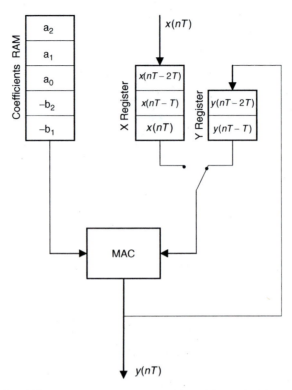

FIGURE 6.4   Simulating a second-order recursive filter of Example 6.1.

TABLE 6.1

Sequence of Operations for Solving a Second-Order Recursive Equation

Step	Operation	C	B	A
0	Clear	—	—	0
1	MA	$-b_1$	$y(nT - T)$	$0 - b_1 y(nT - T)$
2	MA	$-b_2$	$y(nT - 2T)$	$-b_1 y(nT - T) - b_2 y(nT - 2T)$
3	MA	a_o	$x(nT)$	$a_o x(nT) - b_1 \cdot y(nT - T) - b_2 \cdot y(nT - 2T)$

respectively. The operation starts by selecting the final outputs to be multiplied by their corresponding coefficients. The selector is then switched to the left to allow the input samples and their delayed versions to be multiplied by the a's coefficients. Table 6.1 summarizes the necessary steps.

6.3   DESIGN OF FIR FILTERS

The design process of FIR filters follows the same sequence of steps illustrated in the flow chart of Figure 6.1. As before, the specifications of the filter are usually summarized by its tolerance structure. Four approximation techniques for FIR filters will be described here: the windowing method, the frequency sampling method, the optimal method, and the least Pth-norm optimal method. It is worth mentioning that FIR filters are usually realized nonrecursively, as they are open loop systems. The decision for a certain realization topology is application-dependent, as will be demonstrated later. In the following sections, we shall study the main properties of this class of filters.

6.3.1 PROPERTIES OF FINITE IMPULSE RESPONSE FILTERS

FIR filters are described by the following equation:

$$y(nT) = \sum_{k=0}^{N-1} a_k \cdot x(nT - kT)$$

where
- $x(nT)$ = excitation (input)
- $y(nT)$ = response (output)
- N = filter length

Expressing the equation in the Z-domain gives

$$Y(Z) = \sum_{k=0}^{N-1} a_k \cdot Z^{-k} X(Z)$$

Dividing both sides by $X(Z)$, we get

$$H(Z) = \sum_{k=0}^{N-1} a_k Z^{-k} \qquad (6.1)$$

Recall that the transfer function of a certain filter is the Z transform of its impulse response, that is,

$$H(Z) = \sum_{k=0}^{N-1} h(kT) \cdot Z^{-k} \qquad (6.2)$$

Comparing Equations 6.1 and 6.2, we get

$$a_k \equiv h(kT) \quad \text{for } k = 0, 1, 2, \ldots, N - 1$$

From this equality we can conclude that the filter's coefficient $\{a_i\text{'s}\}$ are actually the samples of its impulse response (N samples). The fact that the impulse response has a finite number of samples—that is, of finite duration—explains why such type of filters are called FIR filters. This is in contrast to the other type of filters (IIR) whose impulse response extends indefinitely. FIR filters have some unique features that are summarized in the following sections.

6.3.1.1   Inherent Stability

As FIR filters are mostly realized nonrecursively—that is, as open loop systems (feed forward topologies)—they show no tendency to instability. This statement can be further emphasized by reconsidering Equation 6.2:

$$H(Z) = \sum_{k=0}^{N-1} a_k Z^{-k}$$

$$= \sum_{k=0}^{N-1} a_k Z^{-k} = a_0 + a_1 Z^{-1} + a_2 Z^{-2} + \cdots + a_{N-1} Z^{N-1}$$

which indicates that the transfer function is an $(N-1)$th polynomial; that is, it has $(N-1)$ zeros and no poles, or, like any other open loop system, it has $(N-1)$ poles at the origin.

6.3.1.2   Ease of Implementation

FIR filters are simpler in structure than IIR filters and can be implemented either in software or hardware, as we shall see later.

6.3.1.3   Moderate Finite Word Length Problems

FIR filters are less susceptible to noticeable finite word length errors than IIR filters are.

6.3.1.4   Linear Phase Response

Generally, nonlinear phase response of a system results in a proportional amount of phase and delay distortions. These are measured by the phase delay τ_p and the group delay τ_g, defined as follows:

$$\tau_p = -\frac{\theta(\omega)}{\omega} \tag{6.3}$$

and

$$\tau_g = -\frac{d}{d\omega}\theta(\omega)$$

FIR filters have linear phase response and hence constant group delay characteristics. This specific and unique property of FIR filters is perhaps the most important property. It is further considered as a typical feature of such category of filters. Therefore, FIR filters are the only choice whenever linear phase is in question.

Linear phase and hence constant delay properties of FIR filters are coupled with the inherited symmetry of their impulse responses. Basically, there are two types of symmetry:

- Positive symmetry (symmetrical), where $h(nT) = h(N - n - 1)T$
- Negative symmetry (antisymmetrical), where $h(nT) = -h(N - n - 1)T$

Moreover, there are four possible modes of symmetry, based on whether the filter length N is odd or even, as shown in Figure 6.5. These are:

Positive symmetry
1. N odd
   A typical impulse response for $N = 17$ is illustrated in Figure 6.5a. This response provides constant phase and group delay. It is suitable for all kinds of filter functions.
2. N even
   An impulse response for a filter having length $N = 16$ is depicted in Figure 6.5b. Although it looks similar to the preceding one, it is unsuitable for designing high-pass and band-pass filters.

Negative symmetry
3. N odd
   Due to their antisymmetrical impulse response, this type of filter is capable of providing constant group delay only. Therefore, they are used in the realization of Hilbert transformers[*] and differentiators.[†]
4. N even
   This type is almost the same as type 3.

To have linear phase characteristics, a filter should possess a phase response that satisfies either of the conditions

$$\theta(\omega) = -\alpha\omega \tag{6.4}$$

or

$$= -\alpha\omega + \beta \tag{6.5}$$

in which α and β are constants. The filter transfer function can be expressed as

$$H(Z) = \sum_{n=0}^{N-1} h(nT) \cdot Z^{-n} = \sum_{n=0}^{N-1} h(nT) \cdot e^{-j\omega nT} = \sum_{n=0}^{N-1} h(nT)\{\cos(n\omega T) - j\sin(n\omega T)\}$$

---

[*] A Hilbert transformer is a class of FIR filter that is used mainly in the generation of analytic signals. It is also called a 90-degree phase shifter.

[†] Discrete-time differentiators are also a class of FIR filter that are used, as the name implies, in performing differentiations for discrete-time signals.

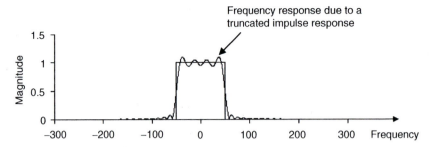

FIGURE 6.8   Gibbs oscillations.

giving

$$h(nT) = \frac{\omega_c T}{\pi} \frac{\sin(n\omega_c T)}{n\omega_c T} \quad n \neq 0$$

$$= \frac{\omega_c T}{\pi} \qquad\qquad n = 0 \qquad\qquad (6.11)$$

This proves that the impulse response of an ideal filter, that has a rectangular magnitude response, is the sinc function that extends from $-\infty$ to $+\infty$. Therefore, to get a finite-length impulse response filter, one has to limit (truncate) such a response to a practically feasible finite length N. A direct way to achieve this, as said before, is to multiply $h(nT)$ by a time function that has a finite length of N samples. Any other samples of the impulse response outside the range $-(N - 1)/2$ to $(N - 1)/2$ are suppressed (attenuated) according to the shape of the used window function. Naturally, the removal of any part of the impulse response through truncation will adversely affect the shape of the reconstructed frequency response, as shown in Figure 6.8. It was found that within the pass band, the resulting filter response experiences some ripples (oscillations) at a frequency and amplitude that are given by the selected window function as well as its length. Such oscillations (known as Gibbs oscillations) are undesirable, because they cause amplitude distortion in the filtered signal. They result due to convolving of the window spectrum with the ideal filter response (multiplying two time-domain signals corresponds to convolving their spectra in the frequency domain). Their amplitude should therefore be maintained within the acceptable limits that are prescribed by the filter's tolerance structure. At the end of the pass band, the response undergoes an overshoot and an almost equal undershoot at the end of the transition band.

The noncausal impulse responses of the four basic filter functions are listed in Table 6.2. To be causal, they should be shifted properly in time, before being used in the design process, as illustrated earlier.

There are many types of windows known in the literature [4]. Some common types are:

1. Rectangular window
2. Bartlett window
3. Von Hann window

TABLE 6.2
Impulse Response of Basic Filter Functions

	Impulse Response $h(nT)$ $n \neq 0$	$n = 0$
LP	$\dfrac{\omega_c T}{\pi} \dfrac{\sin(n\omega_c T)}{n\omega_c T}$	
HP	$-\dfrac{\omega_c T}{\pi} \dfrac{\sin(n\omega_c T)}{n\omega_c T}$	$1 - \dfrac{\omega_c T}{\pi}$
BP	$\dfrac{\omega_{c2} T}{\pi} \dfrac{\sin(n\omega_{c2} T)}{n\omega_{c2} T} - \dfrac{\omega_{c1} T}{\pi} \dfrac{\sin(n\omega_{c1} T)}{n\omega_{c1} T}$	$\dfrac{(\omega_{c2} - \omega_{c1})T}{\pi}$
BS	$\dfrac{\omega_{c1} T}{\pi} \dfrac{\sin(n\omega_{c1} T)}{n\omega_{c1} T} - \dfrac{\omega_{c2} T}{\pi} \dfrac{\sin(n\omega_{c2} T)}{n\omega_{c2} T}$	$1 - \dfrac{(\omega_{c1} - \omega_{c2})T}{\pi}$

4. Hamming window
5. Blackman window
6. Kaiser window
7. Dolph–Chebyshev window
8. Blackman–Harris window
9. Kaiser–Bessel window

6.3.4.1 Rectangular Window

A rectangular window of length N is described by

$$w(nT) = 1 \quad \text{for} \quad -\frac{N-1}{2} < n < \frac{N-1}{2}$$

$$= 0 \quad \text{otherwise}$$

Two main features of any window are of special importance to the filter designer, as they affect to a large extent the performance of the resulting filter. These are:

1. The main lobe width of the spectrum
2. The first side lobe amplitude relative to the main lobe peak amplitude

The frequency spectrum of a rectangular window is plotted in Figure 6.9. It was found that the main lobe width B_m is inversely proportional with the length of the window N.

A filter realized using a rectangular window can provide a maximum stop band attenuation of 21 dB and a transition bandwidth of

$$\Delta\omega' = \frac{5.7}{NT} = \frac{5.7f_s}{N} = \frac{5.7\omega_s}{2\pi N} = \frac{0.9\omega_s}{N}$$

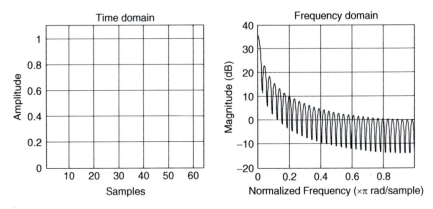

FIGURE 6.9   A rectangular window and its spectrum. (Adapted from MATLAB.)

Let us now define the normalized transition band $\Delta\omega$ as

$$\Delta\omega = \frac{\Delta\omega'}{\omega_s}$$

Thus, for a rectangular window it is given by

$$\Delta\omega = \frac{0.9}{N}$$

The windowed impulse response $h_w(nT)$ is obtained by direct multiplying the window function with the impulse response; that is,

$$h_w(nT) = w(nT) \cdot h(nT) \tag{6.12}$$

This means that the windowed frequency response can correspondingly be obtained by convolving the window function with the frequency response of the filter; that is,

$$H_w(\omega) = W(\omega) * H(\omega) \tag{6.13}$$

As a result of windowing, the filter response shows oscillations (ripples) at both the pass and the stop bands. The amplitudes of such oscillations—Gibbs oscillations—are determined primarily by the first side lobe amplitude of the window spectrum. The ripple ratio (RR), defined as

$$RR\% = \frac{\text{maximum sidelobe amplitude}}{\text{main lobe amplitude}} \times 100$$

which is a good measure for the amplitude of such oscillations. For the case of an 11-sample rectangular window, the RR is

$$RR\% = \frac{2.45}{11} \times 100 = -22.34\%$$

The impulse response, expressed by Equation 6.11 and shown in Figure 6.7b, represents a noncausal filter, as it extends from $-\infty$ to $+\infty$. The same applies to the windowed impulse response, given by Equation 6.12, as it extends from $-(N-1)/2$ to $(N-1)/2$. Therefore, for the filter to be realizable, these functions should be causal. This noncausal/causal transformation of a function can be achieved by shifting it to the right by $(N-1)/2$ samples. Doing so in Equation 6.11, we get the following causal impulse response $h_c(nT)$:

$$h_c(nT) = \frac{\omega_c T}{\pi} \frac{\sin(n-((N-1)/2))\omega_c T}{(n-((N-1)/2))\omega_c T} \qquad n \neq 0$$

$$= \frac{\omega_c T}{\pi} \qquad n = 0$$

(6.14)

Gibbs oscillations, as said before, are a result of convolving the window spectrum with the ideal filter response (brick wall response) as illustrated in Figure 6.10.

It was found that increasing the window length and hence the filter length N results in reducing the amplitude of such oscillations while increasing its frequency. Figure 6.11 depicts the spectra of rectangular windows having different lengths. It is obvious that increasing N causes narrowing of the main lobe width of its spectrum and hence results in sharper peaks in the filter's frequency response. Such sharp and narrow peaks explain why the frequency of Gibbs oscillations increases as N increases, while their amplitudes, on the other hand, decrease proportionally except at the edge of the transition band, where the overshoot and the undershoot amplitudes remain almost unchanged.

EXAMPLE 6.4

Using a rectangular window, design a low-pass filter to have a cut-off frequency at $\omega_c = 1$ rad/s.

The impulse response of an ideal and causal low-pass filter (Figure 6.10b) is given by

$$h_c(nT) = \frac{\omega_c T}{\pi} \frac{\sin n'\omega_c T}{n'\omega_c T}$$

where

$$n' = n - \frac{N-1}{2}$$

To avoid aliasing, choose a sampling frequency that is at least four times the cut-off frequency (Figure 6.12) such that

$$\omega_s = 4\omega_c$$

Now, because

$$\omega_s T = 2\pi$$

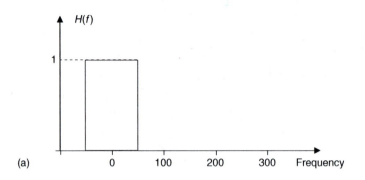

(a)

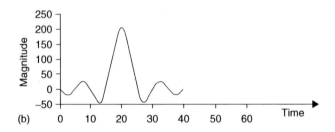

(b)

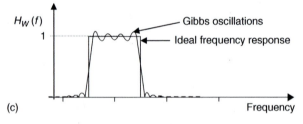

(c)

FIGURE 6.10   Effect of windowing on the magnitude response of an ideal filter. (a) Magnitude response of an ideal filter, (b) windowed impulse response, and (c) windowed magnitude response.

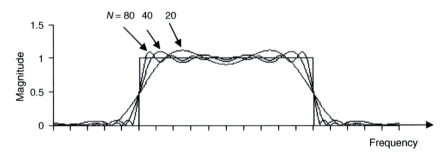

FIGURE 6.11   (See color insert following page 262.) Effect of window length on the response.

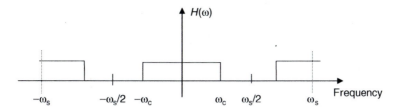

FIGURE 6.12 Magnitude response of an ideal LPF.

then

$$\omega_c T = \pi/2$$

Because no specific requirements are set on the minimum stop band attenuation, we can select arbitrarily a filter length of $N = 11$ (should be an odd number), thus substitution in the earlier equation gives

$$h_c(nT) = \frac{\pi/2}{\pi} \frac{\sin(n - ((11-1)/2))\pi/2}{(n - ((11-1)/2))\pi/2} = \frac{1}{2} \frac{\sin(n-5)\pi/2}{\sin(n-5)\pi/2} \quad \text{for } n \neq 5$$

and

$$= \frac{1}{2} \quad \text{for } n = 5$$

Due to the symmetry of $h(nT)$, it is sufficient to calculate only half the values; that is, for n ranging from 0 to 5, giving

$$h(0) = 0.5 \frac{\sin(-5\pi/2)}{(-5\pi/2)} = \frac{1}{5\pi} = a_0 = a_{10}$$

$$h(T) = 0.5 \frac{\sin(-2\pi)}{(-2\pi)} = 0 = a_1 = a_9$$

$$h(2T) = 0.5 \frac{\sin(-3\pi/2)}{(-3\pi/2)} = -\frac{1}{3\pi} = a_2 = a_8$$

$$h(3T) = 0.5 \frac{\sin(-\pi)}{(-\pi)} = 0 = a_3 = a_7$$

$$h(4T) = 0.5 \frac{\sin(-\pi/2)}{(-\pi/2)} = a_4 = a_6 = \frac{1}{\pi}$$

$$h(5T) = 0.5 = a_5$$

The filter's coefficients have been taken as the impulse response samples, as the window samples are all unity. The filter can be realized in the form shown in Figure 6.13 (transversal form), where 10 unit delays, 7 multipliers, and 1 adder are needed. The impulse response for $N = 11$ is drawn using MATLAB in form of a stem diagram and is illustrated in Figure 6.14.

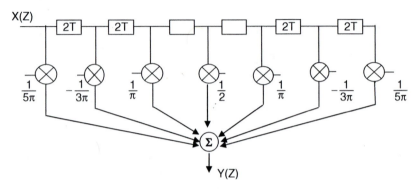

FIGURE 6.13 A realization for the filter of Example 6.4.

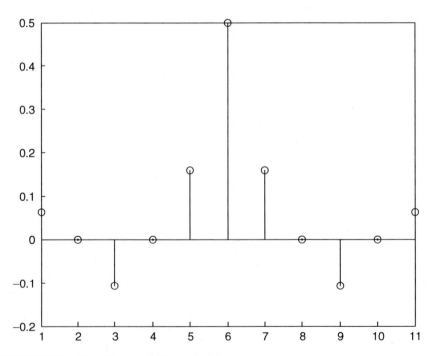

FIGURE 6.14 Stem diagram for Example 6.4.

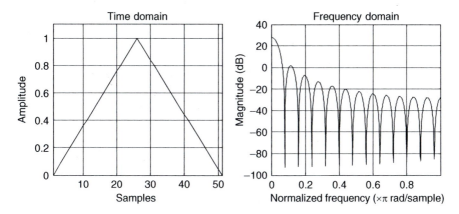

FIGURE 6.15 The Bartlett window, $N = 51$. (Adapted from MATLAB®.)

6.3.4.2 Bartlett Window

This window is described by the following function:

$$w_{Bt}(n) = \frac{n}{(N-1)/2} \qquad 0 \leq n < \frac{N-1}{2}$$

$$= 2 - \frac{n}{(N-1)/2} \qquad \frac{N-1}{2} < n < N-1$$

A plot of this function is shown in Figure 6.15.

A filter realized using a Bartlett window shows a ripple amplitude for the pass band and the stop band of $\delta_p = \delta_r = 0.05$, which gives a pass band attenuation of

$$A_p = 20\log(1 + \delta_p) = 20\log(1 + 0.05) = 0.42 \text{ dB}$$

and a stop band attenuation of

$$A_r = 20\log \delta_r = 20\log 0.05 = -26 \text{ dB}$$

6.3.4.3 Von Hann Window

This window is described by the causal function

$$w_{Hn}(n) = 0.5 - 0.5\cos\left(\frac{2\pi n}{N-1}\right) \quad \text{for } 0 < n < N-1 \tag{6.15}$$

$$= 0 \qquad \qquad \text{otherwise}$$

The Von Hann window, also called the *raised cosine* or sometimes the *hanning window*, is plotted together with its spectrum in Figure 6.16.

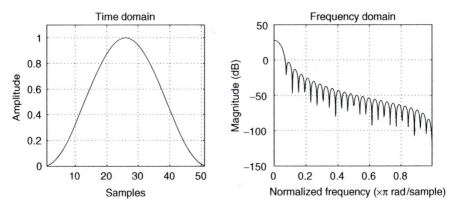

FIGURE 6.16 The Von Hann window. (Adapted from MATLAB®.)

From the window spectrum, one can find out that the associated ripple ratio is 2.62%. This ratio is far better than the ripple ratio of the rectangular window (22.34%). However, this is achieved at the expense of an increased main lobe width (two times wider), which will result in a wider transition band of a realized filter. The attainable ripple amplitudes are

$$\delta_p = \delta_r = 0.0063$$

giving a pass band attenuation of

$$A_p = 20\log(1 + \delta_p) = 0.055 \text{ dB}$$

and a maximum stop band attenuation of

$$A_r = 20 \log \delta_r = -44 \text{ dB}$$

The required filter length is calculated from

$$N = \frac{3.1}{\Delta\omega}$$

A filter realized using the Von Hann window shows the response given in Figure 6.17 for a filter length of $N = 51$. It is clear from the figure that the stop band attenuation is −44 dB, as expected.

6.3.4.4 Hamming Window

This window is derived from the Von Hann window, and is described by

$$w_{\text{Hm}} = 0.54 - 0.46 \cos\left(\frac{2\pi n}{N-1}\right) \quad 0 < n < N - 1$$

$$= 0 \qquad\qquad\qquad\qquad\qquad \text{elsewhere}$$

(6.16)

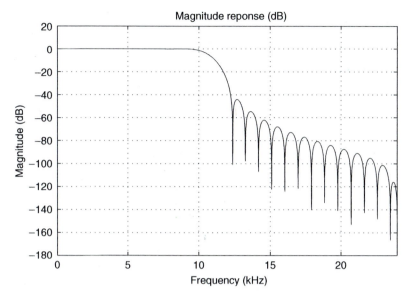

FIGURE 6.17   Magnitude response of an LP FIR filter designed using a Von Hann window.

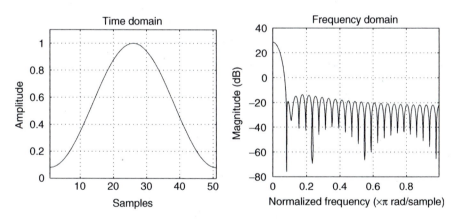

FIGURE 6.18   The Hamming window and its spectrum. (Adapted from MATLAB®.)

A plot of a Hamming window, also called *raised cosine with platform*, is depicted in Figure 6.18, together with its spectrum. The pass band and the rejection band ripples are found to be

$$\delta_p = \delta_r = 0.0022$$

giving a pass band attenuation of

$$A_p = 20\log(1 + 0.0022) = 0.55 \text{ dB}$$

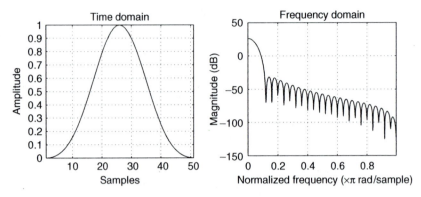

FIGURE 6.19 The Blackman window and its spectrum. (Adapted from MATLAB®.)

and a stop band attenuation of

$$A_r = 20 \log 0.0022 = -53 \text{ dB}$$

which represents an improvement over the attainable stop band attenuation of previous windows of 9 dB.

6.3.4.5 Blackman Window

This window is described by the function

$$w_B(n) = 0.42 - 0.5 \cos\left(\frac{2\pi n}{N-1}\right) + 0.08 \cos\left(\frac{4\pi n}{N-1}\right) \quad 0 \le n < N-1$$

$$= 0 \qquad\qquad\qquad\qquad\qquad\qquad\qquad \text{elsewhere}$$

(6.17)

Figure 6.19 illustrates the window function and its spectrum.

Compared to the previously discussed window functions, the corresponding ripple amplitudes here are greatly reduced to be

$$\delta_p = \delta_r = 0.0002$$

leading to an improved pass band and stop band attenuations of

$$A_p = 20 \log(1 + 0.0002) = 0.0017 \text{ dB}$$

and

$$A_r = 20 \log 0.0002 = -74 \text{ dB}$$

6.3.4.6 Kaiser Window

In this class of windows, a ripple control parameter α is introduced, whose value is selected according to the required stop band attenuation. The magnitude of such parameter determines, in turn, the resulting ripple ratio RR and the main lobe width B_m. It was found that α is directly proportional with B_m and inversely proportional with the ripple ratio RR. Because both values are contradicting, one has to select α that satisfies the specified stop band attenuation. The value of α is determined through one of the following empirical formulas:

$$\begin{aligned} \alpha &= 0 & A_r &\leq 21\,\text{dB}\\ &= 0.5842(A_r - 21)^{0.4} + 0.07886(A_r - 21) & 21 &< A_r < 50\,\text{dB}\\ &= 0.1102(A_r - 8.7) & A_r &\geq 50\,\text{dB} \end{aligned}$$

The window function is described by

$$w_k(n) = \frac{J_o(\beta)}{J_o(\alpha)} \quad 0 \leq n \leq N-1$$

$$= 0 \qquad \text{otherwise} \tag{6.18}$$

where
- $J_o(x)$ = zero-order modified Bessel function of the first kind
- β = function of α and the index n

It is given by

$$\beta(n) = \alpha\sqrt{1 - \left(\frac{2n'}{N-1}\right)^2}$$

in which

$$n' = n - \frac{N-1}{2} \quad 0 < n < N-1$$

The required filter length can be calculated from

$$N \geq \frac{0.9222}{\Delta\omega} + 1 \qquad A_r \leq 21 \tag{6.19}$$

$$= \frac{A_r - 7.95}{14.36\,\Delta\omega} + 1 \qquad A_r > 21 \tag{6.20}$$

where $\Delta\omega$, as before, is the normalized transition bandwidth given by

$$\Delta\omega = \frac{\omega_r - \omega_p}{\omega_s}$$

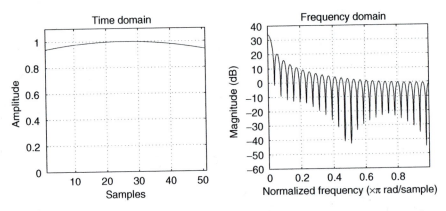

FIGURE 6.20   The Kaiser window and its spectrum, $N = 37$. (Adapted from MATLAB®.)

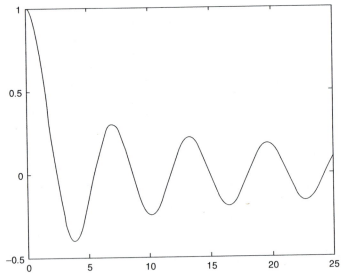

FIGURE 6.21   Modified Bessel function of the zero-order, first kind.

The stop band attenuation A_r is usually estimated using the relationship

$$A_r = -20\log\{\min(\delta_p, \delta_r)\}$$

The Kaiser window and its spectrum are illustrated in Figure 6.20. A plot of Bessel function $J_o(x)$ is shown in Figure 6.21.

Bessel functions can be approximated by the power series [16]

$$J_o(x) = 1 + \sum_{m=1}^{M}\left[\frac{(x/2)^m}{m!}\right]^2$$

for M ranging between 15 and 25.

TABLE 6.3
Summary of the Main Features of Some Usable Window Functions

Window	Window Function	A_p/dB	A_r/dB	N
Rectangular	$w(n) = 1$	0.7416	21	$0.9/\Delta\omega$
Von Hann	$w_{Hn}(n) = 0.5 - 0.5\cos\left(\dfrac{2\pi n}{N-1}\right)$	0.0546	44	$3.1/\Delta\omega$
Hamming	$w_{Hm} = 0.54 - 0.46\cos\left(\dfrac{2\pi n}{N-1}\right)$	0.0194	53	$3.3/\Delta\omega$
Blackman	$w_B(n) = 0.42 - 0.5\cos\left(\dfrac{2\pi n}{N-1}\right) + 0.08\cos\left(\dfrac{4\pi n}{N-1}\right)$	0.0017	74	$5.5/\Delta\omega$
Kaiser	$w_k = \dfrac{J_o(\beta)}{J_o(\alpha)}$	—	>50	$\dfrac{A-7.95}{14.36\,\Delta\omega}$

A Bessel function table for x ranging from 0 to 10 is given in Appendix A2. Table 6.3 summarizes the most important features of some usable window functions. It is repeated in Appendix A4.

The design process of an FIR filter using the window method can be summarized in the following steps:

1. Get the sampled impulse response $h(nT)$ of the desired filter function, as given in Table 6.2.
2. Shift the noncausal impulse response to the right by $(N-1)/2$ samples to get the causal one $h_c(nT)$.
3. Calculate the ripple parameters δ_p and δ_r, then choose the least of them; that is,

$$\delta_{min} = min(\delta_p, \delta_r)$$

4. Calculate the new stop band attenuation $A_r = 20\log\delta_{min}$.
5. Choose a proper causal window function $w(nT)$ from Table 6.3 that satisfies the requirements as given by the tolerance structure. Main criterion is the minimum acceptable stop band attenuation A_r.
6. Estimate the required filter length from Table 6.3, based on the specified normalized transition bandwidth.
7. Calculate the window coefficients (samples) $w(nT)$.
8. Calculate the filter coefficients a_n by multiplying each sample of $h_c(nT)$ with its corresponding one of $w(nT)$ as follows:

$$a_n = h_w(nT) = h_c(nT) \cdot w(nT)$$

It is to be noted here that due to the symmetry in the filter coefficients, it is sufficient to calculate only the first $N/2 + 1$ coefficients. Also, due to smearing effects resulting

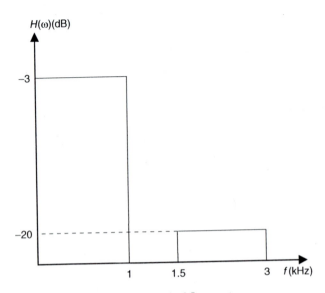

FIGURE 6.22   Tolerance structure for Example 6.5.

from windowing, the obtained cut-off frequency is shifted slightly. An approximate value can be obtained from the relation

$$\omega_c = \frac{\omega_p + \omega_r}{2} = \omega_p + \frac{\Delta\omega'}{2}$$

However, to reduce such effect, one can select a larger value for N, the filter length, than required, as it is inversely proportional with the transition bandwidth ($\Delta\omega' \alpha\ 1/N$). The following examples will help illustrate the method.

EXAMPLE 6.5

Using a proper window function, design a low-pass FIR filter to satisfy the shown tolerance structure (Figure 6.22).

Solution

From the tolerance structure, we get

$$f_c = 1\ \text{kHz}$$

$$f_r = 1.5\ \text{kHz}$$

$$f_s = 3 \times 2 = 6\ \text{kHz}$$

$$\Delta\omega = (1.5 - 1)/6 = 1/12$$

Because $A_r = -20$ dB, we can select a rectangular window. The required length N can be determined from

$$N = \frac{0.9}{\Delta f} = 0.9 \times 12 \approx 11$$

$$\omega_c T = \frac{2\pi}{f_s} = \frac{2\pi}{6} = \frac{\pi}{3}$$

$$\frac{N-1}{2} = 5$$

The causal impulse response of a low-pass filter of length 11, as before, is given by

$$h_c(nT) = \frac{\omega_c T}{\pi} \frac{\sin(n-5)^{\pi/3}}{(n-5)^{\pi/3}}$$

$$= \frac{1}{3} \frac{\sin(n-5)^{\pi/3}}{(n-5)^{\pi/3}} \qquad n \neq 5$$

$$= \frac{1}{3} \qquad\qquad n = 5$$

$$= \frac{1}{3} \frac{\sin(n-5)^{\pi/3}}{(n-5)^{\pi/3}} \qquad n \neq 5$$

$$= \frac{1}{3} \qquad\qquad n = 5$$

The coefficients of the filter are as follows:

$$h(0) = a_0 = a_{10} = -0.055$$

$$h(T) = a_1 = a_9 = -0.0689$$

$$h(2T) = a_2 = a_8 = 0$$

$$h(3T) = a_3 = a_7 = 0.1378$$

$$h(4T) = a_4 = a_6 = 0.2756$$

$$h(5T) = a_5 = 0.333$$

Plotting these values gives the stem diagram shown in Figure 6.23.

EXAMPLE 6.6

Using an appropriate window function, design a low-pass FIR filter to meet the specifications shown in Figure 6.24.

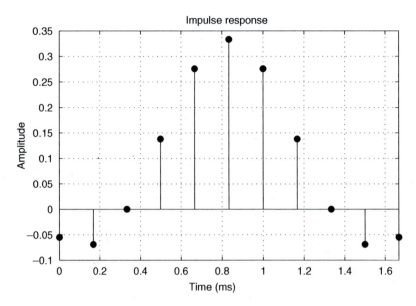

FIGURE 6.23 Results of Example 6.5.

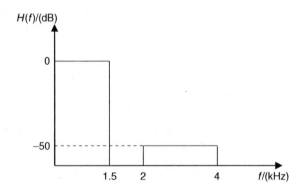

FIGURE 6.24 Tolerance structure for Example 6.6.

Solution

The filter operates at a sampling frequency of

$$f_s = 4 \times 2 = 8 \text{ kHz}$$

and has a normalized transition bandwidth of

$$\Delta\omega = \frac{2 \cdot \pi (f_r - f_p)}{2 \cdot \pi \cdot f_s} = \frac{2 - 1.5}{8} = 0.0625$$

The cut-off frequency can be approximated to

$$f_c = \frac{2 + 1.5}{2} = 1.75 \text{ kHz}$$

giving

$$\omega_c T = 2 \times \pi \times 1.75 \times \frac{1}{8} = 0.437\pi \text{ krad/s}$$

The minimum acceptable stop band attenuation, as given by the tolerance structure, is $A_r = 50$ dB; thus, from Table 6.3, it is evident that Hamming window is an appropriate choice. The required filter length is thus

$$N = \frac{3.3}{\Delta\omega} = \frac{3.3}{0.0625} = 52.3 \approx 53$$

The causal window function from Equation 6.16 is

$$w_{Hm} = 0.54 - 0.46 \cos\left(\frac{2\pi n}{52}\right)$$

The causal impulse response for an LP filter ($N = 53$) is given by

$$h_c(nT) = 0.4375 \frac{\sin(n - 26)0.4375\pi}{(n - 26)0.4375\pi} \qquad n \neq 26$$

$$= 0.4375 \qquad\qquad\qquad n = 26$$

MATLAB Assistance

The following MATLAB commands can be used to get the filters' coefficients, as well as the responses:

```
N = 52;
Wc = 1750/4000;
B = fir1(N, wc, Hamming (N + 1);
fvtool(b)
```

The magnitude, phase, and impulse responses for the filter are illustrated in Figures 6.25a and 6.25b, respectively. The phase shift is seen to be linear throughout the pass band, giving rise to constant group delay (Figure 6.26).

EXAMPLE 6.7

Using a Kaiser window, design a low-pass filter to have the tolerance structure shown in Figure 6.26.

Solution

From the figure, we get the following sampling frequency:

$$f_s = 2 \times 5 = 10 \text{ kHz}$$

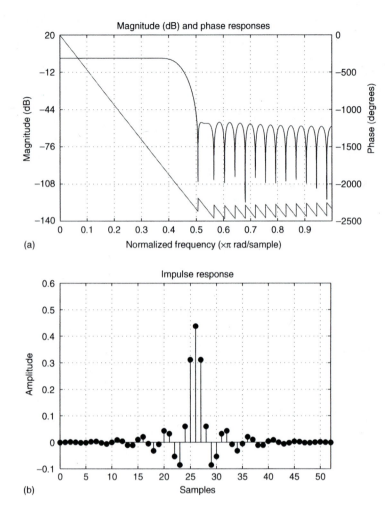

(a)

(b)

FIGURE 6.25 Results of Example 6.6. (a) Magnitude and phase response, (b) stem diagram.

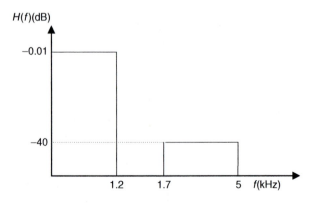

FIGURE 6.26 Tolerance structure for Example 6.7.

The approximate cut-off frequency $f_c = \dfrac{1.2 + 1.7}{2} = 1.45 \text{ kHz};$

then

$$\omega_c T = 2 \times \pi \times 1.45 \times 10^3 \times 10^{-4} = 0.91106 \text{ radians}$$

The ripple amplitudes δ_p and δ_r can be calculated as follows:

$$\delta_p = [10^{0.05A_p} - 1] = 0.00115$$

$$\delta_r = [10^{-0.05A_r}] = 0.01$$

To insure the fulfillment of the attenuation requirements, we should use the smallest one of them—that is,

$$\delta = \min(\delta_p, \delta_r) = 0.00115$$

giving a new stop band attenuation of

$$A_r = 20 \log 0.00115 = -58.8 \text{ dB}$$

which is more than the specified stop band attenuation.

The ripple parameter $\alpha = 0.1102(58.8 - 8.7) = 5.52102$. The required filter length is then

$$N \ge \frac{58.8 - 7.95}{14.36(0.05)} \ge 71$$

$$\beta(n, \alpha) = 5.52102 \sqrt{1 - \left(\frac{n - 35}{35}\right)^2} \qquad \text{for } n = 0:70$$

The window coefficients are calculated from

$$w_k(n, \alpha) = \frac{J_o(\beta)}{J_o(\alpha)}$$

The filter's impulse response is given by

$$h_c(nT) = \frac{\omega_c T}{\pi} \frac{\sin[(n - 35)\omega_c T]}{(n - 35)\omega_c T} = \frac{0.91106}{\pi} \frac{\sin[(n - 35)0.91106]}{(n - 35)0.91106} \qquad n \ne 35$$

$$= \frac{0.9116}{\pi} \qquad n = 35$$

The filter's coefficients are calculated from

$$a_n = h_c(nT) \cdot w_k(n, \alpha)$$

MATLAB Assistance

The following instructions can be used

```
M = 70;
Wc = 1450/5000;
alpha = 5.52102;
b = fir1(M, wc, Kaiser((M + 1),alpha));
fvtool(b)
```

The magnitude, phase responses, and impulse response are plotted in Figure 6.27.
It is clear that the stop band attenuation, the cut-off frequency, and the linear phase
conditions are all met. (Actually, the attenuation exceeds the requirement, which is
recommended.)

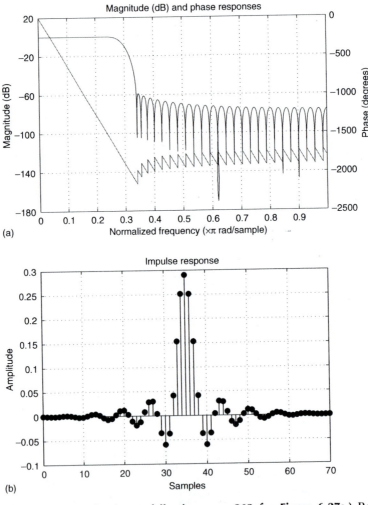

FIGURE 6.27 (See color insert following page 262 for Figure 6.27a.) Results of
Example 6.7. (a) Magnitude and phase response, (b) stem diagram.

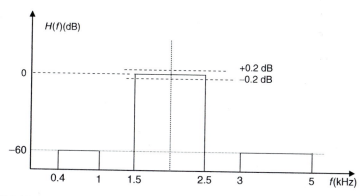

FIGURE 6.28 Tolerance structure of Example 6.8.

Example 6.8

Design a minimum-length FIR filter to satisfy the requirements given by the tolerance structure shown in Figure 6.28.

Solution

From the given diagram, it is clear that the required filter is a band-pass one.
 The sampling frequency is

$$f_s = 5 \times 2 = 10 \text{ kHz}$$

The sampling period $T = 1/f_s = 10^{-4}$ s.
 The normalized transition bands are

$$\Delta f_1 = \frac{1.5 - 1}{10} = 0.05$$

$$\Delta f_2 = \frac{3 - 2.5}{10} = 0.05$$

The approximate cut-off frequencies on both sides of the response are calculated from

$$f_{c1} = 1 + \frac{1.5 - 1}{2} = 1.25 \text{ kHz} \quad \text{and} \quad f_{c2} = 2.5 + \frac{3 - 2.5}{2} = 2.75 \text{ kHz}$$

giving

$$\omega_{c1} T = 2\pi f_{c1} = 2\pi \times 1.25 \times 10^3 \times 10^{-4} = 0.25\pi \text{ radians}$$

and from

$$\omega_{c2} T = 2\pi f_{c2} = 2\pi \times 2.75 \times 10^3 \times 10^{-4} = 0.55\pi \text{ radians}$$

The ripple amplitude in the pass and stop bands are estimated as follows:

$$\delta_p = (10^{0.05A_p} - 1) = (10^{0.01} - 1) = 0.02329$$

$$\delta_r = (10^{0.05A_r}) = 10^{-3} = 0.001$$

and, as before,

$$\delta_r = \min(\delta_p, \delta_r) = 0.001$$

A look at Table 6.3 reveals that two windows come in question:

1. A Blackman window with a length of

$$N_B = \frac{5.5}{\Delta f} = \frac{5.5}{0.05} = 110$$

2. A Kaiser window, whose length is given by

$$N_K \geq \frac{60 - 7.95}{14.36 \times 0.05} \geq 72.49 \approx 73$$

Because a minimum length filter is required, the second option is the best choice. Generally, as a rule of thumb, whenever a minimum-length filter is required, a Kaiser window should be used.

The ripple parameter α for the given attenuation is calculated from

$$\alpha = 0.1102(60 - 8.7) = 5.65$$

giving

$$J_o(5.65) = 0.043$$

and

$$\beta(n, \alpha) = 5.65\sqrt{1 - \left(\frac{n - 36}{36}\right)^2}$$

The filter impulse response, as given in Table 6.2, is

$$h_c(nT) = \frac{\omega_{c2} - \omega_{c1}}{\pi}\left[\frac{\sin \omega_{c2}T(n - 36)}{\omega_{c2}T(n - 36)} - \frac{\sin \omega_{c1}T(n - 36)}{\omega_{c1}T(n - 36)}\right]$$

MATLAB Assistance

For a band-pass filter, the following instructions can be used:

```
w = [wc1, wc2];
B = fir1(N,w, Kaiser(N + 1,alpha));
fvtool(b)
```

The obtained responses are depicted in Figure 6.29.

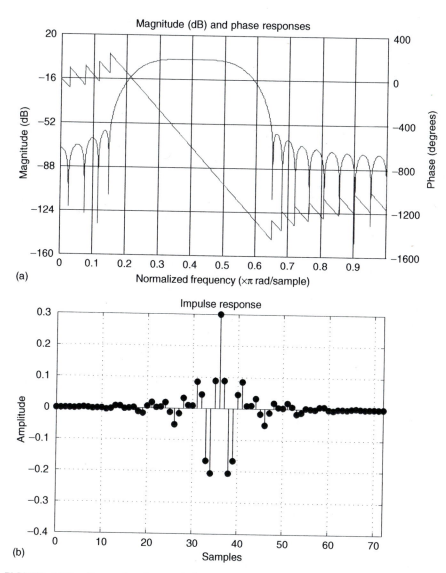

FIGURE 6.29 (See color insert following page 262 for Figure 6.29a.) Results of Example 6.8. (a) Magnitude and phase response, (b) stem diagram.

It is clear from the plots that stop band attenuations of more than 60 dB have been achieved, which meets the requirements.

6.3.5 OPTIMAL METHOD

As discussed before, windowing the impulse response of a filter results in ripples in the amplitude response that reach a peak (overshoot) near the transition band edges. The shape and amplitude of such peaks are window-type dependent. In some applications, however, the resulting amplitude distortion is unacceptable. The optimal equiripple FIR filter design due to Parks and McClellan provides FIR filters that represent the best approximation to the desired frequency response in a Chebyshev sense. The amplitude response is enforced to oscillate between two predetermined limits within the pass band and the stop band. As depicted in Figure 6.26, ripples in the pass band vary between $\pm\delta_p$, while in the stop band they vary between 0 and $+\delta_r$. Filters having the property of equiripple pass band and stop band responses are called *optimum filters*.

The designed filter should have a response $H(\omega)$ that minimizes the error function $E(\omega)$ defined as

$$E(\omega) = W_b[G(\omega) - H(\omega)] \qquad (6.21)$$

where
- $G(\omega)$ = ideal (rectangular) filter response
- W_b = weight assigned to each band reflecting their relative importance as described by the stop band and the pass band ripples

That is,

$$\frac{W_p}{W_r} = \frac{\delta_r}{\delta_p} \qquad (6.22)$$

The filters' coefficients should be selected such that the condition

$$\min[\max|E(\omega)|]$$

remains satisfied over both bands. The resulting response will be a unique weighted Chebyshev approximation to $G(\omega)$ over the interval $0-\pi$.

The locations of the maxima and minima on the frequency scale are called the *extrema* (marked with solid dots in Figure 6.30). To achieve linear phase response, there will be $r + 1$ or $r + 2$ such extrema, where

$$r = \frac{N+1}{2} \quad \text{for odd positive symmetry or}$$

$$= \frac{N}{2} \quad \text{for even positive symmetry}$$

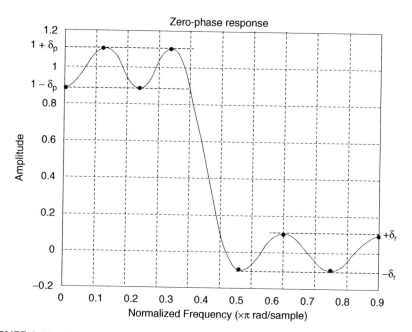

FIGURE 6.30 Equiripple response of an FIR filter (the dark dots indicate the extrema).

The design process involves the determination of these extremal frequencies from which the magnitude response and hence the impulse response are easily obtained. An iterative computational technique that is based on the Remez-exchange algorithm [16] is usually used. The required filter length is determined through the empirical formulas given in Equations 6.23 and 6.24 for low-pass and band-pass filters, respectively.

6.3.5.1 Low-Pass Filter Length

$$N_{LP} = D_{LP}(\delta_P, \delta_r) - f_{LP}(\delta_P, \delta_r) + 1 \tag{6.23}$$

where

$$D_{LP}(\delta_P, \delta_r) = \frac{\log \delta_r \left[a_1 (\log \delta_P)^2 + a_2 \log \delta_P + a_3 \right] + \left[a_4 (\log \delta_P)^2 + a_5 \log \delta_P + a_6 \right]}{\Delta f}$$

and

$$f_{LP} = (11.01217 + 0.51244[\log \delta_P - \log \delta_r]) \Delta f$$

Δf is the normalized transition bandwidth with

$$a_1 = 5.309E - 3 \qquad a_2 = 7.114E - 2 \qquad a_3 = -4.761E - 1$$

$$a_4 = -2.66E - 3 \qquad a_5 = -5.941E - 1 \qquad a_6 = -4.278E - 1$$

6.3.5.2 Band-Pass Filter Length

$$N_{BP} = D_{BP}(\delta_P, \delta_r) + f_{BP}(\delta_P, \delta_r) + 1 \qquad (6.24)$$

where

$$D_{BP}(\delta_P, \delta_r) = \frac{\log \delta_r \left[b_1 (\log \delta_P)^2 + b_2 \log \delta_P + b_3 \right] + \left[b_4 (\log \delta_P)^2 + b_5 \log \delta_P + b_6 \right]}{\Delta f}$$

and

$$f_{BP} = \left(-14.6 \log \left(\frac{\delta_P}{\delta_r} \right) - 16.9 \right) \Delta f$$

with

$$b_1 = 0.01201 \qquad b_2 = 0.09664 \qquad b_3 = -0.51325$$

$$b_4 = 0.00203 \qquad b_5 = -0.5705 \qquad b_6 = -0.44314$$

The approximation steps using the optimal method can be summarized as follows:

1. Specify f_p, f_r, f_s, and the ratio δ_p/δ_r.
2. Assign weights (W_p, W_r) to both bands.
3. Estimate a suitable filter length.
4. Evaluate the optimal set of extremal frequencies using the Remez-exchange algorithm.
5. Using the extremal frequencies, find the frequency response.
6. Get the impulse response samples.

EXAMPLE 6.9

Using MATLAB, design an optimal FIR filter to have the shown tolerance structure, then plot its responses (Figure 6.31).
Consider a filter length 32. The following MATLAB instructions can be used:

```
N = 32;                          % Filter order
A = [.2  .2   1   1  .2   .2];   % Amplitude vector
F = [0  .2  .4  .7  .9  1];      % Frequency vector
W = [1   5   1];                 % Weight vector
B = remez(N, F , A , W);
fvtool(B)
```

The resulting magnitude, phase, and impulse responses are depicted in Figure 6.32.

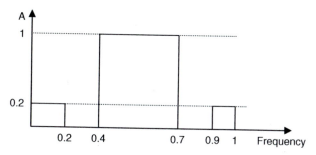

FIGURE 6.31   Tolerance structure of Example 6.9.

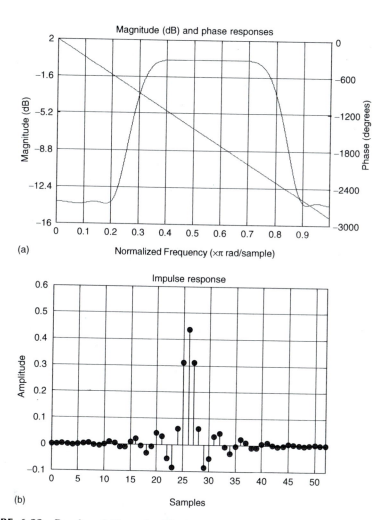

(a)

(b)

FIGURE 6.32   Results of Example 6.9. (a) Magnitude and phase response, (b) stem diagram.

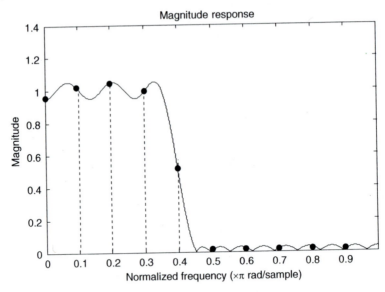

FIGURE 6.33 Sampling the magnitude response of a filter.

6.3.6 Frequency Sampling Method

To design an FIR filter, employing this method for a given prescribed magnitude response, the next steps that should be followed are:

1. Sample the magnitude response at enough points.
2. Take the inverse Fourier transform for the samples.
3. The resulting time samples are the samples of the filter's impulse response $h(nT)$ which are the required filter's coefficients a_n's, as $a_n = h(nT)$.

Needless to say that better result—that is, a filter whose response shows minimum deviations from the desired one—can be obtained by taking as many samples of the frequency response as possible. Figure 6.33 illustrates how the magnitude response is sampled.

EXAMPLE 6.10

Using the frequency sampling method, redesign the filter described in Example 6.9.

Solution

A MATLAB "m-file" can be composed as follows:

```
N = 22;
F = [0   0.1   0.2 0.8 0.9 1];
A = [.2 .2 1 1 .2 .2];
B = fir2(N, F, A);
hfvt = fvtool(B);
```

The obtained responses are illustrated in Figure 6.34.

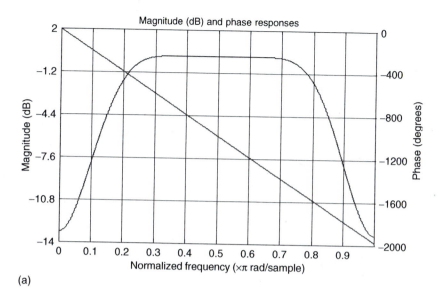

(a)

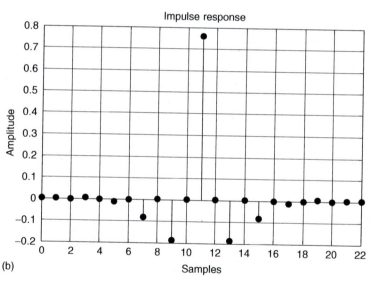

(b)

FIGURE 6.34 Results of Example 6.10. (a) Magnitude and phase response, (b) impulse response.

6.3.7 Least Pth-Norm Optimal Method

This design method delivers optimal FIR filters in the least Pth-norm sense. The resulting filters do not necessarily possess linear-phase property (the impulse response shows no symmetry). The MATLAB command b = firlpnorm (N, F, EDGES, A, W) provides an FIR filter that has an order N, representing the best approximation to the magnitude response described by the two vectors F and A, in the least Pth-norm sense, where P can vary between $P_{init} = 2$ and $P_{final} = 128$. The vector EDGES gives

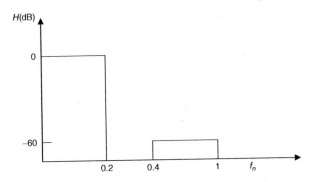

FIGURE 6.35   Tolerance structure of Example 6.11.

the band-edge frequencies for multiband designs. The resulting filter can be checked using the command freqz. The vectors F and A, describing the tolerance structure, should have the same length. The vector W has the same length of F, while A is used as a weight for the error. It defines how much emphasis should be put on minimizing the error in the vicinity of each frequency point relative to the other points. The function firlpnorm may include the parameter P, describing the minimum and the maximum values of P used in the algorithm.

EXAMPLE 6.11

Design an FIR filter to satisfy the tolerance structure shown in Figure 6.35 using the least Pth-norm optimal method. Let $N = 30$.

The following MATLAB instructions can be used to design the filter and to get its responses:

```
N = 30;
F = [0 0.2 0.4 1];
E = F;
A = [1 1 0 0];
W = [1 1 10 10];
b = firlpnorm(N,F,E,A,W);
fvtool(b)
```

The nonlinearity of the phase response over the pass band is evident from Figure 6.36b.

6.3.8   REALIZATION TOPOLOGIES FOR FIR FILTERS

After approximation, the next step is the realization of the obtained transfer function in form of a network containing multipliers, unit delays, and adders. A main criterion in selecting a specific structure is the number of components used in the implementation. Of special interest is the number of multipliers and delays. In view of hardware complexity, multipliers among other components are the most complex and hence the most time-consuming element. If the number of unit delays in a filter network is made equal to the filter's order, the resulting realization is called a *canonic* realization. In the following sections, the different possible realizations will be discussed, along with the required number of components for each realization.

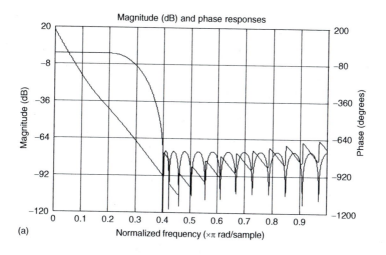

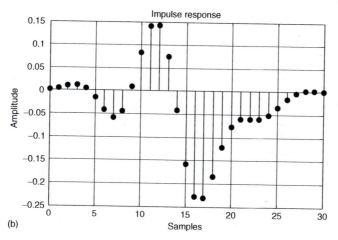

FIGURE 6.36 (See color insert following page 262 for Figure 6.36a.) Results of Example 6.11. (a) Magnitude and phase response, (b) stem diagram.

6.3.8.1 Transversal Structure

This is the most straightforward realization, as it directly implements the transfer function. The resulting structure is in the form of a tapped delay line. There are three possible variants of such structure: the direct form, the nested direct form, and the cascaded form.

6.3.8.1.1 The Direct Form

As the name implies, this topology realizes *directly* the transfer function

$$H(Z) = \sum_{k=0}^{N-1} a_k Z^{-k} = a_o + a_1 Z^{-1} + a_2 Z^{-2} + \cdots + a_{N-1} Z^{N-1}$$

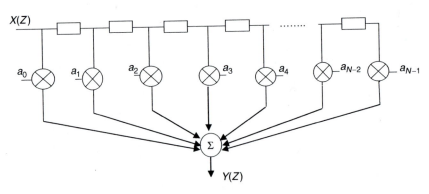

FIGURE 6.37   The direct transversal FIR filter topology.

As shown in Figure 6.37, $(N - 1)$ delays are connected in a chain with N transversal branches (taps) containing multipliers, representing the filter coefficients. The result of multiplications are summed up in an adder to provide the output $Y(Z)$. The main feature of such structure is its simplicity and ease of implementation.

The required hardware comprises:

$$
\begin{array}{ll}
N - 1 & \text{delays} \\
1 & \text{adder} \\
N & \text{multipliers}
\end{array}
$$

Delay elements are usually realized employing shift registers. When implemented using a DSP processor, the following computational operations are required:

$$
\begin{array}{ll}
N & \text{multiplications} \\
N & \text{memory cycles for fetching the coefficients} \\
N - 1 & \text{memory cycles to store input samples} \\
N - 1 & \text{additions}
\end{array}
$$

The direct realization topology is usually known as the *delay, multiply, add* (DMA) topology, which represents the sequence of operations.

6.3.8.1.2 The Cascaded-Direct Form

A second structure could be obtained by factorizing the transfer function into multiplicative second-order terms. The transfer function can then be represented as

$$H(Z) = \prod_{i=1}^{M} H_i(Z)$$

where

$$M = \frac{N - 1}{2}$$

indicating that the filter can be realized as a cascade of second-order sections (*biquads*), as we did in the realization of analog active filters in Chapter 3. A realization in this form is depicted in Figure 6.38. Such a structure has the appealing advantage of being less sensitive to coefficients errors and quantization noise. However, its use is not common, as it is not realizable through most of the known DSP processors.

It is evident from the figure that the components count is as follows:

$$
\begin{aligned}
N + M - 1 \quad &\text{multipliers} \\
N - 1 \quad &\text{delays} \\
M \quad &\text{adders}
\end{aligned}
$$

6.3.8.1.3   The Nested Direct Form

Rewriting the transfer function in the form:

$$Y(Z) = a_o + Z^{-1}[a_1 + Z^{-1}[a_2 + \cdots + Z^{-1}[a_{N-3} + Z^{-1}[a_{N-2} + Z^{-1}[a_{N-1}]]\cdots]$$

We get a different form for realization, as illustrated in Figure 6.39.

The required hardware for a filter of length N is:

$$
\begin{aligned}
N \quad &\text{multipliers} \\
N - 1 \quad &\text{adders} \\
N - 1 \quad &\text{delays}
\end{aligned}
$$

Due to its architecture, this topology is usually known as the *multiply, add, delay* (MAD) topology.

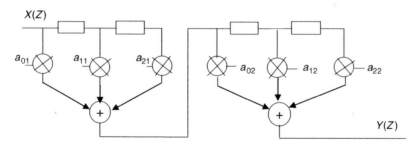

FIGURE 6.38   The cascaded direct realization for FIR filters.

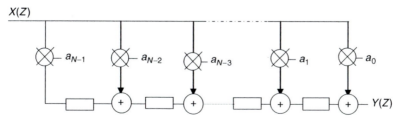

FIGURE 6.39   The nested direct form.

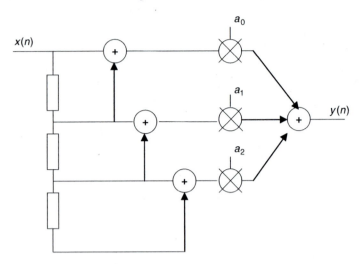

FIGURE 6.40 The DAM topology.

6.3.8.1.4 *The Feedforward Structure*
As an example, consider the third-order filter whose transfer function has the form

$$H(Z) = (1 + Z^{-1})(a_o + a_1 Z^{-1} + a_2 Z^{-2})$$

Such topology is usually known as *delay, add, multiply* (DAM) topology, shown in Figure 6.40. For a filter of length N the following components are needed

$$
\begin{array}{ll}
N - 1 & \text{delays} \\
N + 1 & \text{adder} \\
N - 1 & \text{multiplier}
\end{array}
$$

6.3.8.1.5 *Linear Phase Structure*
An FIR filter has linear phase response and hence a symmetrical impulse response and a symmetrical set of coefficients. This might suggest a saving in the number of multipliers. A modification to the direct transversal structure should therefore be introduced, whereby similar coefficients are made to share a common multiplier in the realization. Figure 6.41 illustrates diagrammatically the concept.

The required components count reduces to

$$
\begin{array}{ll}
\dfrac{N-1}{2} + 1 & \text{multipliers} \\[2mm]
\dfrac{N-1}{2} & \text{adders} \\[2mm]
N - 1 & \text{delays}
\end{array}
$$

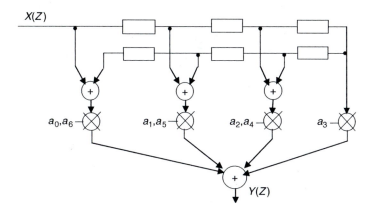

FIGURE 6.41 Linear phase structure for a sixth-order FIR filter.

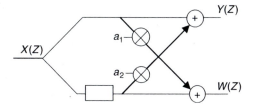

FIGURE 6.42 A first-order lattice FIR filter section.

6.3.8.2 Lattice (Transpose) Structure

A first-order lattice section is shown in Figure 6.42. Two different outputs $Y(Z)$ and $W(Z)$ are obtainable, giving rise to two transfer functions $H(Z)$ and $G(Z)$. Lattice structure is suitable for some specific applications, such as speech processing and linear prediction applications (see Chapter 10). The two outputs $Y(Z)$ and $W(Z)$ are derived as follows:

$$Y(Z) = X(Z) + a_1 Z^{-1} X(Z) = X(Z)\{1 + a_1 Z^{-1}\} \qquad (6.25)$$

$$W(Z) = a_1 X(Z) + Z^{-1} X(Z) = X(Z)\{a_1 + Z^{-1}\} \qquad (6.26)$$

Giving the transfer functions

$$H(Z) = \frac{Y(Z)}{X(Z)} = 1 + a_1 Z^{-1} \qquad (6.27)$$

and

$$G(Z) = \frac{W(Z)}{X(Z)} = a_1 + Z^{-1}$$

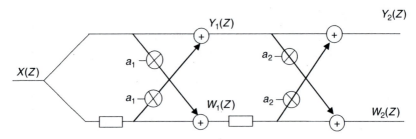

FIGURE 6.43   A second-order lattice filter.

Higher-order sections can be obtained by cascading as many first-order sections as are required. Figure 6.43 illustrates a second-order section.

To find the transfer functions $H_2(Z)$ and $G_2(Z)$, we write

$$Y_2(Z) = Y_1(Z) + a_2 Z^{-1}W_1(Z)$$

$$W_2(Z) = a_2Y_1(Z) + Z^{-1}W_1(Z)$$

Substituting from Equations 6.25 and 6.26 in the previous equations for $Y_1(Z)$ and $W_1(Z)$, we get

$$Y_2(Z) = (1 + a_1Z^{-1}) X(Z) + a_2 Z^{-1}(a_1 + Z^{-1}) X(Z)$$

and

$$W_2(Z) = a_2(1 + a_1Z^{-1}) X(Z) + (a_1 + Z^{-1}) X(Z)$$

giving

$$H_2(Z) = Y_2(Z)/X(Z) = 1 + (a_1 + a_1a_2)Z^{-1} + a_2Z^{-2} \qquad (6.28)$$

and

$$G_2(Z) = W_2(Z)/X(Z) = a_2 + (a_1 + a_1a_2)Z^{-1} + Z^{-2} \qquad (6.29)$$

Recalling that the general form of a second-order transfer function of an FIR filter has the form

$$H(Z) = a_o^* + a_1^* Z^{-1} + a_2^* Z^{-2} \qquad (6.30)$$

Thus, by comparing coefficients in Equations 6.28 and 6.30, we can conclude that

$$a_o^* = 1$$

$$a_1^* = a_1(1 + a_2)$$

and

$$a_2^* = a_2$$

EXAMPLE 6.12

Realize in lattice form the FIR filter described by

$$H(Z) = 2 - 3.6Z^{-1} + 6Z^{-2}$$

Solution

$H(Z)$ can be rewritten as

$$H(Z) = 2[1 - 1.8Z^{-1} + 3Z^{-2}]$$

Comparing coefficients with those of Equations 6.28 and 6.29, we get

$$a_1^* = a_1(1 + a_2) = -1.8$$

$$a_2^* = a_2 = 3$$

giving

$$a_1 = -1.8/(1 + 3) = -0.45$$

The factor 2 in $H(Z)$ is taken into account by including a multiplier at the input, as shown in Figure 6.44.

6.3.8.3 Fast Convolution

The time-domain output of a digital filter, $y(nT)$, is given by

$$y(nT) = h(nT) * x(nT)$$

That is, it is the result of convolving the impulse response with the input sequence $x(nT)$. In the frequency domain, this corresponds to multiplying the input $X(Z)$ by the transfer function $H(Z)$. Thus, a rather different realization technique can be suggested:

1. Transform both of the sampled impulse response $h(nT)$ and the input sequence $x(nT)$ to the frequency domain.
2. Multiply the transformed sequences $H(Z)$ and $X(Z)$ to get $Y(Z)$.
3. Inverse transform $Y(Z)$ to get the sequence $y(nT)$.

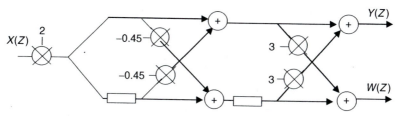

FIGURE 6.44 Lattice realization for the filter of Example 6.9.

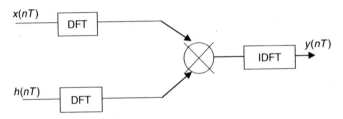

FIGURE 6.45 Fast convolution topology.

Transforming a time sequence into the frequency domain and vice versa is performed using the DFT (which will be described in Chapter 7). A schematic diagram
describing the concept is given in Figure 6.45. This method has the limitation that
both of $h(nT)$ and the set of input data samples $x(nT)$ should have the same length. If
it is not the case, dummy zeros (augmenting zeros) have to be added.

6.3.8.4   Realization Using MATLAB®

MATLAB can be employed to implement FIR filters in form of cascaded second-
order section using the following commands:

```
num = [a₀  a₁  a₂ ... a_{N-1}];
den = [1   0 ... 0];
[z,p,k]  = tf2zp(num,den);
sos = zp2sos(z,p,k)
```

with k being a multiplier (representing the overall gain).
    The result will be the coefficients of each section separately.

EXERCISE 6.2

Realize the following filter in form of second-order sections using MATLAB:

$$H(Z)= 1 + 2Z^{-1} + 3Z^{-2} + 4Z^{-3} + 3Z^{-4} + 2Z^{-5} + Z^{-6}$$

Hint:

The coefficients and hence the multipliers of each second-order section (here three
cascaded sections) can be obtained using the following MATLAB instructions:

```
num = [1 2 3 4 3 2 1];
den = [1 0 0 0 0 0 0];
[z,p,k]  = tf2zp(num,den);
sos = zp2sos(z,p,k)
```

6.4   DESIGN OF IIR FILTERS

Filters of this type are described by the following difference equation:

$$y(nT) = \sum_{i=0}^{N} a_i x(nT - iT) - \sum_{i=1}^{N} b_i y(nT - iT) \qquad (6.31)$$

where

- N is the order of the filter
- a_i's and b_i's are the filter's coefficients

It is easy to conclude from Equation 6.31 that the present output $y(nT)$ is made up of a weighted sum of the present input and N past inputs together with a weighted sum of N past outputs.

Taking the Z transform, we get

$$Y(Z) = X(Z) \sum_{i=0}^{N} a_i Z^{-i} - Y(Z) \sum_{i=1}^{N} b_i Z^{-i}$$

$$Y(Z) \left\{ 1 + \sum_{i=1}^{N} b_i Z^{-i} \right\} = X(Z) \sum_{i=0}^{N} a_i Z^{-i}$$

Giving for the transfer function

$$H(Z) = \frac{Y(Z)}{X(Z)} = \frac{\sum_{i=0}^{N} a_i Z^{-i}}{1 + \sum_{i=1}^{N} b_i Z^{-i}} \tag{6.32}$$

which is a rational function of two polynomials $N(Z)$ and $D(Z)$. The zeros of $N(Z)$ and $D(Z)$ are the zeros and poles of $H(Z)$, respectively. The goal of the design process, as we already know from FIR filters, is to find a suitable order N for the filter and its coefficients a_i's and b_i's. Throughout the design process, the vast amount of knowledge and experience gained over the past five decades in the field of analog filter design mentioned earlier are put to use.

IIR filters are the only choice wherever selectivity—that is, magnitude response, rather than phase response—is the main concern. This is due to the fact that selectivity is usually achieved at relatively low filter orders, as compared to the required orders of FIR filters for the same selectivity. This means that not only smaller components count but also higher throughput rates are possible, allowing higher frequency signals to be processed. From the describing Equation 6.31, it is clear that IIR filters are designed recursively; that is, as closed loop systems. Therefore, they are prone to oscillations. Thus, it is fundamental before proceeding with the design process to test the stability of the filter.

6.4.1 Testing the Stability of IIR Filters

Stability of IIR filters is an important design issue, as unstable filters are useless. Stability in essence means that a bounded input to a filter should yield bounded output. As IIR filters are realized recursively—that is, in feedback loops—they could thus be:

1. Unstable
2. Marginally stable
3. Stable

From our previous study of analog filters, we already know that stable filters have their poles on the left hand side in the Laplace domain. Thus, due to the relationship between the S-plane and the Z-domain, poles of stable filters should lie inside the unit circle in the Z-plane; that is,

$$|P_i| < 1$$

(6.33)

On the other hand, unstable and marginally stable filters have their poles either outside the unit circle or on its circumference; that is,

$$|P_i| \geq 1$$

Determining the exact pole locations of a filter's transfer function is thus the main goal of any test of stability. This would be an easy job if the denominator's polynomial could be factorized into multiplicative first-order terms, as given in Equation 6.34:

$$H(Z) = \frac{N(Z)}{(Z - P_1)(Z - P_2)(Z - P_3)\ldots(Z - P_N)}$$

(6.34)

where the P_i's represent the pole locations. Thus if condition 6.33 is found to be satisfied, then the filter of concern is definitely stable. The following example demonstrates the process.

EXAMPLE 6.13

Test the stability of the following transfer function:

$$H(Z) = \frac{1 - Z^{-1} + Z^{-2}}{1 - Z^{-1} + 1.5}$$

Solution

H(Z) can be written in the form

$$H(Z) = \frac{Z^2 - Z + 1}{Z^2 - Z + 1.5}$$

The denominator's polynomial D(Z) is

$$D(Z) = Z^2 - Z + 1.5$$

Its zeros are located at

$$Z_{1,2} = \frac{1}{2} \pm j\sqrt{2.5}$$

and plotted in Figure 6.46.
  Because the imaginary part $= \sqrt{2.5} = 1.12$, that is >1.
  Therefore, the filter is definitely *unstable*.

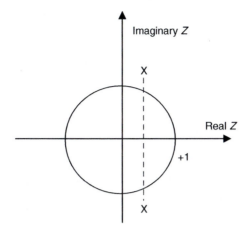

FIGURE 6.46 Pole plot of Example 6.13.

It is clear that for higher-order filters, getting the roots of the denominator's polynomial will not be an easy job; therefore, a systematic procedure for testing the stability of higher-order IIR filters is needed. One direct algorithm is the *Jury criterion* (to be described shortly). The method is programmable and gives precise decisions and requires simple calculations.

6.4.1.1 Jury Criterion

Assume a given transfer function of the form

$$H(Z) = \frac{N(Z)}{D(Z)} = \frac{N(Z)}{\sum_{i=0}^{N} b_i Z^{N-i}} \quad b_o > 0$$

Considering the denominators polynomial $D(Z)$ with positive orders of Z, the filter could be stable if *all* the following *necessary but insufficient* conditions are satisfied:

i. $|b_o| > |b_N|$

ii. $D(1) > 0$ (6.35)

iii. $(-1)^N D(-1) > 0$

On the other hand, the filter is definitely unstable if any *one* of these conditions is not met. However, stability is not fully guaranteed, even if they are all met. Therefore, to insure stability, with all conditions in 6.37 satisfied, one has to construct Table 6.4 as follows.

The first row contains the denominator coefficients starting from b_o. The next row elements are the same coefficients, yet with a reversed order.

TABLE 6.4

Jury Criterion Table

Row			Coefficients		
1	b_0	b_1	b_2		b_N
2	b_N	b_{N-1}	b_{N-2}		b_0
3	c_0	c_1	c_2		c_{N-1}
4	c_{N-1}	c_{N-2}	c_{N-3}		c_0
5	d_0	d_1	d_2	d_{N-2}	
6	d_{N-2}	d_{N-3}	d_{N-4}	d_0	
7			....		
8			....		
⋮					
$2N-3$	r_0	r_1	r_2		

The "c's" coefficients of the third row are obtained from Equation 6.36a:

$$c_i = \begin{vmatrix} b_0 & b_{N-i} \\ b_N & b_i \end{vmatrix} \quad \text{with } i = 0, 1, 2, ..., N-1 \tag{6.36a}$$

Again, the following condition is checked:

$$|c_0| > |c_{N-1}|$$

If it is violated, then the filter is unstable and the test should be terminated; otherwise, it continues by writing the elements of the fourth row; namely, the c's coefficients, yet in a reversed order. Another set of coefficients are then to be deduced from the third and fourth rows, according to Equation 6.36b:

$$d_i = \begin{vmatrix} c_0 & c_{N-1-i} \\ c_{N-1} & c_i \end{vmatrix} \quad \text{with } i = 1, 2, 3, ..., N-2 \tag{6.36b}$$

The condition $|d_0| > |d_{N-2}|$ should be satisfied for the filter to be stable.

The process continues until the row number $2N-3$ is reached, where only three elements remain; namely, r_0, r_1, and r_2. The filter is definitely stable if the coefficients of the last row—row number $2N-3$—satisfy the condition

$$|r_0| > |r_2| \tag{6.37}$$

EXAMPLE 6.14

Check the stability of the filter described by

$$H(Z) = \frac{Z^3 + 3Z^2 - Z + 1}{Z^4 + 3Z^3 + Z - 4}$$

Solution

The denominator polynomial

$$D(Z) = Z^4 + 3Z^3 + 0Z^2 + Z - 4$$

from which $b_o = 1, b_4 = -4$.

Now because $|b_o| < |b_N|$, then the filter is *unstable*.

MATLAB Check

The following commands can be used to find the pole location:

```
den = [1 3 0 1 -4];
num = [1 3 -1 1];
[z,p,k] = tf2zp(num,den)
zplane(num,den);grid
```

$$\text{poles} = -3.2168, -0.3528 + 1.1062i, -0.3528 - 1.1062i, 0.9223$$

Because some poles are off the unit circle, the filter is definitely unstable.

EXAMPLE 6.15

Using the Jury criterion, check the stability of the filter described by

$$H(Z) = \frac{2Z^{-1} + 2}{3 + 2Z^{-1} - Z^{-2} + Z^{-3} + Z^{-4}}$$

Then verify your answer using MATLAB.

Solution

We start first by multiplying both sides of $H(Z)$ by the highest power—that is, Z^4—giving

$$H(Z) = \frac{2Z^3 + 2Z^4}{3Z^4 + 2Z^3 - Z^2 + Z + 1}$$

That is,

$$D(Z) = 3Z^4 + 2Z^3 - Z^2 + Z + 1$$

TABLE 6.5

Example 6.15

Row	Coefficients				
1	3	2	−1	1	1
2	1	1	−1	2	3
3	8	7	−2	1	
4	1	−2	7	8	
5	63	58	−23		

The three necessary conditions are first tested:

i. $|b_o| = 3 > |b_4| = 1$ is satisfied; then we proceed to the second condition.

ii. $D(1) = 3 + 2 - 1 + 1 + 1 = 6 > 0$; that is satisfied also.

iii. $(-1)^4 D(-1) = 3 - 2 - 1 + 1 = 2 > 0$, which is also satisfied.

Therefore we should construct the Jury table with a maximum of $2N - 3 = 5$ rows (Table 6.5), where Equation 6.36 has been used in the calculations. Now because

$$|r_o| = 63 > |r_2| = 23$$

then the filter is *stable*.

MATLAB Check

```
den = [3 2 -1 1 1];
num = [2 2];
[z,p,k] = tf2zp(num,den)
zplane(num,den);grid
```

giving

$$z = -1$$

$$p = 0.4658 + 0.5834i, 0.4658 - 0.5834i, -1.0000, -0.5982$$

From the resulting plot, shown in Figure 6.47, it is easy to determine that all poles are located inside the unit circle (the pole $P = -1$ is being cancelled by a zero at the same frequency), and hence the filter is *stable*.

EXERCISE 6.3

Use MATLAB to test the stability of the filter described by

$$H(Z) = \frac{1 + Z^2}{1 + Z + 2Z^2 + 3Z^3 + 4Z^4 + 5Z^5}$$

Repeat if the sign of the second-order term in the denominator polynomial is reversed.

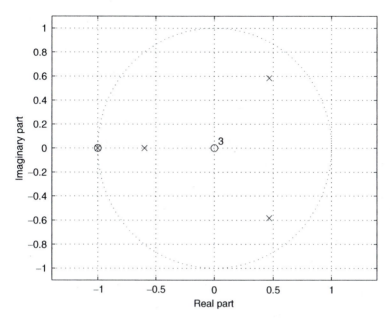

FIGURE 6.47 Zero-pole plot for the filter of Example 6.15.

6.4.2 REALIZATION TOPOLOGIES FOR IIR FILTERS

In this design step, the transfer function, after being tested for stability, is then turned into a network using the three basic building blocks: unit delays, adders, and multipliers. The way these components are arranged to realize the transfer function is very important, as it determines not only the required hardware and thus the cost, but also probable errors and the processing speed. This last property is important, in that it puts a limit on the maximum possible sampling rate and hence the bandwidth of the signals to be processed by the filter. Of great importance is the number of multipliers and the unit delays. The term "canonic," used before in analog LC circuits, is employed here also to denote that the number of the used delays is equal to the order of the filter. There are several possible realization form "topologies" for IIR filters; however, we are going to consider the most commonly used of them.

6.4.2.1 Direct Form

Consider the Nth order transfer function

$$H(Z) = \frac{Y(Z)}{X(Z)} = \frac{N(Z)}{D(Z)}$$

where

$$N(Z) = \sum_{i=0}^{N} a_i Z^{-i}$$

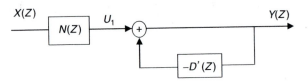

FIGURE 6.48 Direct realization of a recursive filter.

Let

$$D(Z) = 1 + D'(Z)$$

where

$$D'(Z) = \sum_{i=1}^{N} b_i Z^{-i}$$

Substituting in the previous equation, we get

$$H(Z) = \frac{N(Z)}{1 + D'(Z)} \equiv N(Z) \cdot \frac{1}{1 + D'(Z)} \equiv H_1(Z) \cdot H_2(Z) \qquad (6.38)$$

One possible realization of Equation 6.38 is shown in Figure 6.48. This means that the process of realization can be divided into the realization of two individual and simpler transfer functions:

1. $N(Z)$, which is realized as a feed forward function (open loop)
2. $D'(Z)$, which is included in a feedback loop (closed loop)

The output U_1 of the function $N(Z)$ can be expressed as

$$U_1 = N(Z) \cdot X(Z) = \left[\sum_{i=0}^{N} a_i Z^{-i} \right] \cdot X(Z)$$

Now taking the first term out of the summation, we can write

$$U_1 = \left[a_o + \sum_{i=1}^{N} a_i Z^{-i} \right] \cdot X(Z) = [a_o + N_o] \cdot X(Z)$$

Then taking one delay out of the function N_o we get

$$= \left[a_o + Z^{-1} \sum_{i=1}^{N} a_i Z^{-i+1} \right] X(Z) \qquad (6.39)$$

$$= [a_o + Z^{-1} \cdot N_1(Z)] \cdot X(Z)$$

where $N_1(Z)$ is a transfer function whose order is less by one compared to $N(Z)$. Equation 6.39 can alternatively be written in the form

$$U_1 = [a_o + N_1(Z) \cdot Z^{-1}] \cdot X(Z) \qquad (6.40)$$

This suggests that there are two possible configurations for $N(Z)$ based on the position of the delay represented by the term Z^{-1}, as illustrated in Figure 6.49.

Proceeding in this way, we can write either

$$N_1(Z) = a_1 + Z^{-1} \cdot N_2(Z)$$

or

$$= a_1 + N_2(Z) \cdot Z^{-1}$$

which indicates that there are again two possible realizations for $N_1(Z)$ and hence four realizations for $N(Z)$. Repeating the previous procedure results in a successive reduction of the order of $N(Z)$ until no further reduction is possible. Although there are 2^N configurations, only one configuration is usually repeated. Figure 6.50 depicts the result using configuration I.

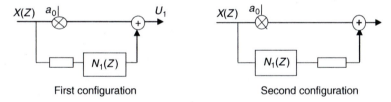

First configuration   Second configuration

FIGURE 6.49   Possible realization topologies for $N(Z)$.

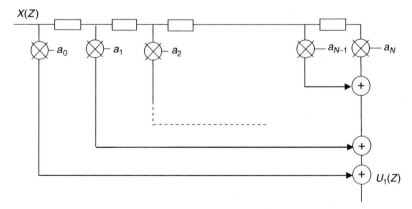

FIGURE 6.50   One possible realization for the transfer function $H_1(Z)$.

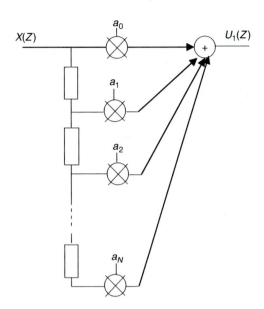

FIGURE 6.51   An alternative realization topology for the transfer function $H_1(Z)$.

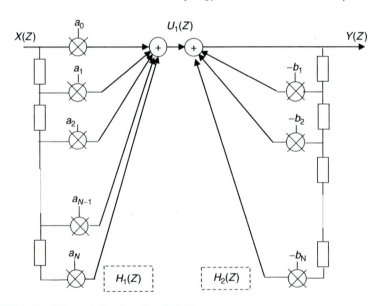

FIGURE 6.52   Direct realization of an IIR filter.

Figure 6.51 illustrates the same configuration but with the delays rotated 90° and using only one adder.

Using the same technique of successive reduction of order in the realization of the second multiplicative term $H_2(Z)$ of Equation 6.38, we get the configuration illustrated in Figure 6.52. It is evident from the figure that $2N$ delays are needed to realize an Nth-order transfer function.

6.4.2.2 Direct Canonic Realization

The realization just described contains N redundant delays. Thus, it is necessary, from the engineering view point, to eliminate them. Recalling that

$$H_1(Z) \cdot H_2(Z) = H_2(Z) \cdot H_1(Z)$$

which suggests that it is possible to exchange the positions of the two transfer functions in the cascade. As shown in Figure 6.53b, if we do that, the two delay lines get closer, such that the same sample is flowing in both of them. This means that we can save one line of them and join the points A with A', B with B', and so on, giving what is called *canonic* realization. A *canonic realization* is defined as

A realization for a transfer function that contains a number of delays that is equal to the highest order of its denominator's polynomial.

Thus by exchanging the position of $H_1(Z)$ and $H_2(Z)$ in the cascade, as shown in Figure 6.53, we end up with the canonic realization shown in Figure 6.54, where only N delays are needed for the implementation of an Nth-order transfer function.

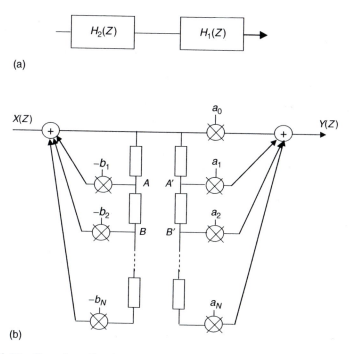

FIGURE 6.53 Canonic realization of an IIR filter. (a) Swapping the blocks in the cascade, (b) deriving a canonic realization for an IIR filter

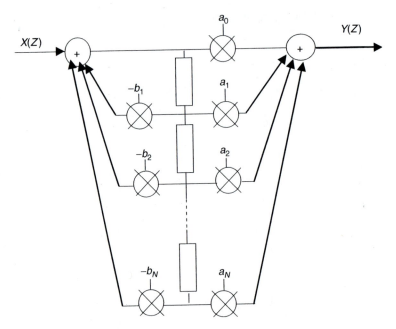

FIGURE 6.54   Direct canonic realization of an Nth-order IIR filter.

6.4.2.3 Cascaded Realization

An Nth-order transfer function $H(Z)$, if written in the form

$$H(Z) = \prod_{i=1}^{M} H_i(Z) = H_1(Z) \cdot H_2(Z) \cdot H_3(Z) \cdots H_M(Z) \qquad (6.41)$$

can be realized in form of M cascaded lower-ordered filters usually chosen to be canonic second-order sections, as depicted in Figure 6.55.

The transfer function $H_i(Z)$ can be expressed as follows:

$$H_i(Z) = \frac{a_{oi} + a_{1i}Z^{-1} + a_{2i}Z^{-2}}{1 + b_{1i}Z^{-1} + b_{2i}Z^{-2}} \quad \text{with } i = 1 : M \qquad (6.42)$$

$$M = \frac{N}{2} \quad \text{for } N \text{ even}$$

$$= \frac{N+1}{2} \quad \text{for } N \text{ odd}$$

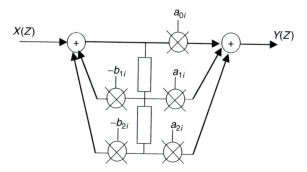

FIGURE 6.55 A standard second-order direct-canonic section.

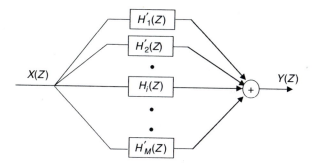

FIGURE 6.56 Parallel realization of IIR filters.

6.4.2.4 Parallel Realization

This is a rather popular topology, in which a number of M second-order sections are paralleled. This can be made possible if the transfer function is expressed in the form

$$H(Z) = \sum_{i=1}^{M} H_i(Z) = H_1'(Z) + H_2'(Z) + \cdots + H_M'(Z)$$

with

$$M = \frac{N}{2} \quad \text{for } N \text{ even}$$

and

$$= \frac{N+1}{2} \quad \text{for } N \text{ odd}$$

The functions $H_i'(Z)$ are usually realized as second-order standard direct canonic sections. The resulting topology is depicted in Figure 6.56.

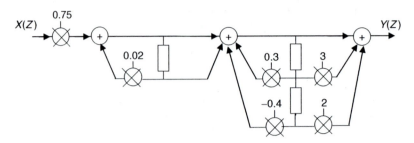

FIGURE 6.57　Realization of H_1 in Example 6.16.

Dividing the transfer function into M second-order sections is usually done by using partial fraction expansion (PFE), as described in Appendix A3.

EXAMPLE 6.16

Realize the filters described by the following transfer functions:

$$H_1 = 0.75 \frac{1 + Z^{-1}}{1 - 0.02Z^{-1}} \cdot \frac{1 + 3Z^{-1} + 2Z^{-2}}{1 - 0.3Z^{-1} + 0.4Z^{-2}}$$

and

$$H_{II} = \frac{1 - 0.6Z^{-1}}{1 - 0.3Z^{-1} + 0.4Z^{-2}} + \frac{4}{1 - 0.02Z^{-1}} - 7.5$$

A look at the transfer functions reveals that H_1 can better be realized as a cascade of two filters: a first-order filter, preceded by a multiplier of 0.75, and a second-order section. The realization is shown in Figure 6.57. The second filter H_{II} can be realized as a parallel combination of three branches: a multiplier, a first-order section, and a second-order section. The realization is depicted in Figure 6.58.

6.4.2.5　Lattice Topology

In contrast to other realization topologies, the unit section in this type is characterized, as illustrated in Figure 6.59, by a single input and two outputs for the case of a second-order section. This unique structure finds applications in speech processing at large, and specifically in linear prediction coding.

The output $Y_2(Z)$ is given by

$$Y_2(Z) = Y_1(Z) + a_2 Z^{-1} Y_2(Z)$$

$$Y_2(Z)\{1 - a_2 Z^{-1}\} = Y_1(Z)$$

(6.43)

and

$$Y_1(Z) = X(Z) + a_2 W_1(Z) \cdot Z^{-1}$$

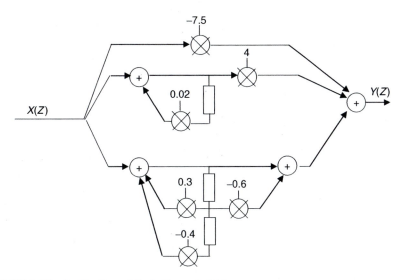

FIGURE 6.58   Realization of H_{II} in Example 6.16.

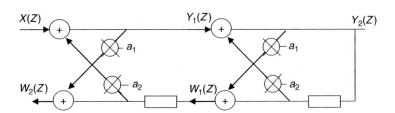

FIGURE 6.59   A second-order lattice section.

where

$$W_1(Z) = (Z^{-1} + a_1)Y_2(Z)$$

Substituting for W_1 in Y_1, we get

$$Y_1(Z) = X(Z) + a_2 Z^{-1}(Z^{-1} + a_1)Y_2(Z)$$

Substituting from 6.42 for Y_1 in the previous equation yields

$$Y_2(Z)\{1 - a_2 Z^{-1}\} = X(Z) + a_2 Z^{-1}(Z^{-1} + a_1)Y_2(Z)$$

$$Y_2(Z)\{1 - a_2 Z^{-1} - a_1 a_2 Z^{-1} - a_2 Z^{-2}\} = X(Z)$$

$$H(Z) = \frac{Y_2(Z)}{X(Z)} = \frac{1}{1 - a_2(1 + a_1)Z^{-1} - a_2 Z^{-2}}$$

(6.44)

The second output is given by

$$W_2(Z) = Z^{-1}W_1(Z) + a_1Y_1(Z)$$

$$= Z^{-1}[a_1 + Z^{-1}]Y_2(Z) + a_1X(Z) + a_1a_2Z^{-1}(a_1 + Z^{-1})Y_2(Z)$$

$$= Y_2(Z)\{Z^{-2} + a_1Z^{-1} + a_1a_2Z^{-2} + a_1^2a_2Z^{-1}\} + a_1X(Z)$$

$$= \frac{X(Z)}{1 - a_2(1 + a_1)Z^{-1} - a_2Z^{-2}}\{a_1(1 + a_1a_2)Z^{-1}$$

$$+ (1 + a_1a_2)Z^{-2}\} + a_1X(Z)$$

$$= \left[\frac{(1 + a_1a_2)\{a_1Z^{-1} + Z^{-2}\}}{1 - a_2(1 + a_1)Z^{-1} - a_2Z^{-2}} + a_1\right] \cdot X(Z)$$

giving

$$H_2(Z) = \frac{W_2(Z)}{X(Z)} = \frac{a_1 + a_1(1 - a_2)Z^{-1} + Z^{-2}}{1 - a_2(1 + a_1)Z^{-1} - a_2Z^{-2}} \qquad (6.45)$$

6.4.3 APPROXIMATIONS FOR IIR FILTERS

The outcome of the approximation step for IIR filters is two sets of coefficients {a} and {b} of the filter's describing equation:

$$y(nT) = \sum_{i=0}^{N} a_i x(nT - iT) - \sum_{i=1}^{N} b_i y(nT - iT)$$

Taking the Z-transform, we can write

$$Y(Z) = \sum_{i=0}^{N} a_i Z^{-i} X(Z) - \sum_{i=1}^{N} b_i Z^{-i} Y(Z)$$

$$Y(Z)\left\{1 + \sum_{i=1}^{N} b_i Z^{-i}\right\} = \sum_{i=0}^{N} a_i Z^{-i} X(Z)$$

giving the transfer function

$$H(Z) = \frac{Y(Z)}{X(Z)} = \frac{\sum_{i=0}^{N} a_i Z^{-i}}{1 + \sum_{i=1}^{N} b_i Z^{-i}} = \frac{a_o + a_1Z^{-1} + a_2Z^{-2} + \cdots + a_NZ^{-N}}{1 + b_1Z^{-1} + b_2Z^{-2} + \cdots + b_NZ^{-N}}$$

We are going to consider the following approximation techniques:

1. The Invariant Impulse Response method that is based on designing a filter whose sampled impulse response is derived from the impulse response of an equivalent analog filter.
2. The BZT method, where $H(S)$, the transfer function of an equivalent analog filter, is mapped into the Z-domain to get the transfer function $H(Z)$.
3. The least Pth-norm method.

6.4.4 INVARIANT IMPULSE RESPONSE METHOD

This method is based on designing a digital filter that has a sampled impulse response similar to the sampled impulse response of an equivalent analog filter. Consider the impulse response $h(t)$ of a simple RC filter, shown in Figure 6.60. To design a digital filter that has the same response, we first sample $h(t)$ at a high enough sampling frequency f_s. Mathematically, this could be expressed by replacing t by nT where $T = 1/f_s$ and n extends from 0 to ∞. The method seeks a filter whose impulse response samples coincide with those of an equivalent RC filter. The obtained discrete-time impulse response $h(nT)$ is then Z-transformed to get $H(Z)$. However, it is not always guaranteed that the process will deliver a filter whose frequency response is similar to the required one. This is to be expected, due to the large number of responses that can include the sampling points. Thus, certain deviations from the requirements are to be expected. It is obvious that reducing T by increasing the sampling frequency f_s would minimize the possible deviation from the required response, yielding a better approximation. The method can be summarized in the following steps:

1. Get $H(S)$, the continuous-time transfer function of an analog filter that satisfies the prescribed magnitude response.
2. Inverse Laplace transform $H(S)$ to get the impulse response $h(t)$.
3. Derive a discrete-time version of $h(t)$ by replacing t by the discrete time nT.
4. Apply the Z-transform to $h(nT)$ to get $H(Z)$.

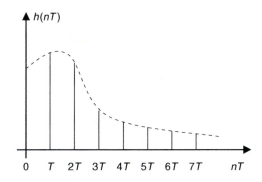

FIGURE 6.60   Sampled impulse response of an IIR filter.

It should be mentioned that this method is suitable only for the design of band-limited functions such those of LPF and BPF. They are not suitable for HPF and BSF. Also, to prevent aliasing errors, the sampling frequency should be high enough. Design data for the filters are given by their poles and zeros distribution or through their tolerance structures, as we shall see in the following examples.

EXAMPLE 6.17

Using the invariant impulse method, design an IIR filter that has the pole-zero distribution shown in Figure 6.61.

Solution

From the given pole-zero distribution, we can construct the transfer function $H(S)$ of the filter as follows:

$$H(S) = \frac{S + \alpha}{(S + \alpha - j\omega_o)(S + \alpha + j\omega_o)}$$

Using PFE (Appendix A3) we get

$$H(S) = \frac{1/2}{(S + \alpha) + j\omega_o} + \frac{1/2}{(S + \alpha) - j\omega_o}$$

The impulse response can then be obtained by taking the inverse Laplace transform for the previous expression, giving

$$h(t) = L^{-1}H(S) = 1/2\left[e^{-(\alpha + j\omega_o)t} + e^{-(\alpha - j\omega_o)t}\right]u_{-1}$$

where u_{-1} is a unit step function (just added to indicate causality). It is described by

$$u_{-1} = 0 \qquad t < 0$$
$$= 1 \qquad t \geq 0$$

The discrete-time impulse response $h(nT)$ is then obtained by replacing t by nT as follows:

$$h(nT) = 1/2\left[e^{-(\alpha + j\omega_o)nT} + e^{-(\alpha - j\omega_o)nT}\right]$$

Taking the Z-transform, we get

$$H(Z) = Z\{h(nT)\} = \frac{1/2}{1 - e^{-(\alpha + j\omega_o)T}Z^{-1}} + \frac{1/2}{1 - e^{-(\alpha - j\omega_o)T}Z^{-1}}$$

$$= \frac{1 - e^{-\alpha T}\cos\omega_o T\, Z^{-1}}{1 - 2e^{-\alpha T}\cos\omega_o T\, Z^{-1} + e^{-2\alpha T}Z^{-2}} \qquad (6.46a)$$

which is equivalent to

$$H(Z) = \frac{a_o + a_1 Z^{-1}}{1 + b_1 Z^{-1} + b_2 Z^{-2}} \qquad (6.46b)$$

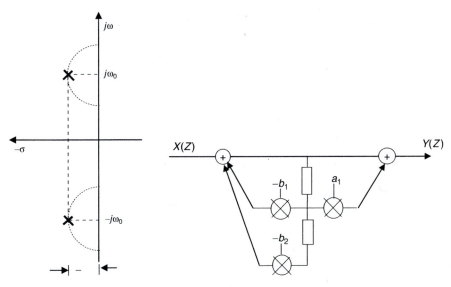

FIGURE 6.61 Pole-zero plot for Example 6.17.

FIGURE 6.62 Filter realization of Example 6.17.

Comparing corresponding coefficients in Equations 6.46a and 6.46b, we get

$$a_o = 1 \quad a_1 = -e^{-\alpha T} \cos\omega_o T$$

and

$$b_1 = -2e^{-\alpha T} \cos\omega_o T \quad b_2 = e^{-2\alpha T}$$

A canonic realization for the filter is given in Figure 6.62.

Alternatively, we can rewrite the transfer function $H(S)$ in the form

$$|H(S)| = \frac{S + \alpha}{(S + \alpha)^2 + \omega_o^2}$$

The impulse response $h(t)$ can directly be obtained by inverse transforming $H(S)$ without performing PFE. Using the Laplace transform table, we find out that

$$h(t) = e^{-\alpha t} \cos\omega_o t$$

then

$$h(nT) = e^{-n\alpha T} \cos n\omega_o$$

Using Z-transform tables, we can get directly

$$Z\{h(nT)\} = H(Z) = \frac{1 - e^{-aT} \cos\omega_o T Z^{-1}}{1 - 2e^{-aT} \cos\omega_o T Z^{-1} + e^{-2aT} Z^{-2}}$$

which is the same transfer function obtained earlier.

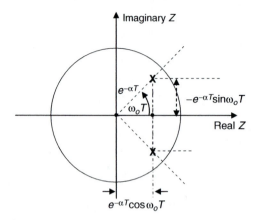

FIGURE 6.63 Mapped pole-zero distribution of Example 6.17.

To find the pole-zero distribution in the Z-domain, we multiply both sides of $H(Z)$ by Z^2, giving

$$H(Z) = \frac{Z^2 - e^{-\alpha T}\cos\omega_o T \cdot Z}{Z^2 - 2e^{-\alpha T}\cos\omega_o T Z + e^{-\alpha T}}$$

Factorization of $H(Z)$ gives

$$H(Z) = \frac{Z(Z - e^{-\alpha T}\cos\omega_o T)}{[Z - e^{-\alpha T}(\cos\omega_o T + j\sin\omega_o T)][Z - e^{-\alpha T}(\cos\omega_o T - j\sin\omega_o T)]}$$

from which it is easy to conclude that $H(Z)$ has two zeros located at

$$Z_1 = 0 \quad \text{and} \quad Z_2 = e^{-\alpha T}\cos\omega_o T$$

and two poles at

$$P_{1,2} = e^{-\alpha T}\cos\omega_o T \pm je^{-\alpha T}\sin\omega_o T$$

Figure 6.63 illustrates the new mapped poles and zeros in the Z-plane.

EXAMPLE 6.18

Design a normalized third-order Butterworth low-pass filter as a recursive filter, using the IIR method.

Solution

The poles of a normalized third-order Butterworth filter are distributed along the circumference of a unit circle, as shown in Figure 6.64.
    The poles are located at

$$P_{1,3} = -\alpha \pm j\beta$$
$$P_2 = -\alpha_o$$

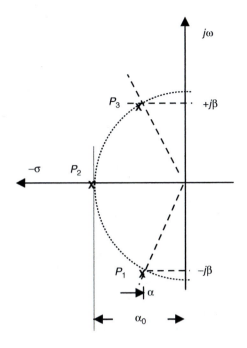

FIGURE 6.64 Pole-zero plot of Example 6.18.

The corresponding analog filter transfer function is given by

$$H(S) = \frac{H_o}{(S + \alpha_o)(S + \alpha + j\beta)(S + \alpha - j\beta)}$$

where H_o is selected such that

$$H(0) = 1$$

giving

$$H_o = \alpha_o(\alpha^2 + \beta^2)$$

From the geometry of the plot, we find that

$$\alpha_o = 1$$

$$\alpha = \sin 30 = \frac{1}{2} \quad \text{and} \quad \beta = \cos 30 = \frac{\sqrt{3}}{2}$$

such that

$$H_o = 1 \cdot \left(\frac{1}{4} + \frac{3}{4}\right) = 1$$

Applying PFE to $H(S)$, we get

$$H(S) = \left[\frac{1}{S+\alpha_o} - \frac{S+(2\alpha-\alpha_o)}{(S+\alpha)^2 + \beta^2}\right] = \left[\frac{1}{(S+\alpha_o)} - \frac{S+\alpha}{(S+\alpha)^2+\beta^2} - \frac{\alpha-\alpha_o}{(S+\alpha)^2+\beta^2}\right]$$

The impulse response $h(t)$ is obtained by applying the inverse Laplace transform as follows:

$$h(t) = L^{-1}H(S) = \left[e^{-\alpha_o t} - e^{-\alpha t}\cos\beta t - \frac{\alpha-\alpha_o}{\beta}e^{-\alpha t}\sin\beta t\right]u_{-1}$$

Substituting for t by nT gives

$$h(nT) = \left[e^{-\alpha_o nT} - e^{-\alpha nT}\cos\beta nT - \frac{\alpha-\alpha_o}{\beta}e^{-\alpha nT}\cos\beta nT\right]$$

Taking the Z-transform, we get

$$H(Z) = Z\{h(nT)\}$$

$$= \left[\frac{Z}{Z-e^{-\alpha_o T}} - \frac{Z^2 - Z(e^{-\alpha T}\cos\beta T)}{Z^2 - (2e^{-\alpha T}\cos\beta T)Z + e^{-2\alpha T}}\right.$$

$$\left. - \frac{\alpha-\alpha_o}{\beta}\frac{(e^{-\alpha T}\sin\beta T)Z}{Z^2 - (2e^{-\alpha T}\cos\beta T)Z + e^{-2\alpha T}}\right]$$

Substituting for the constants and dividing by the highest order of Z gives

$$H(Z) = \frac{1}{1-e^{-T}Z^{-1}} + \frac{-1+e^{-T/2}\cos\left(\sqrt{3/2}\right)T\cdot Z^{-1}}{1-2Z^{-1}e^{-T/2}\cos\left(\sqrt{3/2}\right)T + Z^{-2}e^{-T}}$$

$$+ \frac{1}{\sqrt{3}}\frac{e^{-T/2}\sin\left(\sqrt{3/2}\right)T\cdot Z^{-1}}{1-2Z^{-1}e^{T/2}\cos\left(\sqrt{3/2}\right)T + e^{-T}Z^{-2}} \qquad (6.47)$$

The filter can be realized in either one of the forms discussed in the following sections.

6.4.4.1 Direct Canonic Realization

For a direct canonic realization, the transfer function should take the form

$$H_1(Z) \cdot H_2(Z) = H_2(Z) \cdot H_1(Z)$$

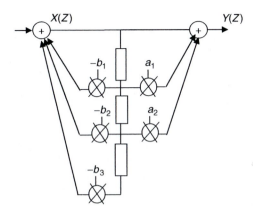

FIGURE 6.65   Direct canonic realization of Example 6.18.

Thus by exchanging the position of $H_1(Z)$ and $H_2(Z)$ in the cascade, as shown in Figure 6.64, the two delay lines get closer, such that one line of them could be saved. This results in what is called *canonic* realization, shown in Figure 6.65. In this realization, only N delays are needed for the implementation of an Nth-order transfer function.

$$H(Z) = \frac{a_o + a_1 Z^{-1} + a_2 Z^{-2} + a_3 Z^{-3}}{1 + b_1 Z^{-1} + b_2 Z^{-2} + b_3 Z^{-3}}$$

Comparing coefficients, we get (Figure 6.65)

$$a_o = 0$$

$$a_1 = e^{-T} + e^{-T/2}\left(\frac{1}{\sqrt{3}} \sin \frac{\sqrt{3}}{2} T - \cos \frac{\sqrt{3}}{2} T \right)$$

$$a_2 = e^{-T} + e^{-(3T/2)T}\left(\frac{1}{\sqrt{3}} \sin \frac{\sqrt{3}}{2} T + \cos \frac{\sqrt{3}}{2} T \right)$$

6.4.4.2 Parallel Realization

Combining the last two terms in Equation 6.47, we get

$$H(Z) = \frac{1}{1 - e^{-T}Z^{-1}} + \frac{-1 + e^{T/2}\left(\cos\left(\sqrt{3}/2\right)T + \left(\sqrt{3}/2\right)\sin\left(\sqrt{3}/2\right)T \right)Z^{-1}}{1 - 2e^{-(T/2)}\cos\left(\sqrt{3}/2\right)TZ^{-1} + e^{-T}Z^{-2}}$$

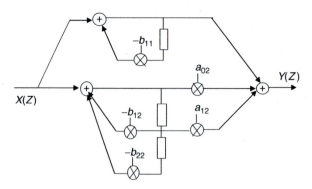

FIGURE 6.66 Parallel realization for Example 6.18.

which is similar to

$$H(Z) = \frac{1}{1 + b_{11}Z^{-1}} + \frac{a_{o2} + a_{12}Z^{-1}}{1 + b_{12}Z^{-1} + b_{22}Z^{-2}}$$

That is, it represents a parallel connection of a first-order section with a second-order one, as diagramed in Figure 6.66.

6.4.4.3 Cascaded Realization

Combining the three terms of Equation 6.47, we get a transfer function in the form

$$H(Z) = \frac{Z^{-1}(a_o + a_1Z^{-1})}{(1 - e^{-T}Z^{-1})\left(1 - 2e^{-(T/2)}\cos\left(\sqrt{3}/2\right)T\ Z^{-1} + e^{-T}Z^{-2}\right)} = \frac{A \cdot B}{C \cdot D}$$

in which

$$a_o = e^{-T} + e^{-(T/2)}\left(\frac{1}{\sqrt{3}}\sin\frac{\sqrt{3}}{2}T - \cos\frac{\sqrt{3}}{2}T\right)$$

and

$$a_1 = e^{-(T/2)} - e^{-(3/2T)}\left(\frac{1}{\sqrt{3}}\sin\frac{\sqrt{3}}{2}T + \cos\frac{\sqrt{3}}{2}T\right)$$

The previous equation can be written as

$$H(Z) = \frac{A}{C} \cdot \frac{B}{D} \quad \text{or} \quad = \frac{B}{C} \cdot \frac{A}{D}$$

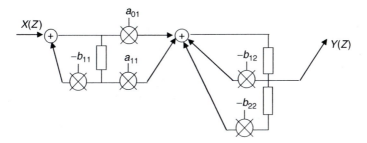

FIGURE 6.67   Cascaded realization of Example 6.18, first configuration.

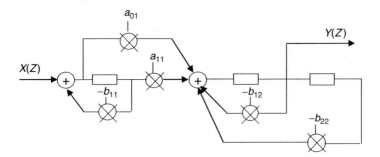

FIGURE 6.68   Cascaded realization of Example 6.18, second configuration.

thus

$$H(Z) = \frac{a_o + a_1 Z^{-1}}{1 - e^{-T} Z^{-1}} \cdot \frac{Z^{-1}}{1 - 2e^{-(T/2)} \cos\left(\sqrt{3}/2\right) T \cdot Z^{-1} + e^{-T} Z^{-2}}$$

$$\equiv \frac{a_{o1} + a_{11} Z^{-1}}{1 + b_{11} Z^{-1}} \cdot \frac{a_{12} Z^{-1}}{1 + b_{12} Z^{-1} + b_{22} Z^{-2}}$$

Comparing coefficients, we get

$$a_{o1} = a_o, \quad a_{11} = a_1, \quad a_{12} = 1$$

$$b_{11} = -e^{-T}, \quad b_{12} = -2e^{-(T/2)} \cos\frac{\sqrt{3}}{2} T, \quad \text{and} \quad b_{22} = e^{-T}$$

A direct canonic realization of this transfer function is given in Figure 6.67. The same circuit is redrawn differently in Figure 6.68.

It is interesting to find the corresponding pole-zero distribution in the Z-plane. This can be obtained by making the substitution $Z = e^{-ST}$. A schematic diagram for the distribution is illustrated in Figure 6.69 and redrawn exactly in Figure 6.70b.

Recalling that

$$S = \sigma + j\omega$$

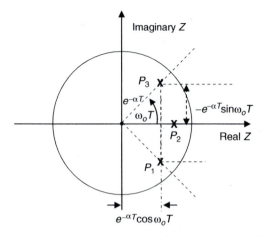

FIGURE 6.69   Mapped pole-zero distribution of Example 6.18.

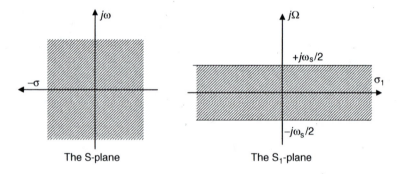

FIGURE 6.70   Mapping of the S-plane into the S_1-plane.

then

$$P_{1,3} = e^{(-\alpha \pm j\beta)T} = e^{-\alpha T}[\cos \beta T \pm j \sin \beta T]$$

$$P_2 = e^{-\alpha_o T}, \alpha_o = 1$$

6.4.5   BZT

The amount of calculations needed in designing a filter using the IIR method is sometimes lengthy, as we have to transform the analog transfer function into the time domain before transforming it to the Z-domain. Tables are required to be available for both transforms. A sufficient training with PFE is also necessary. Furthermore, the resulting filter response could deviate from the desired one. Due to these reasons, this method is seldom used in practice. A rather more efficient and appealing method is the BZT, in which direct mapping into the Z-domain is achieved. Through this transform, the entire S-plane is mapped into the region between two horizontal and parallel lines in a new frequency domain, the S_1 domain, we shall call it the digital frequency to distinguish it from the analog frequency S. The two parallel lines are separated by a distance that corresponds to the sampling frequency ω_s as illustrated in Figure 6.70.

The mapping equation is

$$S = \frac{2}{T} \tanh \frac{S_1 T}{2} \tag{6.48}$$

It follows that any band-unlimited function in the S-plane will be band-limited in the S_1-plane. Substituting for tanh x by its exponential form, Equation 6.48 becomes

$$S = \frac{2}{T} \frac{e^{S_1 T/2} - e^{(-S_1 T)/2}}{e^{S_1 T/2} + e^{(-S_1 T)/2}} \tag{6.49}$$

Then substituting for the analog frequency S by $j\omega$ and for the digital frequency S_1 by $j\Omega$, we get

$$j\omega = \frac{2}{T} \frac{e^{j\Omega T/2} - e^{(-j\Omega T/2)}}{e^{j\Omega T/2} + e^{-j\Omega T/2}} = \frac{2}{T} j \tan \frac{\Omega T}{2}$$

leading to

$$\omega = \frac{2}{T} \tan \frac{\Omega T}{2} \tag{6.50}$$

or

$$\Omega = \frac{2}{T} \tan^{-1} \frac{\omega T}{2}$$

Multiplying both sides of Equation 6.49 by $e^{ST/2}$ and substituting for $Z = e^{S_1 T}$, we get

$$S = \frac{2}{T} \frac{1 - Z^{-1}}{1 + Z^{-1}} \tag{6.51}$$

which is the BZT equation. This equation enables direct mapping from the S-domain directly to the Z-domain without need for transforming to the time domain first.

6.4.5.1 Warping Effect

Plotting Equation 6.48 reveals that the relation between the analog frequency ω and the digital frequency Ω is linear only for small values of Ω and assumes increasing slopes as it becomes larger. Thus a virtual multiband filter that has equipass bands, as shown in Figure 6.71, would result after transformation in a nonuniform multiband filter; that is, a frequency distortion or *warping*. This might suggest that to avoid such phenomenon, some kind of countermeasure (*predistortion*) should be taken before

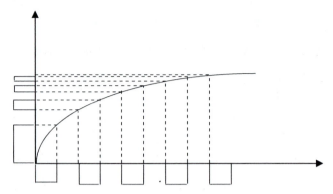

FIGURE 6.71 The warping effect.

applying the transform. The process is called *prewarping*. It is described by Equation 6.50 the "prewarping equation":

$$\omega_i = \frac{2}{T} \tan \frac{\Omega_i T}{2}$$

The nonlinearity of this relationship should explain the resulting frequency distortion (warping).

The design process using the BZT can be summarized in the following steps:

1. Prewarp all given frequencies Ω_i's using Equation 6.50.
2. Use the prewarped poles and zeros frequencies to construct the transfer function.
3. Apply the BZT to get $H(Z)$ as follows:

$$H(Z) = H(S)\Big|_{S = \frac{2}{T} \frac{1-z^{-1}}{1+z^{-1}}}$$

EXAMPLE 6.19

Design, using the BZT, a first-order LPF to have a cut-off frequency at 100 Hz, assuming a sampling frequency of 1000 Hz.

Solution

Given $\Omega_c = 2 \times \pi \times 100 = 628$ rad/s with $T = 1/f_s = 0.001$,

thus

$$\Omega_c T = 0.628$$

Prewarping Ω_c gives

$$\omega_c = \frac{2}{T} \tan \frac{\Omega_c T}{2} = \frac{2}{T} \tan 0.314 = 0.3247$$

The transfer function of a normalized, first-order LPF is

$$H_n(Z) = \frac{1}{S+1}$$

To denormalize $H(Z)$, we substitute S by (S/ω_c), giving

$$H_d(Z) = \frac{\omega_c}{S+\omega_c}$$

Applying the BZT using the formula

$$S = \frac{2}{T}\frac{1-Z^{-1}}{1+Z^{-1}}$$

we get

$$H(Z) = \frac{2/T \tan \Omega_c T/2}{2/T(1-Z^{-1}/1+Z^{-1}) + 2/T \tan \Omega_c T/2}$$

Dividing both sides by $2/T \tan(\Omega_c T/2)$, we get

$$H(Z) = \frac{1}{(\cot \Omega_c T/2)(1-Z^{-1}/1+Z^{-1}) + 1}$$

From this, it is easy to conclude that the multiplier $2/T$ in the prewarping and the BZT formulas could be dropped, and the denormalization and Z-transform steps could be accomplished in one single step by substituting for S by

$$S = \frac{1}{\omega_c}\frac{1-Z^{-1}}{1+Z^{-1}}$$

in the normalized transfer function.

For our first-order LP filter, substitution gives (see Figure 6.72)

$$H(Z) = \frac{1}{(1/0.3247)(1-Z^{-1}/1+Z^{-1}) + 1} = \frac{0.3247(1+Z^{-1})}{1.3247 - 0.6752Z^{-1}}$$

$$= 0.245\frac{1+Z^{-1}}{1-0.5097Z^{-1}}$$

Simple analysis of the numerator and denominators reveals that the filter has a zero at $Z = -1$ and a pole at $Z = 0.5097$.

EXERCISE 6.4

Repeat the previous problem for a normalized first-order HP filter, giving the associated pole-zero plot. (Compare the location of the zero in both cases.)

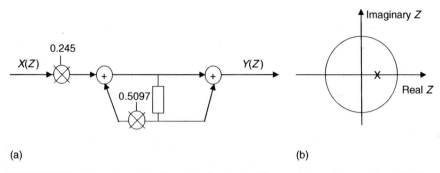

(a)   (b)

FIGURE 6.72 Canonic realization of a first-order filter and its pole-zero plot. (a) Filter realization of Example 6.19, (b) pole-zero plot.

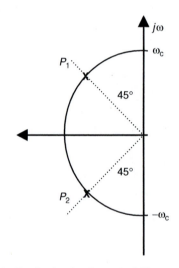

FIGURE 6.73 Pole distribution for Example 6.20.

EXAMPLE 6.20

Using the BZT method, design a normalized second-order Butterworth LP filter.

Solution

Given $\Omega_c = 1$ rad/s, and assuming a sampling frequency of 1 Hz, prewarping Ω_c gives

$$\omega_c = \tan\frac{\Omega_c T}{2} = \tan\frac{1}{2} = 0.5463 \text{ rad/s}$$

The poles P_1 and P_2, as illustrated in Figure 6.73, are located at

$$P_{1,2} = -0.5463(\sin 45 \pm j \cos 45) = -0.386 \pm j0.386$$

$$= \sigma_P \pm j\omega_P$$

Thus, the transfer function $H(S)$ can be formulated, as before, to be

$$H(S) = \frac{H_o}{S^2 + 2\sigma_p + (\sigma_p^2 + \omega_p^2)} = \frac{H_o}{S^2 + 2\sigma_p + \omega_c^2} = \frac{H_o}{S^2 + 0.7726S + 0.298}$$

with $H_o = 0.298$.

Applying the BZT, we get

$$H(Z) = \frac{0.298[1 + Z^{-1}]}{[1 - Z^{-1}]^2 + 0.7726[1 + Z^{-1}][1 - Z^{-1}] + 0.298[1 + Z^{-1}]^2}$$

$$= \frac{0.298[1 + 2Z^{-1} + Z^{-2}]}{0.298[1 + 2Z^{-1} + Z^{-2}] + 0.7726[1 - Z^{-2}] + [1 - 2Z^{-1} + Z^{-2}]}$$

$$= \frac{0.298[1 + 2Z^{-1} + Z^{-2}]}{2.0706 - 1.404Z^{-1} + 0.5254Z^{-2}}$$

$$= \frac{0.1439[1 + 2Z^{-1} + Z^{-2}]}{[1 - 0.678Z^{-1} + 0.2537Z^{-2}]}$$

or alternatively, starting with the normalized second-order Butterworth transfer function

$$H(S) = \frac{1}{S^2 + \sqrt{2}S + 1}$$

then substituting for S by

$$S = \frac{1}{\omega_c} \frac{1 - Z^{-1}}{1 + Z^{-1}} = \cot\left(\frac{1}{2}\right) \cdot \frac{1 - Z^{-1}}{1 + Z^{-1}} = 1.83 \frac{1 - Z^{-1}}{1 + Z^{-1}}$$

giving

$$H(Z) = \frac{[1 + Z^{-1}]^2}{(1.83)^2[1 - Z^{-1}]^2 + \sqrt{2}(1.83)[1 - Z^{-1}][1 + Z^{-1}] + [1 + Z^{-1}]^2}$$

$$= \frac{0.1439[1 + 2Z^{-1} + Z^{-2}]}{[1 - 0.678Z^{-1} + 0.2537Z^{-2}]}$$

A canonic realization for the filter is shown in Figure 6.74.

EXERCISE 6.5

Repeat the previous problem for a cut-off frequency $f_c = 100$ Hz and a sampling frequency of 10 times f_c. Plot the pole-zero distribution.

$$\left(\text{Answer: } H(Z) = \frac{0.067[1 + 2Z^{-1} + Z^{-2}]}{[1 - 1.143Z^{-1} - 0.865Z^{-2}]}\right)$$

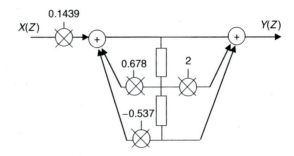

FIGURE 6.74   Filter realization of Example 6.20.

EXAMPLE 6.21

Design a third-order Butterworth LPF to have a cut-off frequency at 4 kHz using the BZT method and assuming a sampling frequency of 10 kHz.

Solution

Given $f_c = 4$ kHz and $T = 1/f_s = 10^{-4}$ s, thus $\Omega_c T = 2 \times \pi \times 4 \times 10^3 \times 10^{-4} = 2.513$ rad.
   Prewarping gives

$$\omega_c = \tan \frac{\Omega_c T}{2} = \tan 1.2565 = 3.0762 \text{ rad/s}$$

The pole frequencies can be obtained from the pole-zero distribution of a normalized third-order Butterworth filter, shown in Figure 6.58, giving

$$P_{1,3} = -\alpha \pm j\beta = -0.5 \pm j0.866$$

and

$$P_2 = -\alpha_o = -1$$

The transfer function $H(S)$ can be expressed as

$$H(S) = \frac{H_o}{(S + \alpha_o)[S^2 + 2\alpha S + (\alpha^2 + \beta^2)]} = \frac{1}{(S + 1)[S^2 + 2S + 1]}$$

Substituting for

$$S = \frac{1}{\omega_c} \frac{1 - Z^{-1}}{1 + Z^{-1}} = 0.3249 \frac{1 - Z^{-1}}{1 + Z^{-1}} = \delta \frac{1 - Z^{-1}}{1 + Z^{-1}}$$

gives

$$H(Z) = \frac{[1 + Z^{-1}][1 + Z^{-1}]^2}{(\delta[1 - Z^{-1}] + [1 + Z^{-1}])\delta^2[1 - Z^{-1}]^2 + 2\delta[1 - Z^{-1}][1 + Z^{-1}] + [1 + Z^{-1}]^2}$$

$$= \frac{[1 + Z^{-1}][1 + Z^{-1}]^2}{[(1 + \delta) + (1 - \delta)Z^{-1}][(1 + \delta + \delta^2) + (2 - 2\delta^2)Z^{-1} + (1 - \delta + \delta^2)Z^{-2}]}$$

$$= \frac{[1 + Z^{-1}][1 + Z^{-1}]^2}{[1.3249 + 0.6751Z^{-1}][1.4305 + 1.789Z^{-1} + 0.7806Z^{-2}]}$$

$$= \frac{0.52763[1 + Z^{-1}][1 + Z^{-1}]^2}{[1 + 0.5095Z^{-1}][1 + 1.2506Z^{-1} + 0.5457Z^{-2}]}$$

$$= 0.52763 \quad \frac{[1 + Z^{-1}]}{[1 + 0.5095Z^{-1}]} \times \frac{[1 + 2Z^{-1} + Z^{-2}]}{[1 + 1.2506Z^{-1} + 0.5457Z^{-2}]}$$

which could be realized as a cascade of a first-order canonic section cascaded with a second-order canonic one. Figure 6.75 illustrates such a realization.

MATLAB Check

The fdatool is best suited in this case, because the order is given. Thus in the command window, write fdatool. In the GUI, select the filter specifications and watch the plot. Select the option "full view analysis" to watch the responses clearly. However, the following MATLAB commands can be used to design the filter:

```
[z,p,k] = butter(3, .8);
[num,den] = zp2tf(z,p,k);
sos=tf2sos(num,den)
fvtool(num,den)
```

giving

$$\text{sos} = \begin{array}{llll} (a_{1i}) & 0.5276 & 0.5276 & 0 \\ (a_{2i}) & 1.0000 & 2.0000 & 1.0000 \end{array} \qquad \begin{array}{llll} (b_{1i}) & 1.0000 & 0.5095 & 0 \\ (b_{2i}) & 1.0000 & 1.2505 & 0.5457 \end{array}$$

which is consistent with the previous results.

EXAMPLE 6.22

Using the BZT method, design a Butterworth IIR filter to satisfy the tolerance structure shown in Figure 6.76.

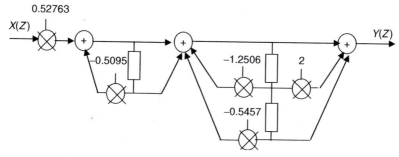

FIGURE 6.75   Canonic realization of the filter in Example 6.21.

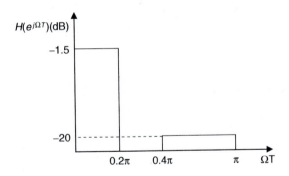

FIGURE 6.76 Tolerance structure for Example 6.22.

Solution

We usually start solving such problems by first calculating the ripple factors ε_p, ε_r as follows:

$$\varepsilon_p = \sqrt{10^{0.1A_p} - 1} = \sqrt{10^{0.15} - 1} = 0.6423$$

$$\varepsilon_r = \sqrt{10^{0.1A_r} - 1} = \sqrt{10^2 - 1} = 9.949$$

Prewarping the frequencies Ω_p and Ω_r gives

$$\omega_p = \tan\frac{\Omega_p T}{2} = \tan\frac{0.2\pi}{2} = 0.315$$

$$\omega_r = \tan\frac{\Omega_r T}{2} = \tan 0.2\pi = 0.6368$$

The required filter order is calculated from

$$N_B \geq \frac{\log(\varepsilon_r / \varepsilon_p)}{\log(\omega_r / \omega_p)} = \frac{\log 15.49}{\log 2.02} = 03.89 \approx 4$$

The cut-off frequency can be determined from (Figure 6.77)

$$\omega_c = \varepsilon_p^{-(1/N)} \omega_p = 0.3518$$

From the shown pole distribution, we can get the normalized pole frequencies at

$$P_{1,4} = -[\sin 22.5 \pm j\cos 22.5] = -[0.38266 \pm j0.92387]$$

$$P_{2,3} = -[\sin 67.5 \pm j\cos 67.5] = -[0.92387 \pm j0.38266]$$

The transfer function of the normalized filter can then be formulated as

$$H(S) = \frac{H_o}{(S^2 + 0.76532S + 1)(S^2 + 1.84776S + 1)}$$

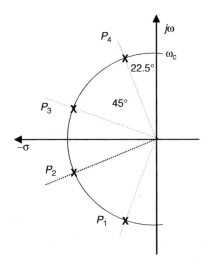

FIGURE 6.77   Pole distribution for Example 6.22.

Now to apply the BZT, we put

$$S = \frac{1}{\omega_c} \frac{1 - Z^{-1}}{1 + Z^{-1}} = 2.7106 \frac{1 - Z^{-1}}{1 + Z^{-1}}$$

giving

$$H(Z) = \frac{[1 + Z^{-1}]^2}{[\delta^2(1 - Z^{-1})^2 + 2\alpha_1\delta(1 - Z^{-2}) + (1 + Z^{-1})^2]}$$

$$\times \frac{[1 + Z^{-1}]^2}{[\delta^2(1 - Z^{-1})^2 + 2\alpha_2\delta(1 - Z^{-2}) + (1 + Z^{-1})^2]}$$

$$= \frac{[1 + Z^{-1}]^4}{(1 + 2\alpha_1\delta + \delta^2) + 2(1 - \delta^2)Z^{-1} + (1 - 2\alpha_1\delta + \delta^2)Z^{-2}}$$

$$\times \frac{1}{(1 + 2\alpha_2\delta + \delta^2) + 2(1 - \delta^2)Z^{-1} + (1 - 2\alpha_2\delta + \delta^2)Z^{-2}}$$

$$= \frac{[1 + Z^{-1}]^4}{[(10.42135) - 12.6947Z^{-1} + 6.27335Z^{-2}][8.84815 - 12.694Z^{-1} + 3.33935Z^{-2}]}$$

$$= \frac{0.0108[1 + Z^{-1}]^4}{[1 - 1.218Z^{-1} + 0.6019Z^{-2}][1 - 1.4346Z^{-1} + 0.3774Z^{-2}]}$$

which could be divided into two second-order transfer functions as follows:

$$H(Z) = 0.0108 \frac{[1 + 2Z^{-1} + Z^{-2}]}{[1 - 1.218Z^{-1} + 0.6019Z^{-2}]} \cdot \frac{[1 + 2Z^{-1} + Z^{-2}]}{[1 - 1.4346Z^{-1} + 0.3774Z^{-2}]}$$

The filter can be realized in form of two cascaded second-order canonic sections.

MATLAB Check

```
[N, wc] = buttord (.2,.4,-1.5, -20);   % estimating the order
[num, den] = butter (N, wc);           % Butterworth transfer function
[z, p, k] = tf2zp (num,den);           % zeros and poles locations
sos=zp2sos(z,p,k);                     % realization in second order
                                         sections
fvtool (num, den)                      % plot the magnitude, phase and
                                         impulse responses
```

EXAMPLE 6.23

Design a 0.5 dB Chebyshev filter to satisfy the shown tolerance structure using the BZT method and a sampling frequency of 1 kHz (Figure 6.78).

Solution

Prewarping all frequencies neglecting the constant term $2/T$, as discussed earlier, gives

$$\omega_p = \tan\frac{\Omega_p T}{2} = \tan\frac{2 \times \pi \times 100 \times 10^{-3}}{2} = 0.325$$

Similarly

$$\omega_r = \tan 0.3\pi = 1.376$$

The ripple factors are obtained from

$$\varepsilon_p = \sqrt{10^{0.1A_p} - 1} = \sqrt{10^{0.05} - 1} = 0.3493$$

$$\varepsilon_r = \sqrt{10^{0.1A_r} - 1} = \sqrt{10^2 - 1} = 9.9498$$

The required filter order is calculated from

$$N_{\text{Cheby}} \geq \frac{\cosh^{-1} \varepsilon_r/\varepsilon_p}{\cosh^{-1} \omega_r/\omega_p} \geq \frac{4}{1.76} \geq 2.29 \approx 3$$

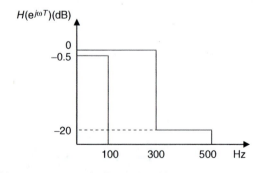

FIGURE 6.78 Tolerance structure for Example 6.23.

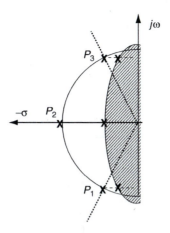

FIGURE 6.79   Pole distribution of the filter in Example 6.23.

The Butterworth poles are separated, as before, by an angle:

$$\theta = 180/3 = 60°$$

and are located at

$$P_{1,3} = -\omega_p[\sin \varphi_1 \pm j \cos \varphi_1]$$
$$P_2 = -\omega_p \sin \varphi_2$$

It was found that Chebyshev poles are distributed along an ellipse (shown hatched in Figure 6.79). The Chebyshev poles compression factor γ can be calculated from the equation

$$\gamma = \frac{1}{N}\sinh^{-1}\frac{1}{\varepsilon_p} = \frac{1}{3}\sinh^{-1}\frac{1}{0.3493} = 0.59138$$

The Chebyshev poles, as illustrated in Figure 6.79, are located at

$$P_{1,3} = -\omega_p[\sinh \gamma \sin \varphi_1 \pm j \cosh \gamma \cos \varphi_1] = -0.1018 \pm j0.32$$

$$P_2 = -\omega_p \sinh \gamma = -0.20359$$

The rest is left as an exercise for the reader. The frequency response of the filter is shown in Figure 6.80.

MATLAB Check

To check your results, use the following MATLAB instructions:

```
[num,den] = cheby1(3,0.5,0.2);
fvtool(num,den);
```

Or, alternatively, use the `fdatool` as before.

Hints:

• To view the pass band ripples, change the analysis parameters from the analysis menu and select magnitude
• cheby1 (N, Ap, fp' = 2fp/fs)

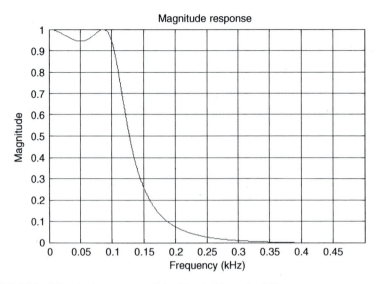

FIGURE 6.80   Magnitude response of the filter in Example 6.23.

6.4.5.2   Designing Other Types of Filters

HPF, BPF, and BSF functions can be derived from LPF through the transformations tabulated in this section. The design procedures for any filter function other than the LPF using the BZT can be summarized as follows:

1. Prewarp all critical frequencies of the required digital filter.
2. Construct the transfer function of a corresponding LPF.
3. Transform the normalized LP transfer function into the required filter function, using Table 6.5.
4. Apply the BZT to get $H(Z)$.
5. Determine the filter coefficients from the transformed transfer function.

Table 6.6 lists the basic filter transformations from a normalized LP filter into denormalized LPF, HPF, BPF, and BSF in which

$$W = \omega_2 - \omega_1 \text{ is the pass band/stop band of a BPF/BSF, respectively.}$$
$$\omega_o = \sqrt{\omega_1 \omega_2} \text{ is the filter's center frequency.}$$

where ω_1 and ω_2 are the 3-dB frequencies of the filter.

EXAMPLE 6.24

Design a Butterworth HP filter to meet the specifications given by the tolerance structure depicted in Figure 6.81.

TABLE 6.6
Basic Filter Transformations

Type of Transformation	Transformation Equation
LP→HP	$\bar{S} = \dfrac{\omega_c}{S}$
LP→BP	$\bar{S} = \dfrac{S^2 + \omega_o^2}{WS}$
LP→BS	$\bar{S} = \dfrac{WS}{S^2 + \omega_o^2}$

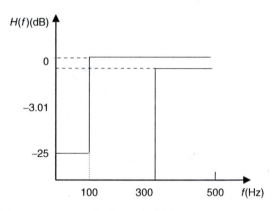

FIGURE 6.81 Tolerance structure for Example 6.24.

Solution

From the figure, the sampling frequency $f_s = 500 \times 2 = 1$ kHz. The sampling period $T = 10^{-3}$ s:

$$f_p = f_c = 300 \text{ Hz (at } -3 \text{ dB)}$$

$$f_r = 100 \text{ Hz}$$

Prewarping gives

$$\omega_p = \tan\frac{\Omega_p T}{2} = \tan\frac{2 \times \pi \times 300 \times 10^{-3}}{2} = \tan 0.3\pi = 1.37638$$

$$\omega_r = \tan\frac{\Omega_r T}{2} = \tan\frac{2\pi \times 100 \times 10^{-3}}{2} = \tan 0.1\pi = 0.325$$

The ripple factors are

$$\varepsilon_p = \sqrt{10^{0.301} - 1} = 1$$

$$\varepsilon_r = \sqrt{10^{2.5} - 1} = 17.75$$

The required filter order is obtained from

$$N_{HP} \geq \frac{\log(\varepsilon_r/\varepsilon_p)}{\log(\omega_p/\omega_r)} \geq \frac{\log 1.2493}{\log 4.235} \geq 1.992 \approx 2$$

The normalized second-order analog LP Butterworth filter is given by

$$H_{LP}(S) = \frac{1}{S^2 + \sqrt{2}S + 1}$$

From Table 6.5, for transformation from LP to HP, we substitute

$$\overline{S} = \frac{1}{S}$$

in the previous equation, to get

$$H_{HP}(S) = \frac{1}{(1/S)^2 + \sqrt{2}(1/S) + 1} = \frac{S^2}{1 + \sqrt{2}S + S^2}$$

Now, to apply the BZT, we substitute for

$$S = \frac{1}{\omega_p}\frac{1 - Z^{-1}}{1 + Z^{-1}}$$

in the previous equation, giving

$$H(Z) = \frac{(1 - Z^{-1})^2}{(1 - Z^{-1})^2 + \sqrt{2}\omega_p(1 - Z^{-2}) + \omega_p^2(1 + Z^{-1})^2}$$

$$= \frac{1 - 2Z^{-1} + Z^{-2}}{\left(1 + \sqrt{2}\omega_p + \omega_p^2\right) + Z^{-1}(2\omega_p^2 - 2) + Z^{-2}\left(1 - \sqrt{2}\omega_p + \omega_p^2\right)}$$

$$= 0.2065 \frac{1 - 2Z^{-1} + Z^{-2}}{1 + 0.369Z^{-1} + 0.1958Z^{-2}}$$

A realization in the direct form for the filter is shown in Figure 6.82.

MATLAB Check

```
[N,wc] = buttord(.6, .2, -3.02, -25);    % wp > wr both are
                                          normalized to ws/2

[num,den] = butter (N,wc,'high');
[z,p,k] = tf2zp (num,den);
sos = zp2sos(z,p)
fvtool (num,den)
```

giving

numerator's coefficients:	0.2079	−0.4157	0.2079
denominator's coefficients:	1.0000	0.3635	0.1950

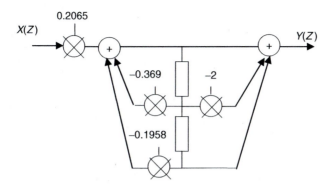

FIGURE 6.82   Canonic realization for the filter in Example 6.24.

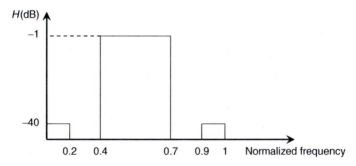

FIGURE 6.83   Tolerance structure for Example 6.25.

EXAMPLE 6.25

Using MATLAB, design a Chebyshev BPF to meet the specifications described by the shown tolerance structure (Figure 6.83).

The following MATLAB m-file can be used to design the filter:

```
wp = [.4, .7];
ws = [.2, .9];
Rp = 1; Rs = -40;
[N,wn] = cheb1ord(wp, ws, Rp, Rs) [a,b] = cheby1(N, Rp, wn);
[z,p,k] = tf2zp(a,b)
sos = zp2sos(z,p,k)
fvtool(a,b)
```

Hint:

In the filter visualization menus choose Analysis then analysis parameters and magnitude. The obtained results are listed here, and the frequency response is illustrated in Figure 6.84.

$$N = 4$$

$$wn = [0.4000 \quad 0.7000]$$

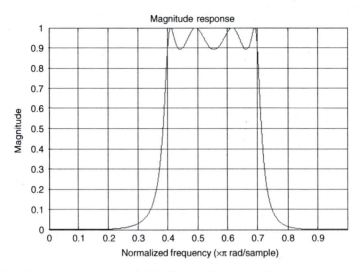

FIGURE 6.84 Magnitude response of the filter in Example 6.25.

$$z = 1.0001 \qquad\qquad -1.0001$$

$$1.0000 + 0.0001i \quad -1.0000 + 0.0001i$$

$$1.0000 - 0.0001i \quad -1.0000 - 0.0001i$$

$$0.9999 \qquad\qquad -0.9999$$

$$p = -0.5544 + 0.7703i$$

$$-0.5544 - 0.7703i$$

$$0.2854 + 0.8961i$$

$$0.2854 - 0.8961i$$

$$-0.3204 + 0.7899i$$

$$-0.3204 - 0.7899i$$

$$0.0298 + 0.8414i$$

$$0.0298 - 0.8414i$$

$$k = 0.0084$$

The coefficients of the filter, assuming a realization in form of four cascaded second-order sections, are as follows:

0.0084	−0.0167	0.0084	1.0000	−0.0597	0.7088
1.0000	2.0000	1.0000	1.0000	0.6409	0.7266
1.0000	−2.0000	1.0000	1.0000	−0.5709	0.8845
1.0000	2.0000	1.0000	1.0000	1.1088	0.9007

6.4.6 LEAST PTH-NORM OPTIMAL IIR FILTER DESIGN METHOD

The least Pth-norm algorithm delivers an optimal IIR filter. It operates directly in the Z-domain, without the need to bilinear-transform the transfer function. Furthermore, it allows the design of filters having arbitrary magnitude responses. The numerator and the denominator polynomials can have different degrees ($N < M$). This is in addition to designing the traditional filter functions (LP, HP, BP, and BS filters). The designed optimal filter is described by the transfer function h_{opt}, defined as

$$h_{opt} = \min\left(\sum_{i=0}^{N}|H_i - G_i|^p\right)^{1/p}$$

where

   H_i = ideal response at the ith frequency
   G_i = response to be optimized at the same frequency

   To design a filter using this method, the following MATLAB instructions are useful:

```
N = Numerator order
M = Denominator order
F = [0 f_p f_r 1];              % Frequency vector
E = F;                          % Frequency edges
A = [1 1 0 0];                  % Magnitude vector (LP filter)
[b,a] = iirlpnorm(N,M,F,E,A);
[z,p,k] = tf2zp(b,a)
sos = zp2sos(z,p,k)
v = fvtool(b,a)
```

EXAMPLE 6.26

Using the least Pth-norm method, design an LPF to satisfy the tolerance structure shown in Figure 6.85.

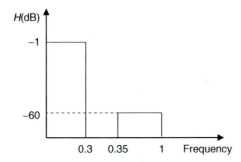

FIGURE 6.85   Tolerance structure for Example 6.26.

Solution

Assuming an order of 8, the following MATLAB file can be used to deliver the detailed design (Figure 6.86):

```
N = 8;                          % Numerator order
M = N;                          % Denominator order
F = [0 0.3 0.35 1];             % Frequency vector
E = F;                          % Frequency edges
A = [1 1 0 0];                  % Magnitude vector
[a,b] = iirlpnorm(N,M,E,F,A);
[z,p,k] = tf2zp(a,b)
sos = zp2sos(z,p,k)
fvtool(a,b)
```

Results:

$z =$

$-0.5631 + 0.5569i$

$-0.5631 - 0.5569i$

$0.0707 + 0.9961i$

$0.0707 - 0.9961i$

$0.3617 + 0.9320i$

$0.3617 - 0.9320i$

$0.4455 + 0.8952i$

$0.4455 - 0.8952i$

$p =$

$0.5476 + 0.8001i$

$0.5476 - 0.8001i$

$0.5372 + 0.7070i$

$0.5372 - 0.7070i$

$0.5384 + 0.5184i$

$0.5384 - 0.5184i$

$0.5437 + 0.1966i$

$0.5437 - 0.1966i$

$k = 0.0099$

$sos =$

0.0099	0.0112	0.0062	1.0000	−1.0874	0.3343
1.0000	−0.1415	0.9972	1.0000	−1.0768	0.5587
1.0000	−0.7233	0.9994	1.0000	−1.0744	0.7884
1.0000	−0.8910	0.9999	1.0000	−1.0951	0.9400

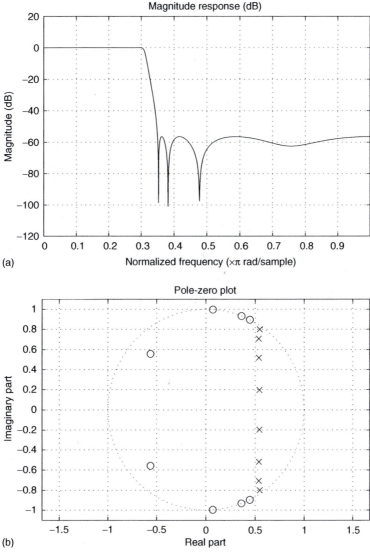

FIGURE 6.86 **(See color insert following page 262 for Figure 6.86a.)** MATLAB results for the filter in Example 6.26. (a) Magnitude response, (b) pole-zero plot.

6.5 EFFECT OF FINITE WORD LENGTH ARITHMETIC

Practical digital filters are realized using finite word length arithmetic. Thus, all coefficients, numbers, and the sum and products of two numbers should be quantized (truncated) to the nominal word length dictated by the front end ADC. IIR filters suffer much from finite word length effects, as they can lead to instability. The resulting errors degrade the performance of the designed filter. Therefore, it is imperative to check their effects before implementing a filter. Increasing the word length can help reduce such effects, but because this is reflected in a corresponding increase in the cost, the optimum word length should be determined. The degradation in performance depends on:

1. Word length
2. Type of arithmetic (fixed-point or floating-point)
3. Method of quantizing filters' coefficients
4. Filter topology
5. Method of implementation (hardware or software)

Before we select the proper corrective actions to minimize their effects, let us first study the different sources of errors caused by finite word length. In the following sections, we shall discuss such sources and their potential remedies.

6.5.1   QUANTIZATION NOISE DUE TO ADCs

As described in Chapter 4, due to the discrete-time nature of the process of quantizing a continuous input signal, errors are to be expected. This type of error is irreversible and is called *quantization noise*.

Remedies:

1. Increasing word length.
2. Applying multirate techniques (which will be discussed in Chapter 8).

6.5.2   ERRORS DUE TO COEFFICIENTS QUANTIZATION

Poles and zeros of a filter's transfer function are sensitive to any changes in the design values of the coefficients. Slight changes in these values are accompanied by a corresponding shift in the poles and zeros locations. This shift becomes critical in narrow-band filters, where the poles are nearest to the circumference of the unit circle. Driving any pole outside this circle causes oscillations. Furthermore, such a shift does affect the shape of the filter's frequency response, in that shifts in the pole locations causes a shift in the pass band response, so the response could exceed the limits set by the tolerance structure. Shifts in the zeros affect the stop band characteristics.

Remedy:

Choosing the optimum word length.

6.5.3   ERRORS DUE TO OVERFLOW

Adding two large numbers of the same sign can cause overflow. In two's complement arithmetic, the number of bits necessary to represent the sum can exceed the design word length. This might enforce the output to change its sign, such that large positive numbers can be changed into large negative ones and vice versa.

Remedy:

Downscale the inputs to keep the output within limits.

6.5.4   ERRORS DUE TO PRODUCT ROUND-OFF

The product of an n-bit input and an n-bit coefficient is definitely a number of length $2n$-bits. For this result to be further processed, it should be truncated to assume a word length of n-bit again. As result of such round-off errors, the SNR of the output reduces; in addition, the filter can either break into small-scale oscillations or produce a constant output without any input.

Remedy:

Modern DSP processors support double word length accumulation by allowing the product ($2n$-bit length) to be stored in a $2n$-bit product register, then adding at a word length of $2n$ bits before performing quantization.

6.6   SUMMARY

In this chapter, the design and realization of both types of digital filters (FIR and IIR) were described. After presenting important filter specifications employing the tolerance structure, the steps involved in the design process of digital filters were discussed. The issue of hardware vs. software realization was then presented, and an example of a dedicated processor system was given. The design process of FIR filters was presented illustrating their excellent phase characteristics. Popular approximation techniques were given, with special emphasis on the windowing method. Several windows were described, along with their important features. The optimal, frequency sampling, and the least Pth-norm methods were then described and applied using MATLAB. The design problem of IIR filters was then handled, and a standard and systematic test of stability (the Jury criterion) was described. Approximation techniques such as the Invariant Impulse Response method (IIR) and the BZT method were presented. In such techniques, use was made of the already acquired knowledge of analog filter design from Chapter 3. The undesired warping effects that are a side effect of the BZT, together with a method to overcome such effects, were then described. The least Pth-norm method was then introduced as an alternative to the BZT method.

6.7   REVIEW QUESTIONS

1. Digital filters are classified according to their impulse response. What are these types? How are they realized and implemented?
2. Sketch a block diagram for a processor system that is dedicated for software realization of a digital filter, stating the function of each block.
3. Explain why the response of a discrete-time system is repeated around multiples of the sampling frequency.
4. What is meant by "warping effect"? How is it possible to avoid this effect?
5. What is a window function? Why and when is it used? Give examples of some usable window functions.
6. What is the criterion for selecting a window function?
7. State why it is necessary in some applications to have a linear phase response.
8. Sketch the pole-zero distribution of a fifth-order Chebyshev HPF.
9. Prove that delaying a certain variable by n sampling periods in the time domain is equivalent to multiplying it by Z^{-n} in the frequency domain.
10. What is meant by "canonic realization"? Sketch a canonic realization for a fifth-order IIR filter. Compare the number of the used components in a canonic and noncanonic realizations.

11. Show how you can get the impulse response of a certain FIR filter, given its transfer function.
12. Show why FIR filters possess linear phase and hence constant group delay characteristics.

6.8   PROBLEMS

1. For the filter described by

$$y(nT) = 3x(nT) - 2x(nT - T) - x(nT - 2T) - 2y(nT - T) - y(nT - 2T)$$

a. Calculate its response for the given excitation.
b. Deduce its transfer function $H(Z)$ giving its type and order.
c. Test its stability.
d. Sketch one possible realization topology.

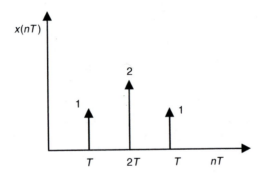

2. The given transfer function represents the filter shown in the figure of Problem 6.1. Find the values of the constants A and B, then suggest a new realization.

$$H(Z) = \frac{Z(3 + 9Z)}{(Z + 0.2)(Z + 0.4)}$$

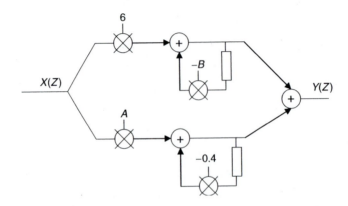

3. Give an expression for the transfer function $H(Z)$ of the filter shown in the following figure, stating its type and order. Find its sampled impulse response, as well as the excitation sequence $x(nT)$ that provides the shown output sequence $y(nT)$.

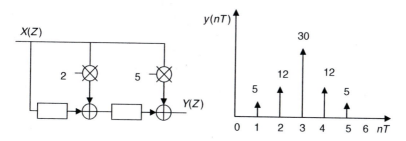

4. The following difference equation describes a digital filter:

$$y(nT) = x(nT) + 4x(nT - 2T) + 3x(nT - 4T) - 0.7y(nT - 2T) - 0.1y(nT - 4T)$$

   a. State its type and order.
   b. Derive its transfer function $H(Z)$.
   c. Sketch the first few samples of its impulse response.
   d. Sketch its pole-zero distribution in the Z-plane, then check its stability.
   e. Give a canonic realization for the filter.
5. The following difference equation describes a digital filter:

$$y(nT) = 4x(nT) - 3x(nT - T) - 2x(nT - 2T) + x(nT - 3T) - y(nT - T)$$

   a. Determine its type and order.
   b. Deduce its transfer function $H(Z)$.
   c. Find its poles.
   d. Check its stability.
   e. Deduce its output in response to the shown excitation.

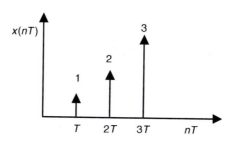

6. Deduce the overall transfer function $H_o(Z)$ and the impulse response $h_o(nT)$ of the given filter system, if the individual impulse responses of its elements are as shown in the following figure.

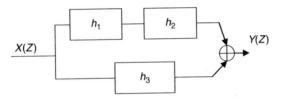

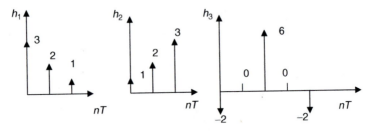

(a) Filter system

(b) Individual impulse responses

7. Deduce the transfer function $H(Z)$ of the filter in the following figure. Determine its type and order. Give suitable values for its multipliers to have a constant phase and group delay. Sketch the resulting impulse response.

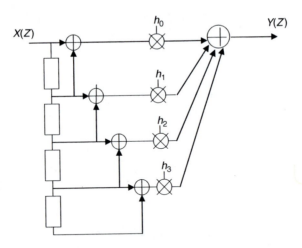

8. A digital filter is described by the following difference equation:

$$y(nT) = x(nT) - 2x(nT - T) - 3x(nT - 2T) + 4x(nT - 3T) - y(nT - T)$$

a. Determine its type and order.
b. Deduce its transfer function $H(Z)$.
c. Find its poles.
d. Check its stability.
e. Deduce its output in response to the shown excitation.
f. Verify your results using MATLAB.

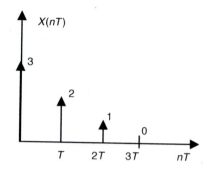

9. For the filter described by

$$y(nT) = 4x(nT) + 3x(nT - T) + 2x(nT - 2T) + x(nT - 3T)$$
$$+ 2x(nT - 4T) + 3x(nT - 5T) + 4x(nT - 6T)$$

   a. Find $H(Z)$.
   b. Calculate its response to the shown excitation.
   c. Prove that it is a constant delay filter.
   d. Plot its pole-zero distribution in the Z-plane using MATLAB.
   e. Draw a circuit diagram for the filter.

10. Give an expression for the transfer function $H(Z)$ of the filter shown in the following figure, stating its type and order. Find its sampled impulse response, as well as its response to the shown excitation.

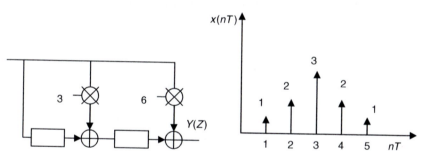

11. Derive $H(Z)$ of the filter shown in the following figure, then reconfigure it in the standard canonic form. Compare between the two configurations, considering the number of multipliers.

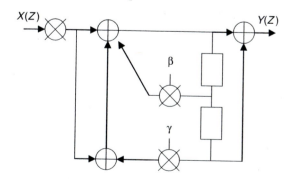

12. Calculate the response $Y(Z)$ of the filter described by

$$y(nT) = 4x(nT) + 3x(nT - T) + 2x(nT - 2T) + x(nT - 3T)$$

$$+ 2x(nT - 4T) + 3x(nT - 5T) + 4x(nT - 6T)$$

to the excitation shown in the following figure, then derive its transfer function $H(Z)$ as well as its impulse response.

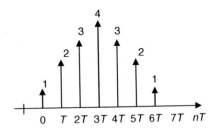

13. Derive $H(Z)$ for the shown filter, then reconfigure it in the standard canonic form.

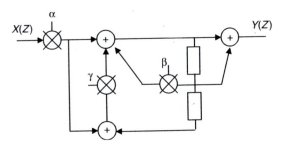

14. The following difference equation describes a digital filter:

$$y(nT) = x(nT) + 4x(nT - 2T) + 3x(nT - 4T)$$
$$- 0.7y(nT - 2T) - 0.1y(nT - 4T)$$

a. State its type and order.
b. Derive its transfer function $H(Z)$.
c. Sketch the first few samples of its impulse response.
d. Sketch its pole-zero distribution in the Z-plane.
e. Give a canonic realization for the filter.
f. Check its stability using MATLAB.

15. Derive $H(Z)$ for the filter shown in the following figure, then plot its pole(s) and zero(s) in the Z-plane. What is the range of β that ensures stability?

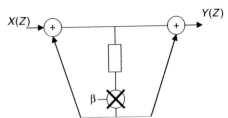

16. Find the range of θ over which the filter described by

$$H(Z) = \frac{1 - Z^{-1}}{1 - \tan\theta \cdot Z^{-1} - 0.4226Z^{-2}}$$

remains stable. Illustrate your answer using sketches, and then draw a canonic realization for the filter.

17. For the filter described by

$$y(nT) = 3x(nT) - 2x(nT - T) - x(nT - 2T) - 2y(nT - T) - y(nT - 2T)$$

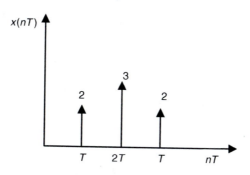

a. Calculate its response for the given excitation.
b. Deduce its transfer function $H(Z)$.
c. Find its type and order.
d. Test its stability using MATLAB.
e. Sketch one possible realization topology.

18. The following difference equation describes a digital filter:

$$y(nT) = 6x(nT) + 5x(nT - T) - 2x(nT - 2T) + x(nT - 3T) - y(nT - T)$$

a. Deduce its transfer function $H(Z)$.
b. Find its poles.
c. Check its stability.
d. Deduce its output in response to the shown excitation.

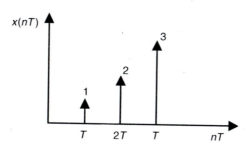

19. Using a suitable window function, design an FIR filter to satisfy the tolerance structure shown in the following figure. Verify your design using MATLAB.

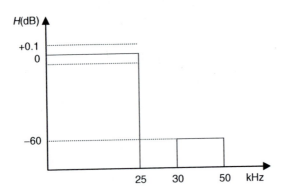

20. Using a suitable window function, design an FIR filter to satisfy the tolerance structure shown in the following figure. Verify your design using MATLAB. Redesign the filter using the least Pth-norm method.

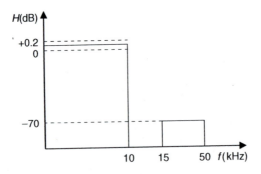

21. Using a suitable window function, design a linear phase filter to satisfy the tolerance structure shown in the following figure. Use MATLAB to prove the correctness of your design.

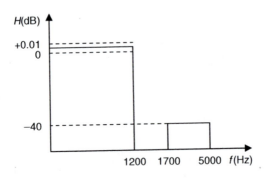

22. Using a suitable window function, design an FIR filter to satisfy the tolerance structure shown in the following figure. Verify your design employing MATLAB.

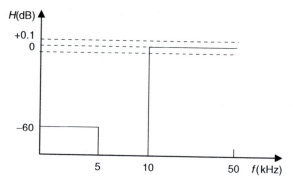

23. Using a suitable window function, design a linear phase filter to satisfy the tolerance structure shown in the following figure. Repeat the design using MATLAB and compare both results.

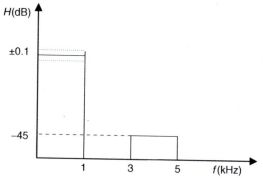

24. Using an appropriate window function, design an FIR LPF to provide an attenuation of 60 dB/octave. The maximum frequency of signals is limited to 500 Hz. To avoid aliasing errors, all signals are oversampled at 10 kHz.

25. Using the least Pth-norm method, design a Butterworth filter that satisfies the tolerance structure shown in the following figure.

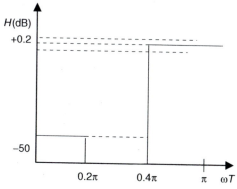

26. Using the optimal method, design a second-order IIR filter to satisfy the requirements given in the tolerance structure shown in the following figure, and then sketch its pole-zero distribution in the Z-plane.

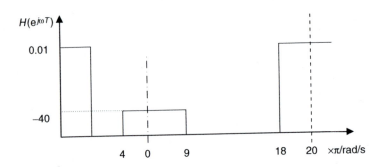

27. The following figure shows the tolerance structure of a certain BPF. Using the frequency sampling method, design the filter, then suggest a canonical realization.

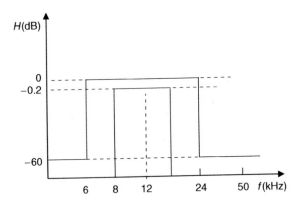

28. Using the BZT method, design a BSF to satisfy the tolerance structure shown in the following figure, then realize the filter in a canonical form.

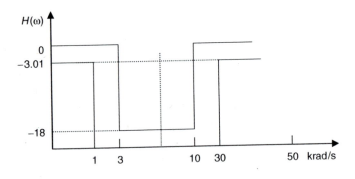

6.9   MATLAB® PROJECT

Use MATLAB to design a filter that satisfies the shown tolerance structure as:

1. An FIR filter employing:
    a. A Kaiser window
    b. The least Pth-norm optimal method
2. An IIR filter employing:
    a. The BZT
    b. The least Pth-norm optimal method

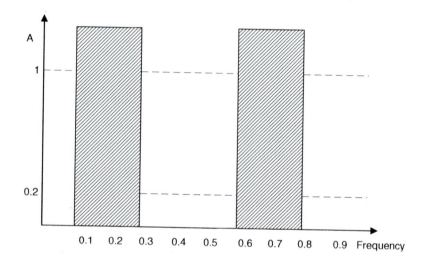

7 Multirate Signal Processing

7.1 INTRODUCTION

Multirate signal processing usually means the handling of a data stream at more than one sampling rate for improving the performance of some signal processing tasks. Two basic operations are encountered here:

1. *Decimation* (*downsampling*) or sampling rate reduction
2. *Interpolation* (*upsampling*) or sampling rate increase

The reasons why the data rate of a certain signal is changed are many, but to mention a few:

- Providing compatibility between a data stream and a processor
- Improving the computational efficiency of a certain processor
- Eliminating the need for high-order antialiasing filters while using simple low-cost ADCs
- Reducing the quantization noise of ADCs through oversampling
- Making it possible to deal with signals of different bandwidths at the same time
- Reducing the storage capacity required to save a data stream
- Reducing the required lengths of narrow-band digital filters by decimating the input data rate
- Reducing transmitting channels occupation

Whatever the reason is for changing the sampling rate either upwards or downwards, it should be possible to bring it back after processing to its original rate without causing any serious errors to the data.

7.2 DECIMATION (SAMPLING RATE REDUCTION)

As said before, decimation or downsampling means reducing the sampling rate of a certain data stream by an integer factor α (the decimation ratio) where $\alpha > 1$. This can simply be achieved by dropping a sample every other sample or dropping a sample every two samples depending on α. However, due to this sampling rate reduction, repeated spectra around multiples of the new sampling rate come closer to each other, such that aliasing between them becomes possible. Therefore, to avoid such an effect, a band-limiting (antialiasing) filter is usually inserted prior to each decimator, in order to limit the base band to a maximum of $f_s/2\alpha$. As depicted in

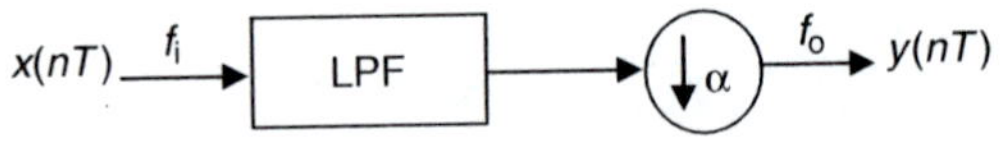

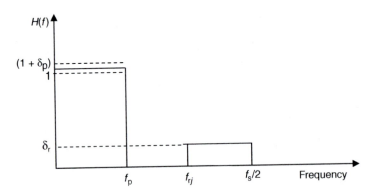

FIGURE 7.1   The decimation process.

FIGURE 7.2   Specifications of an antialiasing filter.

Figure 7.1, the decimator is represented by a circle with a downward arrow, labeled with α.

Decimation can be achieved in several cascaded stages, such that

$$\alpha = f_o/f_i = \alpha_1 \cdot \alpha_2 \cdot \alpha_3 \cdot \alpha_4 \cdots \alpha_n$$

where f_s and f_o are the input and output sampling frequencies, respectively.

Proper design of the required antialiasing filters is fundamental to prevent any deterioration in the SNR of the input stream due to probable overlapping of adjacent spectra. One of the important design data is the stop band edge f_{rj} of the jth filter, given by

$$f_{rj} = f_{oj} - \frac{f_s/2}{\alpha} \tag{7.1}$$

where

f_{rj} = rejection band edge of the jth filter, as shown in Figure 7.2
f_{oj} = output data rate of the jth stage
f_s = input sampling rate

The normalized transition bandwidth Δf_i of the ith filter is a rather important design data. It is calculated from

$$\Delta f_j = \frac{f_p - f_{rj}}{f_s}$$

where f_p is the pass band edge of interest.

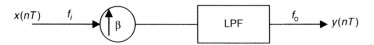

FIGURE 7.3   The process of interpolation.

7.3   INTERPOLATION (SAMPLING RATE EXPANSION)

Interpolation, or upsampling, is the reverse process of decimation. The sampling rate f_s is raised to βf_s, with $\beta > 1$ (an integer) by inserting $\beta - 1$ dummy zeros between every two successive samples. However, to ensure continuity and smoothness of the data stream, an interpolating filter (an LPF) has to be introduced immediately after the interpolator, to limit the signal spectrum to $f_s/2$. The insertion of $\beta - 1$ zeros will result in redistributing the energy among all other samples. Thus, to compensate for this loss in amplitude, a pass band gain of β has to be provided by the filter to the samples. The process is diagrammed in Figure 7.3.

7.4   FRACTIONAL SAMPLING RATE CHANGE

It is evident from the previous discussion that to halve the sampling rate (i.e., $\alpha = 2$), one has to drop every other sample. Accordingly, decimation by a noninteger factor—for example, 0.75—seems to be unrealizable, as it is not possible in practice to drop a noninteger number of samples. However, a solution to this problem is possible if we consider such a fractional rate as a ratio between two integers—for example, 3/4. In this case, decimation should be carried out in two successive stages, according to the following steps:

1. A decimation stage with $\alpha = 4$ that is followed by
2. An interpolation stage with $\beta = 3$

Figure 7.4 illustrates the process diagrammatically. Two scenarios are possible: in (a), the decimator precedes the interpolator, and the sequence is reversed in (b), where the decimator follows the interpolator instead.

It is thus evident from Figure 7.4 that configuration (b), where interpolation is performed first, seems to be better, as the two LPFs could be merged into one filter, thereby saving processing time and cost in addition to preventing aliasing that could be introduced due to decimation. The resulting configuration is depicted in Figure 7.4c.

It is apparent that the performance of the used filters will determine the performance of the designed system. Due to their excellent phase characteristics and ease of implementation, it is a common practice to use FIR filters in the implementation of the antialiasing and interpolating filters (also called antiimaging filters). The windowing and optimal methods, discussed in Chapter 6, are usually used in the design. Figure 7.5 illustrates the tolerance structures of the filters used for decimation and interpolation. As said before, a pass-band gain of β should be provided in the interpolating filter.

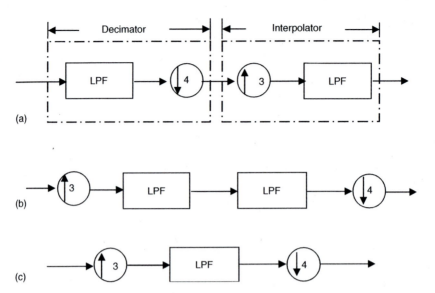

FIGURE 7.4 Possible configurations of a fractional rate change. (a) Performing decimation first, then interpolation, (b) interpolation followed by decimation, (c) merging the antiimaging filter and the antialiasing filter into one filter.

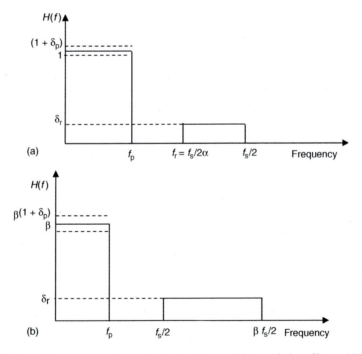

FIGURE 7.5 Tolerance structures for decimation and interpolation filters. (a) Typical specifications of an antialiasing filter (decimation filter), (b) typical specifications of an antiimaging filter (interpolation filter).

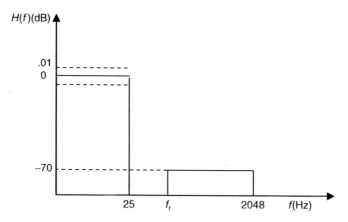

FIGURE 7.6   Tolerance structure for Example 7.1.

EXAMPLE 7.1

Design a single-stage decimator to provide a decimation factor of 64. The input signal is sampled at 4.096 kHz. The specifications of the required antialiasing filter are summarized in the tolerance structure given in Figure 7.6.

Solution

We start by calculating the ripple amplitudes in the pass and stop bands of the filter, like we did before in Chapter 6, as follows:

$$A_p = 0.01 = 20 \log(1 + \delta_p)$$

giving

$$\delta_p = 0.001155$$

and

$$A_r = -70 = 20 \log \delta_r$$

giving

$$\delta_r = 0.000316$$

Because $\delta_r < \delta_p$, the design should be based on A_r; that is,

$$A_r = -70 \text{ dB}$$

Thus, a Blackman window can be used.

The output frequency $= f_o = \dfrac{4096}{64} = 64\,\text{Hz}$

The rejection band edge $= f_r = f_o - \dfrac{f_s}{2 \times \alpha} = 64 - \dfrac{4096}{2 \times 64} = 32\,\text{Hz}$

The normalized transition band $= \Delta f = \dfrac{f_r - f_p}{f_s} = \dfrac{32 - 25}{4096} = 1.708 \times 10^{-3}$

Giving a filter length of

$$N = \frac{5.5}{1.708 \times 10^{-3}} = 3215.9 \approx 3217$$

which is too large! Therefore, it is recommended to perform the decimation in several cascaded stages, as will be discussed in the following sections.

7.5   CASCADED DECIMATORS/INTERPOLATORS

As the required decimation/interpolation ratio increases, the transition bandwidth of the associated filters becomes narrower, giving filter lengths that are impractical to implement. Therefore, to get reasonable filter orders, one can perform the process in m cascaded stages, rather than in a single stage. The required ratio can then be split into several smaller factors $\alpha_1, \alpha_2, \alpha_3$, such that

$$\alpha_o = \alpha_1 \cdot \alpha_2 \cdot \alpha_3 \cdots \alpha_m$$

or

$$\beta_o = \beta_1 \cdot \beta_2 \cdot \beta_3 \cdots \beta_m$$

Besides the achievable relaxation on the filter's "characteristics," the inherent adverse effect of finite word length problems are markedly reduced. Based on the requirements of the specific problem, there is usually an optimum number of the stages in the cascade and optimum distribution of the decimation/interpolation ratios among the stages. As an example, consider all possible decimation ratios of a decimator to achieve an overall ratio of 32. The following cases can come in question:

32×1	single stage
16×2	two stages
8×4	two stages
$4 \times 4 \times 2$	three stages
$4 \times 2 \times 2 \times 2$	four stages

The optimum number of the stages and the decimation/interpolation factors of each stage are not easy to determine. One important criterion is based on selecting the number of stages m that minimizes the computational effort reflected in the number of multiplications per second (MPS) and the storage requirements (SR) for the filter's coefficients, where

$$\text{MPS} = \sum_{j=1}^{m} N_j f_{s_i} \tag{7.2}$$

$$\text{SR} = \sum_{i=1}^{m} N_i \tag{7.3}$$

and N_i is the length of the ith filter.

However, as a general rule, the number of stages rarely exceeds three or four. Thus, one can exhaust all possible sets of factors to find the one that minimizes

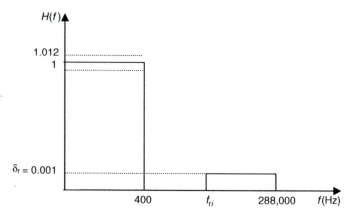

FIGURE 7.7   Tolerance structure of Example 7.2.

either one of Equations 7.2 and 7.3. It is also recommended that the ratios should be descending [16]; that is,

$$\alpha_1 > \alpha_2 > \alpha_3 > \cdots > \alpha_m$$

There is no closed formula that could be used to find the optimal ratio distribution among the stages in a cascade. However, one can use the following empirical formula for a cascade of two decimation stages having the ratios α_1 and α_2 [4].

$$\alpha_{1opt} = \frac{2\alpha}{2 - \Delta f + \left(\sqrt{2\alpha\Delta f}\right)}$$

in which Δf is the normalized transition width.

$$\alpha_{2opt} = \frac{\alpha_o}{\alpha_1}$$

where

$$\alpha_o = \alpha_1 \cdot \alpha_2$$

EXAMPLE 7.2

Design an efficient decimator to reduce the sampling rate of a certain data stream by a factor of 576 to attain a value of 1 kHz. Use the windowing method to design the FIR filters that should satisfy the tolerance structure given in Figure 7.7.

Solution

As before, the minimum ripple amplitude is considered to be

$$\delta = \min(\delta_p, \delta_r) = \min(0.012, 0.001)$$

$$\delta_{rmin} = 0.001$$

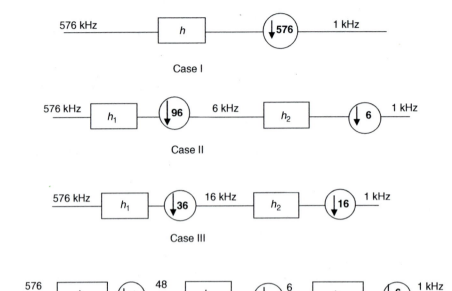

FIGURE 7.8 Some cases of Example 7.2.

giving

$$A_r = -60 \text{ dB}$$

For such attenuation, Blackman and Kaiser windows are possible candidates. However, to get minimum length, we should choose the Kaiser window.

For a stop band attenuation of $A_r > 50$ dB, we have

$$\alpha_k = 0.1102(60 - 8.7) = 5.65326$$

The filter length, as given by Equation 6.20, is

$$N_i = \frac{60 - 7.95}{14.36 \, \Delta f_i} = \frac{3.6247}{\Delta f_i}$$

where Δf_j, as before, is the normalized transition bandwidth of the jth filter.

Now, let us try the following possibilities (Figure 7.8):

Case I. Single stage, $m = 1$:

$$f_r = 1000 - \frac{f_{s_i}}{2\alpha_o} = 1000 - \frac{576,000}{2 \times 576} = 500 \text{ Hz}$$

$$\Delta f_u = \frac{500 - 400}{576,000} = 1.73,611 \times 10^{-4}$$

$$N = \frac{3.6,247}{1.73,611 \times 10^{-4}} = 20,878.2 \approx 20,879$$

which is large and therefore impractical.

The required MPS $= \sum N_i \cdot f_i = 20{,}879 \times 576 \times 10^3 = 12{,}026{,}304 \times 10^3$, and the SR $= \sum N_i = 20{,}879$.

Case II. Two stages, $m = 2$ ($\alpha_1 = 96$ and $\alpha_2 = 6$):

<table>
<tr><td align="center">For the first filter, h_1</td><td align="center">For the second filter, h_2</td></tr>
</table>

$$f_{r_1} = 6000 - \frac{576{,}000}{2 \times 576} = 5500 \qquad f_{r_2} = 1000 - 500 = 500$$

$$\Delta f_1 = \frac{5500 - 400}{576{,}000} = 0.854 \times 10^{-3} \qquad \Delta f_2 = \frac{500 - 400}{6000} = 0.016666$$

giving

$$N_1 = 4245 \quad \text{and} \quad N_2 = 219$$

and

$$\text{MPS} = 576 \times 4245 \times 10^3 + 6 \times 219 \times 10^3 = 2{,}446{,}434 \times 10^3$$

$$\text{SR} = 4245 + 219 = 4464$$

Case III. Two stages, $m = 2$ ($\alpha_1 = 36$ and $\alpha_2 = 16$):

The rejection band edges and the normalized transition bands are

$$f_{r_1} = 16{,}000 - \frac{576{,}000}{2 \times 576} = 15{,}500 \qquad f_{r_2} = 1000 - \frac{57{,}600}{2 \times 576} = 500$$

$$\Delta f_1 = \frac{15{,}500 - 400}{576{,}000} = 26.215 \times 10^{-3} \qquad \Delta f_2 = \frac{500 - 400}{16{,}000} = 6.25 \times 10^{-3}$$

giving

$$N_1 = 139 \quad \text{and} \quad N_2 = 581$$

The corresponding number of multiplications and SR are

$$\text{MPS} = \sum N_i \cdot f_i = (139 \times 576 \times 10^3 + 581 \times 16 \times 10^3) = 89{,}360 \times 10^3$$

and

$$\text{SR} = \sum N_i = 139 + 581 = 720$$

Case IV. Three stages, $m = 3$ ($\alpha_1 = 12$, $\alpha_2 = 8$, and $\alpha_3 = 6$):

The data of the first-stage filter can be calculated as

$$f_{r_1} = 4800 - 500 = 47{,}500 \quad \text{and} \quad \Delta f_1 = \frac{47{,}500 - 400}{576{,}000} = 0.08177$$

giving

$$N_1 = \frac{3.62}{0.08177} = 45$$

The rejection band edge and Δf_2 of the second stage are

$$f_{r_2} = 6000 - 500 = 5500 \quad \text{and} \quad \Delta f_2 = \frac{5500 - 400}{4800} = 0.10625$$

The length of the second-stage filter is then

$$N_2 = \frac{3.62}{0.10625} = 35$$

Repeating the same for the third stage, we get

$$f_{r_3} = 1000 - 500 = 500 \quad \text{and} \quad \Delta f_3 = \frac{500 - 400}{6000} = 0.016666$$

giving

$$N_3 = \frac{3.62}{0.01666} = 219$$

The required calculations are

$$\text{MPS} = \sum N_i \cdot f_i = 45 \times 576 \times 10^3 + 35 \times 48 \times 10^3 + 219 \times 6 \times 10^3 = 28,914 \times 10^3$$

and

$$\text{SR} = \sum N_i = 45 + 35 + 219 = 299$$

Case V. Three stages, m = 3 ($\alpha_1 = 9$, $\alpha_2 = 8$, and $\alpha_3 = 8$):

We get

$$\text{MPS} = 24,600 \times 10^3$$

and

$$\text{SR} = 359$$

Case VI. Four stages, m = 4 ($\alpha_1 = 8$, $\alpha_2 = 6$, $\alpha_3 = 4$, and $\alpha_4 = 3$):

We get

$$\text{MPS} = 20,235 \times 10^3$$

and

$$\text{SR} = 186$$

Case VII. Four stages, m = 4 ($\alpha_1 = 8$, $\alpha_2 = 8$, $\alpha_3 = 3$, and $\alpha_4 = 3$):

We get

$$\text{MPS} = 20,712 \times 10^3$$

TABLE 7.1
Summary of the Results of Example 7.2

Case	m	α_1	α_2	α_3	α_4	N_i	MPS × 10³	SR
I	1	576	—	—	—	20,853	12,026,304	20,879
II	2	96	6			4,239,219	2,446,434	4,464
III	2	36	16	—	—	91,379	89,360	720
IV	3	12	8	6	—	4,535,219	28,914	299
V	3	9	8	8	—	3,533,291	24,600	359
VI	4	8	6	4	3	312,521,109	20,235	186
VII	4	8	8	3	3	313,317,109	20,712	190

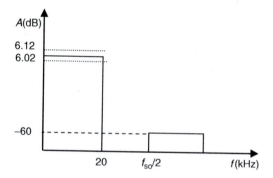

FIGURE 7.9   Tolerance structure of Example 7.3.

and

$$SR = 190$$

The obtained results are summarized in Table 7.1. From the table, we can conclude that Case VI (8/6/4/3) is the *optimum* one.

EXAMPLE 7.3

Design an interpolator that is required to double the rate of a certain audio stream from 44.1 to 88.2 kHz. The filter specifications are depicted in the tolerance structure shown in Figure 7.9.

Solution

The required interpolation ratio $\beta = 2$.

The corresponding pass band gain = $20 \times \log 2 = 6.02$ dB.

The pass band = 20 kHz.

$$A_p = 20\log(1 + \delta_p) = 0.1 \text{ dB}$$

giving

$$\delta_p = 10^{0.05A_p} - 1 = 0.001158$$

and

$$A_r = -20\log \quad \delta_r = -60 \text{ dB}$$

giving

$$\delta_r = 0.001 < \delta_p$$

thus

$$A_r = -60 \text{ dB}$$

The normalized transition bandwidth is then

$$\Delta f = \frac{44.1 - 20}{88.2} = 0.02324$$

Using a Kaiser window, the required filter length is

$$N = \frac{3.6247}{\Delta f} = 154.9 = 155$$

The necessary calculations are

$$\text{MPS} = 88.2 \times 10^3 \times 155 = 15{,}588.2 \times 10^3$$

and

$$\text{SR} = 155$$

7.6 SUMMARY

In several modern applications, data streams must undergo data rate changes for several reasons, such as compatibility, security, improving the computational efficiency of some processors, and so on. The problem was addressed in this chapter, where the two fundamental processes (decimation and interpolation) were described. The need for band-limiting filters (predecimation and postinterpolation) was highlighted. The technique of cascading decimation and interpolating stages was then presented, demonstrating its impact on the order of the required filters. The chapter was enriched with a detailed solution of some examples, in addition to the review questions and problems.

7.7   REVIEW QUESTIONS

1. Why is it sometimes necessary to change the rate of a certain data stream?

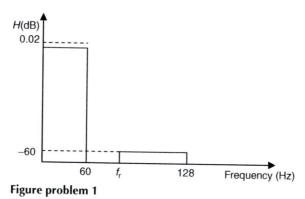

Figure problem 1

2. Why is it necessary to precede decimation stages by a properly designed antialiasing filter?
3. Sketch a tolerance structure for:
   a. An antialiasing filter
   b. An interpolating filter

7.8   PROBLEMS

1. Design a single-stage decimator to reduce the sampling rate of a signal from 256 kHz to 128 Hz. The tolerance structure of the antialiasing filter is shown in the figure shown in problem 1. Comment on the results.
2. Suggest an improved version of the decimator designed in Problem 7.1.
3. Design a two-stage interpolator for a digital audio system to accept an input sampling rate of 44.1 kHz and deliver an output at a sampling rate of 176.4 kHz, given that the band of interest extends from 0 to 16 kHz. Calculate also the MPS and the SR.
4. Design a single-stage decimator to provide a decimation ratio of 4 for a signal that is sampled at 8 kHz. The pass band of the required filter should extend from 0 to 800 Hz, with a maximum ripple of 0.05 dB and a minimum stop band attenuation of −50 dB.
5. Design an efficient decimator to reduce the sampling rate of a certain data stream from 1 to 36 kHz. Allow ripples of 0.01 dB in the pass band and 0.001 dB in the stop band of the required filter, considering the following cases:
   a. A two-stage decimator.
   b. All possible three-stage designs.
   c. All possible four-stage designs.
   d. Find the optimal solution.
6. Design a sampling rate changing system to increase the sampling rate of an audio signal from 44.1 to 110.25 kHz. Try all possible configurations, then find the best one. (*Hint:* Allow a DC gain of β for the interpolation filter.)

8 Discrete-Time Transforms

8.1 INTRODUCTION

Signals are basically time functions that represent changes of a certain physical phenomenon. Traditional signal processing operations—for example, amplification and filtering—are performed on raw signals in the time domain. However, it is sometimes advantageous to process signals while in the frequency domain, where they are viewed from another perspective. Mathematical tools (transforms) are therefore needed to perform mapping of signals from the time to the frequency domain and vice versa. Discrete-time transforms are those transforms that deal with sampled signals. Through such operations, N samples of a certain continuous time signal are transformed into an equal number of discrete frequency components describing the signal spectrum. The resulting components from this *forward transform* are usually complex; that is, they are expressed as magnitude and phase. To be usable, the reverse process of a transform—the *inverse transform*—should be valid; that is, it should be possible to get back a signal in the time domain from its spectrum. Frequency transforms are used extensively in several signal processing applications, such as the following:

1. Data transmission and storage, where it is necessary to reduce (compress) the amount of data to be transmitted and stored.
2. Pattern recognition applications, where a representative and compact set of features are extracted and then classified. Examples are diagnosis of diseases in humans, detection of defects in products or faults in machines, identity verification using biometrics, and so on. Consider, for example, the noise generated due to an internal combustion engine or the sound of a human heart. Any detected changes in the machine noise or the typical heart sound is an indication of disorder and calls for immediate intervention.
3. EKGs, which represent the electrical activity of the heart, are time-domain signals. Any slight change in waveform is a sign of a functional disorder. Such changes might not be observed by inexperienced physicians; therefore, transforming such waveforms into the frequency domain should help pinpoint the disease.
4. De-noising of corrupted signals, where all frequency components having magnitude below a certain chosen threshold are omitted.

Throughout this chapter, we are going to study some commonly used transforms, such as the DFT and its computational algorithm, the FFT; the discrete cosine transform (DCT); the Walsh transform; the Walsh–Hadamard transform; and the Wavelet Transform. Actually, these are just few examples of a long list of known transforms;

each one has its own area of application. Therefore, we are going to study only the most popular ones.

The performance of a certain transform is usually measured by its:

1. Data compression efficiency (the ability to concentrate signal energy at a low frequency)
2. Speed of computation

Usually, signals are represented by infinite time series. However, only data streams of a finite number of digitized terms can be transformed. Therefore, long data streams have to be divided into finite-length records. Selecting a finite number of terms to be transformed at a time means that the infinite sequence of data has been abruptly terminated at both ends (that is, it is as if it were multiplied by a square window of finite duration). This operation has the effect of producing undesirable oscillations "Gibbs oscillations" in the spectrum (Gibbs oscillations were discussed in Chapter 6). To minimize this adverse effect, windows that differ from rectangular windows should be selected.

8.2 DISCRETE FOURIER TRANSFORM (DFT)

The Fourier transform (FT) represents signals in the frequency domain in terms of the real frequency $j\omega$. It can be seen as a special case of the more general Laplace transform known to all electrical engineers. It is also related to the simple Fourier analysis, in that they are both analyzing signals into harmonically related sinusoidal and co-sinusoidal terms. Yet, it is more general, as it can deal with nonperiodic signals.

Let us first define the Fourier series representation of a periodic and continuous time function $f(t)$ of period T_p as

$$f(t) = \sum_{n=0}^{\infty} a_n \cos n\omega_o t + \sum_{n=1}^{\infty} b_n \sin n\omega_o t \qquad (8.1)$$

where a_n and b_n are the amplitudes of the nth harmonic and are given by

$$a_n = \frac{2}{T_p} \int_0^{T_p} f(t) \cos n\omega_o t \, dt \quad n = 1, 2, 3, \ldots$$

$$b_n = \frac{2}{T_p} \int_0^{T_p} f(t) \sin n\omega_o t \, dt \quad n = 1, 2, 3, \ldots$$

in which $\omega_o = 2\pi/T_p$ is the fundamental frequency.

The DC component a_o (average value) is calculated from

$$a_o = \frac{1}{T_p} \int_0^{T_p} f(t) \, dt$$

Equation 8.1 can be rewritten in the compact form as

$$f(t) = \sum_{n=-\infty}^{\infty} c_n e^{jn\omega_0 t} \tag{8.2}$$

from which it is evident that the coefficients c_n's are complex quantities. They are related to the a's and b's through the equation

$$c_n^2 = a_n^2 + b_n^2 \tag{8.3}$$

with the coefficients a's and b's representing their real and imaginary parts, respectively. The corresponding phase angle ϕ_i is given by

$$\phi_i = \tan^{-1} \frac{\text{imaginary part}}{\text{real part}} = \tan^{-1} \frac{\text{Im } c_i}{\text{Re } c_i} = \tan^{-1} \frac{b_i}{a_i}$$

Now if the period T_p is increased indefinitely, the function will not be periodic any more. In this case, a proper tool for transforming the signal to the frequency domain is the FT, defined by the Fourier integral (forward transform) as

$$F(j\omega) = \int_{-\infty}^{\infty} f(t) e^{-j\omega t} \, dt = |G(j\omega)| \, e^{j\phi(\omega)} \tag{8.4}$$

in which

$$|F(j\omega)|^2 = R_e^2 \, F(j\omega) + \text{Im}^2 \, F(j\omega)$$

and

$$\phi(\omega) = \tan^{-1} \frac{\text{Im } F(j\omega)}{\text{Re } F(j\omega)}$$

If the signal $f(t)$ is sampled at a frequency $f_s = 1/T$, then the previous transform will turn into an equivalent discrete form, the DFT, in which:

- The analog frequency ω is replaced by the discrete frequency $k\Omega$.
- The continuous time t is replaced by the sampled time nT.
- The integration is changed into summation of N samples.
- $\Omega = 2\pi/NT$ is the fundamental frequency.

Then Equation 8.4 will take the discrete-time form,

$$F(k\Omega) = \sum_{n=0}^{N-1} f_s(nT) \, e^{-jk\Omega nT} \quad \text{for } k = 0, 1, 2, \ldots, N-1 \tag{8.5}$$

giving the Fourier coefficients $F(k\Omega)$'s.

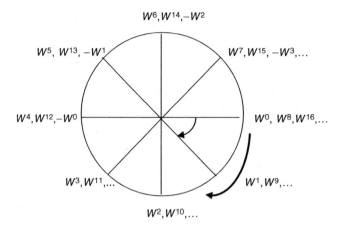

FIGURE 8.1   Twiddle factors for an 8-point DFT.

Now, making the substitution $e^{-j\Omega T} = e^{-j(2\pi/N)} = W_N$, the "twiddle factor," in Equation 8.5, gives

$$F(k\Omega) = \sum_{n=0}^{N-1} f_s(nT)e^{-j(2\pi\, kn/N)}$$

$$= \sum_{n=0}^{N-1} f_s(nT)W_N^{kn} \qquad (8.6)$$

which is the DFT equation. The magnitude of this complex equation is called *the magnitude spectrum*, and its phase represents the phase spectrum. The power spectral density, or simply the power spectrum of the signal, is obtained by squaring this quantity, giving the power dissipated in a 1 Ω resistor. It is easy to conclude that the twiddle factors W's, due to their cyclic nature (as described by Figure 8.1, for the case of an 8-point transform), exhibit the following interesting properties:

$$W_N^i = W_N^{i+N} = W_N^{i+2N}$$

$$W_N^{i+(N/2)} = -W_N^i \qquad (8.7)$$

8.3   PROPERTIES OF THE DISCRETE FOURIER TRANSFORM

The power of the DFT stems from its many useful properties. Knowledge of such properties is fundamental, as it leads to designing powerful algorithms for computing the DFT. The FFT is a good example of efficient algorithms that make use of such properties. In the following sections, we present the most important properties.

8.3.1 LINEARITY

Due to its linearity, it is possible to compute the DFT of composite signals by computing the DFT of its components individually and then applying the superposition principle to get the DFT of the signal. Mathematically, we can write

$$\text{DFT}\{a_1 x_1(nT) + a_2 x_2(nT)\} = a_1 X_1(k) + a_2 X_2(k)$$

where

$$X_i(k) = \text{DFT}\{x_i(nT)\}$$

$$a_i = \text{constant}$$

provided that all sequences have the same lengths; otherwise, dummy zeros should be added to shorter sequences.

8.3.2 SYMMETRY

For an N-point transform, the following relations are valid:

$$\text{Re}\{X(N - K)\} = \text{Re}\{X(K)\}$$

and

$$\text{Im}\{X(N - K)\} = -\text{Im}\{X(K)\}$$

indicating that the magnitude spectrum is symmetrical, while in the phase spectrum, there exists an antisymmetry.

8.3.3 PERIODICITY

The DFT and its inverse deliver periodic sequences of period N, where N is the length of the sequence. This could be referred to as the cyclic nature of the twiddle factor, W_N^{nk}. This implies that

$$X(K) = X(K + N) \quad K = 0, 1, 2,..., N - 1$$

and

$$x(n) = x(n + N) \quad n = 0, 1, 2,..., N - 1$$

To prove this let us find the DFT of the $(K + N)$th component as follows:

$$X(K + N) = \sum_{n=0}^{N-1} x(nT)W^{n(K+N)}$$

From Equation 8.7, since $W^{i+N} = W^i$, we can write

$$X(K + N)\sum_{n=0}^{N-1} x(nT)W^{nK} = X(K)$$

8.3.4 TRANSFORMS OF EVEN AND ODD FUNCTIONS

The transform of the even function $x_e(nT)$ is given by

$$F\{x_e(nT)\} = X_e(k) = \sum_{n=0}^{N-1} x_e(nT)\cos\left(\frac{2\pi kn}{N}\right)$$

Similarly, the transform of the odd function $x_o(nT)$ is given by

$$F\{x_o(nT)\} = X_o(k) = -j\sum_{n=0}^{N-1} x_o(nT)\sin\left(\frac{2\pi kn}{N}\right)$$

where the subscripts e and o denote even and odd functions, respectively.

8.3.5 TRANSFORM OF THE COMPLEX CONJUGATE OF A SEQUENCE

It was found that

$$DFT\{x^*(nT)\} = X^*(N - k)$$

That is, the DFT of the complex conjugate of a sequence is equal to the transform of the reversed sequence. This property is useful especially in evaluating the inverse discrete Fourier transform (IDFT).

8.3.6 CIRCULAR CONVOLUTION

Time or frequency domain convolutions could be easily performed using DFT as follows. Assume that two functions $f_1(nT)$ and $f_2(nT)$ are to be convolved; that is,

$$f(nT) = f_1(nT) * f_2(nT)$$

Using the DFT, we get

$$F_1(K) = DFT\{f_1(nT)\}$$

and

$$F_2(K) = DFT\{f_2(nT)\}$$

thus

$$f(nT) = IDFT\{F_1(K) \cdot F_2(K)\}$$

Circular convolution is useful in several applications, as in digital filtering, where the time-domain output of a certain filter is required. As shown in Figure 8.2, the

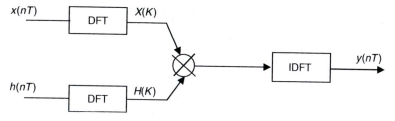

FIGURE 8.2 Filtering in the frequency domain.

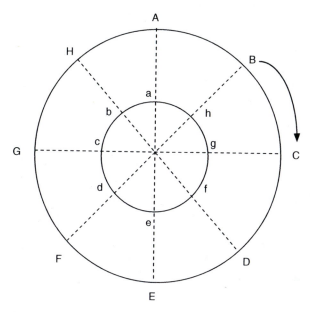

FIGURE 8.3 Circular convolution.

DFT of both the impulse response of the filter and the sampled input signal are first performed. They are then multiplied before taking the IDFT for the product. This is referred to as "filtering in the frequency domain."

However, due to the inherent periodicity of the DFT, the result of the finite-length input signal will be a periodic sequence in the frequency domain. Circular convolution could be represented easily using two concentric circles as shown in Figure 8.3, where the two sequences [S_1: A, B, C, D, E, F, G, H] and [S_2: a, b, c, d, e, f, g, h] are to be convolved.

For the shown setting, every number in the outer circle is multiplied by the corresponding number in the inner circle, giving the first output of the convolution. Then the outer circle is rotated 45° clockwise and each number in the circle is multiplied by the corresponding one in the inner circle to deliver the next output. The process is repeated until the outer circle has been rotated 360°.

8.3.7   Frequency Resolution

The *frequency resolution* Δf of the transform is given by the spacing between its spectral lines (bin spacing); it is given by

$$\Delta f = \frac{f_s}{N} = \frac{1}{NT}$$

where f_s is the sampling frequency and N is the number of samples in the record.

The result of the transform is complex and is usually given in the form of a plot for the magnitude spectrum $|X(K\Omega)|$ (spectral lines separated by Δf) and the phase spectrum $\Phi(K\Omega)$ both at $K = 1, 2, 3, 4, 5, \ldots$; that is, at the frequencies

$$0, \Delta f, 2\Delta f, 3\Delta f, \ldots (N-1)\Delta f$$

8.3.8   Zero Padding

From the previous discussion, it is clear that increasing the number of input samples N to a transform improves its frequency resolution. In some cases, it is necessary to artificially increase N by adding some dummy zeros (augmenting zeros) to the sequence. Such added zeros do not corrupt the sequence, as it might seem; on the contrary, they improve the frequency resolution Δf, thereby giving a better insight onto the spectrum of the input sequence. The process of adding dummy zeros to an input sequence is called *zero padding*.

8.3.9   Windowing

As discussed in Chapter 6, windows are specially designed time functions having finite duration. They are usually employed to shape the ends of the infinite-length data sequence in a form that reduces Gibbs oscillations. The length of a practical data record due to a certain measurement depends naturally on the period of observation and is sometimes lengthy. Truncating a sequence of data would naturally impair the precision of the spectrum resulting from taking the DFT for the record. One serious effect is known as *spectral leakage*. It usually occurs when adjacent spectra overlap, causing irreversible distortion in the spectrum. It might also cause some components to vanish. Because the spectrum of a rectangular window is a sinc function, $\sin x/x$, the resulting spectral lines of the DFT of the input data stream will be modulated by sinc functions. Thus, it is expected that the many side lobes of the repeated spectra around f_s and its multiples will overlap, causing such distortion (aliasing). The extent of this effect depends primarily on the main lobe width of the spectrum of the "virtu-ally" used rectangular window and the amplitudes of its side lobes (a finite-duration or a truncated sequence can be seen, as if it has been windowed by a rectangular win-dow). Spectral leakage could however be minimized, either by incorporating more samples in the record—that is, stretching the rectangular window so as to minimize the width of the main lobe—or, alternatively, by using a finite-length nonrectangular window.

As discussed in Chapter 6, window functions that do not end abruptly are recommended, as they cause minimum aliasing. Examples of popular windows in this respect are Hamming, Kaiser–Bessel, Dolph–Chebyshev, and Blackman–Harris. The selection of the proper window—or "window carpentry"—is a matter of experience. However, a good window is one that has a control parameter for adjusting the width of its main lobe and the amplitude of the side lobes. Besides, there are some other criteria that could be considered when deciding upon a specific window; among them are the resulting equivalent noise bandwidth, processing gain (PG), and main resolution bandwidth.

The *equivalent noise bandwidth* (ENBW) is defined as the bandwidth of an ideal BPF that passes a white noise having a mean square value that is equivalent to that due to the window. Windows that have smaller ENBW are recommended. The PG, on the other hand, is defined as the ratio of the SNR pre- and postwindowing. It depends mainly on the shape of the window.

Due to overlapping of adjacent spectra and the expected crossover, a reduction in the amplitudes of the spectrum off the bin frequencies is easy to detect. The maximum drop in the overall spectrum is referred to as the *scalloping loss* (SL). The drop reaches its maximum (dip) midway between the bins; that is, at the frequencies

$$f_k = (k + 0.5) \cdot \Delta f$$

Window functions other than rectangular ones do contribute less to the SL, as their main lobes are much wider than those of rectangular windows.

8.3.10 OVERLAP CORRELATION

Due to windowing, data at the ends of a sequence are modified; hence they become untrue. To compensate for this effect, windows are overlapped. The data record is divided into small sub-records, each of which contains N samples. A sliding window whose length is longer than N is then used to scan the record. As demonstrated in Figure 8.4, for a 50% overlap, the window length should be 1.5N.

FIGURE 8.4 Data overlapping.

EXAMPLE 8.1

Evaluate and plot the DFT for the sequence {1, 0, 1, 1} that results from sampling a certain speech utterance.

Solution

Because we are dealing with a speech signal, the sampling frequency should be 8 kHz, giving a sampling period of $T = 125$ μs. With $N = 4$, the bin spacing will be

$$\Delta f = \frac{f_s}{N} = \frac{8}{4} = 2\,\text{kHz}$$

The frequency components are given as before by

$$F(k\Omega) = \sum_{n=0}^{N-1} f_s(nT)e^{-jk\Omega nT} \quad \text{for } k = 0, 1, 2, ..., N-1$$

The twiddle factors can easily be calculated to be

$$W_4^0 = 1 = -W_4^2$$

and

$$W_4^1 = -j = -W_4^3$$

giving

$$X(0) = 1 + 1 + 0 + 1 = 3$$
$$X(2) = 1 + 0 - 1 + j = j$$
$$X(4) = 1 + 0 + 1 - 1 = 1$$
$$X(6) = 1 + 0 - 1 - j = -j$$

The results are plotted in Figure 8.5.

8.4 INVERSE FOURIER TRANSFORM

The inverse transform for the DFT is defined as

$$f(nT) = \frac{1}{N}\sum_{K=0}^{N-1} F(K\Omega)W_N^{-nK} \tag{8.8}$$

from which it is evident that it resembles the forward transform formula, except for the reversed sign of the exponent.

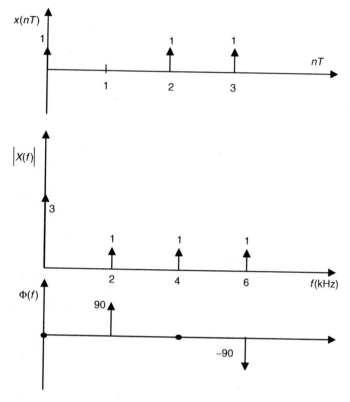

FIGURE 8.5   DFT result of Example 8.1.

EXAMPLE 8.2

Evaluate the inverse DFT for the results obtained in Example 8.1.

Solution

From Equation 8.4, we can write

$$f(nT) = \frac{1}{4}\sum_{K=0}^{3} F(K\Omega)W_4^{-nK} \quad n = 0, 1, 2, 3$$

giving

$$f(0) = \frac{1}{4}[3 + j + 1 - j] = 1$$

$$f(T) = \frac{1}{4}[3 + (j)(j) + 1(-1) + (-j)(-j)] = 0$$

$$f(2T) = \frac{1}{4}[3 - j + 1 + j] = 1$$

$$f(3T) = \frac{1}{4}[3 + (j)(-j) - 1 + (j)(-j)] = 1$$

8.5 FAST FOURIER TRANSFORM (FFT) ALGORITHM

The DFT of a sequence of N samples, studied so far (as described by Equation 8.6) is given by

$$\text{DFT}\{x(nT)\} = X(K\Omega) = \sum_{n=0}^{N-1} x(nT)W_N^{nK} \quad K = 0, 1, 2, \ldots, N-1$$

This means that for each frequency component (bin), N complex multiplications and $(N-1)$ additions are required; thus, for an N point transform, a total of N^2 complex multiplications and $N(N-1)$ additions are to be performed.

It is clear that as N gets larger, the necessary computational effort will increase exponentially. A means for reducing the number of operations is therefore needed. A popular tool in this respect is the FFT. It relies actually on exploring the cyclic property of the twiddle factors or W^i's that allows the elimination of redundant calculations, leaving only $N/2\log_2 N$ complex multiplications needed compared to N^2 multiplications, which means that a savings of $N^2 - N/2\log_2 N$ is achieved. Furthermore, it is evident that as N increases, the amount of savings increases dramatically, which reflects the power of the algorithm. In terms of real-time computations and for a certain sampling period T, the required number of complex multiplications per second (CMPS) is given by

$$\text{CMPS}_{\text{DFT}} = \frac{N}{T}$$

Using the FFT algorithm, it reduces to

$$\text{CMPS}_{\text{FFT}} = \frac{1}{2T}\log_2 N$$

That is, the achievable savings in the size of CMPS becomes more pronounced as N increases. To prove this, consider (for example) the cases of an 8-point transform and a 4096-point transform. In the first case, the percentage saved using the FFT algorithm is only 81.25%; for the second case, it amounts to 99.85% [18].

The algorithm was first introduced by Cooly and Tukey (1965) [19]; since then, it has become an indispensable tool for signal processing engineers. Let us now rewrite Equation 8.6 in matrix form as follows:

$$[X(K)] = \left[W_N^{nk}\right] \cdot [x(n)]$$

or

$$
\begin{bmatrix} X(0) \\ X(1) \\ X(2) \\ \vdots \\ \vdots \end{bmatrix} =
\begin{bmatrix}
W_N^0 & W_N^0 & W_N^0 & W_N^0 & \cdots & W_N^0 \\
W_N^0 & W_N^1 & W_N^2 & W_N^3 & \cdots & W_N^{N-1} \\
W_N^0 & W_N^2 & W_N^4 & W_N^6 & \cdots & W_N^{2(N-1)} \\
\cdots & \cdots & \cdots & \cdots & \cdots & \cdots \\
W_N^0 & W_N^{N-1} & W_N^{2(N-1)} & & \cdots & W_N^{(N-1)^2}
\end{bmatrix}
\begin{bmatrix} x(0) \\ x(1) \\ x(2) \\ \vdots \\ x(N-1) \end{bmatrix} \quad (8.9)
$$

Considering again the cyclic nature of the W's, as given before,

$$W_n^i = W_n^{i+N} = W_N^{i+2N}$$
$$W_N^{i+(N/2)} = -W_N^i$$
$$W_N^2 = W_{N/2}$$

Inspection of the matrix $[W]$, in view of these relations, reveals that it contains a lot of terms that are either equal in magnitude and phase or equal in magnitude and opposite in phase. Therefore, such terms could be calculated only once, resulting in the noticeable reduction in the computations just mentioned.

The basic idea of the algorithm is to perform the computation in m successive (cascaded) stages, where $m = \log_2 N$, provided that N is a multiple of 2; otherwise, a number of dummy zeros have to be added to the input samples (zero padding). In this way, the transform is decomposed into smaller transforms with much simpler calculations. The process of decomposition is called *decimation*. It could be done either in the time domain, by reordering the time samples, or in the frequency domain, by reordering the frequency components. A rather important property of the algorithm is its relatively limited memory requirements, by using the so-called *in-place* calculation. Here the results of a certain stage in the algorithm are used to replace the results of a preceding stage in the same array. This continuous update of the stored data in the memory saves the required storage space, allowing longer records to be processed.

In the *decimation-in-time algorithm* (DIT), the input time sequence of length N is split into even-ordered samples (x_0, x_2, x_4, ...) and odd-ordered samples (x_1, x_3, x_5, ...), each of length $N/2$. Then each of the two groups is further split into even-ordered and odd-ordered groups. The process is repeated m times until we get groups of two samples. This would result at the end in a new order of the input time samples with output frequency components that are ordered in the normal sequence. According to Equation 8.6, this reordering of the input samples should be accompanied by a corresponding reordering of the rows of the matrix $[W]$. The result is that the involved calculations are reduced to either the addition or subtraction of two numbers.

In the *decimation-in-frequency algorithm* (DIF), on the other hand, the opposite takes place, and the algorithm starts with two groups, each containing $N/2$ samples at the input side with normal sequence and ends up at the output side (frequency side) by reordered frequency components in groups of two, as we shall see later.

The mathematical operations involved in each stage can be illustrated by the so-called *butterfly diagram*, shown in Figure 8.6a for a width of 2 (radix-2). This butterfly diagram represents a basic unit in the signal flow graph describing the algorithm. Because the coefficient W is always multiplied by the second input x_2, we can reduce the number of multiplications by taking the multiplier outside the butterfly to take the form shown in Figure 8.6b.

A rule of thumb known as the *bit reversal algorithm* can be used to find the new order of the input time samples in the DIT algorithm or the order of the output

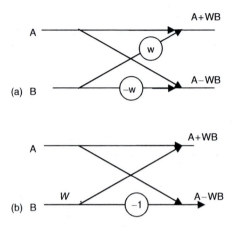

FIGURE 8.6 (a) Basic butterfly diagram, (b) modified butterfly diagram.

TABLE 8.1
Illustrating the Bit Reversal Algorithm

Decimal Index	Binary Index	Bit Reversed Binary Index	Bit Reversed Decimal Index
0	000	000	0
1	001	100	4
2	010	010	2
3	011	110	6
4	100	001	1
5	101	101	5
6	110	011	3
7	111	111	7

frequency components in the DIF algorithm. It can be summarized as follows:

1. Get the binary equivalent for the serially ordered input samples.
2. Take the middle bit as a pivot and replace the MSB's with the LSB's going from the outside to the inside. For even number of bits, consider an imaginary pivot at the middle and reverse bits as above.
3. Find the decimal equivalent of the new reversed binary representations of the orders.

The algorithm is illustrated in Table 8.1 for an 8-point case.

EXAMPLE 8.3

Find the new order of the sample number 12 in 16-point and 32-point sequences, respectively.

Answer

For binary representation of 16 samples, we need 4 bits.

Input index

8 4 2 1   8 4 2 1   Bit reversed index

(12)   ——→   ——→   ——→   (3)
          1 1 0 0   0 0 1 1

For the 32-point sequence, on the other hand, we need 5 bits.

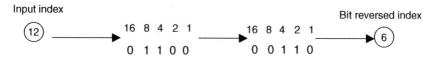

Input index   Bit reversed index

16 8 4 2 1   16 8 4 2 1

(12)   ——→   ——→   ——→   (6)
          0 1 1 0 0   0 0 1 1 0

General Properties of the Butterfly Diagram

- Butterflies of width 1×1 at the input or output of the diagram have inputs or outputs that are in bit reversal order.
- Butterflies having a width of $N/2 \times N/2$ have their inputs in sequence.
- DIT butterflies have their multipliers (weights) W's at their inputs, whereas those in DIF diagrams have their weights at their outputs.
- Butterfly diagrams of the DIF algorithm with inputs in bit reversal sequence are equivalent to those of the DIT diagram with input sequence in order with interchanged inputs and outputs; that is, reversing the direction from left to right.
- Butterfly diagrams of DIF algorithm with inputs in sequence are equivalent to the DIT diagram with inputs in bit reversal order with interchanged inputs and outputs.
- Butterfly diagram of the inverse fast Fourier transform (IFFT) 1 algorithm is equivalent to the DIF1 diagram with a reversed exponent sign of the twiddle factor and outputs divided by N. It is also equivalent to DIT1 with changed direction with the rest as before.

8.5.1 Decimation in-Time (DIT) Algorithm

As mentioned earlier, the DIT algorithm reduces the amount of calculations performed on the time samples. The input sequence of samples is therefore successively split into odd- and even-ordered halves until we get in the limit unit operations, containing only two terms. The order of the samples will accordingly be changed (reshuffled) as we have demonstrated. This could be expressed as

$$X(K\Omega) = \sum_{n=0}^{(N/2)-1} x(2n) W_N^{2nK} \blacktriangleleft + \sum_{n=0}^{(N/2)-1} x(2n+1) W_N^{(2n+1)K}$$

$$\vert\!\!\longleftarrow \text{ N/2 odd terms } \longrightarrow\!\!\vert\!\!\longleftarrow \text{ N/2 even terms } \longrightarrow\!\!\vert$$

Taking W^K out of the summation in the second term, we get

$$X(K\Omega) = \sum_{n=0}^{(N/2)-1} x(2n)\, W_N^{2nK} + W^K \sum_{n=0}^{(N/2)-1} x(2n+1)\, W_N^{2nK} \quad \text{N/2 even terms} \quad (8.10)$$

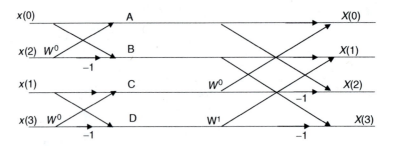

FIGURE 8.7   Butterfly diagram for a 4-point FFT.

Further splitting of the summations into odd and even terms will yield groups of two samples, or "unit butterflies." FFT algorithms are usually represented by their signal flow graphs containing butterfly units of different widths (number of samples included)—for example, 2, 4, 8, and 16 samples. Generally, for an N-point transform, the graph consists of m cascaded butterfly stages, where $m = \log_2 N$. The width of the butterfly in succeeding stages is continuously doubled; thus it starts with 1×1 in the first stage, then 2×2 in the second, and ends up with $N/2 \times N/2$. Also, it is to be noted that butterfly units in the DIT algorithm have their twiddle factors at their input side on the lower half and are associated with a multiplier of -1. Also the twiddle factors are raised to a power that increases gradually among butterflies in the same stage. The power increment for the ith stage is given by

Twiddle factor power increment of the ith stage

$$\Delta_p = \frac{N}{2^i} \quad i \neq 1$$

Also, the butterfly width of the ith stage is

$$BW_i = 2^{i-1}$$

A signal flow graph for a 4-point case is shown in Figure 8.7.

To calculate the output frequency components, we start by estimating the intermediate components A, B, C, and D at the output of the first stage, as follows:

$$A = x(0) + W^0 x(2)$$
$$B = x(0) - W^0 x(2)$$
$$C = x(1) + W^0 x(3)$$
$$D = x(1) - W^0 x(3)$$

The output components can similarly be calculated from

$$X(0) = A + W^0 \quad C \quad X(2) = A - W^0 C$$
$$X(1) = B + W^1 \quad D \quad X(3) = B - W^1 D$$

Recalling that $W^0 = 1$ and $W^1 = -j$, we can write

$$\begin{aligned}
X(0) &= x(0) + x(2) + x(1) + x(3) \\
X(1) &= x(0) - x(2) - jx(1) + jx(3) \\
X(2) &= x(0) + x(2) - x(1) - x(3) \\
X(3) &= x(0) - x(2) + jx(1) - jx(3)
\end{aligned} \qquad (8.11)$$

EXAMPLE 8.4

Calculate the FT for the sequence {1, 0, 0, 0, 0, 0, 0, 1} of a sampled audio signal at a rate of 8 k sample/s using the DIT–FFT algorithm.

Solution

The signal flow graph shown in Figure 8.8 can be used to get the transform. The resulting eight components can be represented by eight "bins" separated by the frequency $f_s/8 = 1$ kHz.

MATLAB® Help

The following MATLAB instructions can be used to check your results:

```
x = [1 0 0 0 0 0 0 1];    % input vector
y = fft(x)                 % Fourier transform FR (complex)
z = abs(y)                 % Absolute values
theta = angle(y);          % angles in rad/s
phi = theta*180/pi         % angles in degress
```

```
yielding
Columns 1 through 6
2.0000 1.7071 + 0.7071i 1.0000 + 1.0000i 0.2929 + 0.7071i 0
0.2929 - 0.7071i
Columns 7 through 8
1.0000 - 1.0000i 1.7071 - 0.7071i
The absolute values are
2.0000 1.8478 1.4142 0.7654 0 0.7654 1.4142 1.8478
with the angles
0 22.5000 45.0000 67.5000 0 -67.5000 -45.0000 -22.5000
```

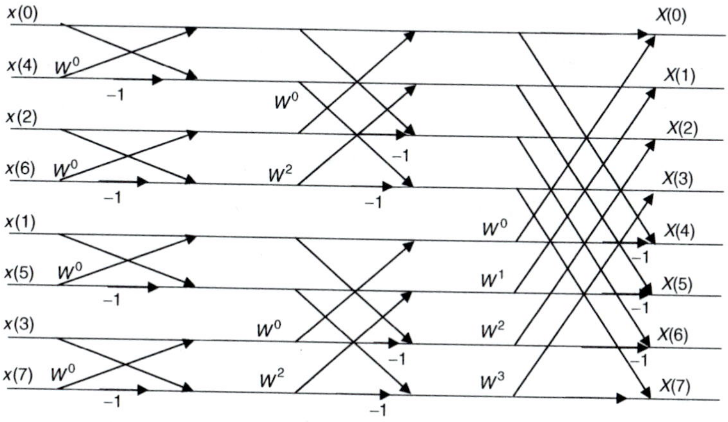

FIGURE 8.8 Butterfly diagram for Example 8.4.

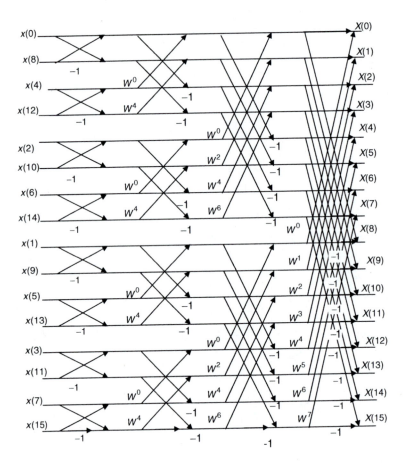

FIGURE 8.9   Signal flow graph for Example 8.5.

EXAMPLE 8.5

Sketch the butterfly diagram for a 16-point DIT–FFT algorithm.

Solution

Applying the bit-reversal algorithm discussed before, using 4 bits to represent the 16 points, we can easily find that the new order of the input samples are as follows:

(0, 8, 4, 12, 2, 10, 6, 14, 1,9, 5, 13, 3, 11, 7, 15)

The signal flow graph shown in Figure 8.9 will consist of m cascaded stages where m = $\log_2 16 = 4$.

EXERCISE 8.1

Calculate the twiddle factors W_{16}^i for the earlier butterfly diagram. Insert the results in a table.

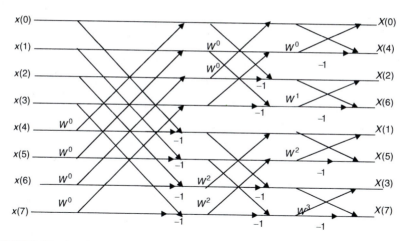

FIGURE 8.10 An alternative DIT–FFT algorithm (DIT_{II}–FFT).

An Alternative DIT–FFT Algorithm (DIT_{II}–FFT)

The same results could be obtained using a modified form of the butterfly diagram, where the input and output sides are exchanged. In such a case, the time samples remain in their ordinary sequence, whereas the output frequency components will be bit reversed; that is, they should be reordered. The resulting signal flow graph is illustrated in Figure 8.10.

EXERCISE 8.2

Find the DFT of the sequence {1, 1, 1, 1, 0, 0, 0} using the described two DIT–FFT algorithms and prove their consistency. Draw the amplitude and phase spectra, assuming a sampling rate of 10 kHz.

8.5.2 DECIMATION-IN-FREQUENCY (DIF) ALGORITHM

The butterfly diagram of the DIF–FFT algorithm is similar, to some extent, to the DIT one just described. The algorithm therefore delivers bit reversed frequency components. On the input side, however, the time samples are applied in their sequential order. Also, the twiddle factors, although following the same rules, are located at the output of each butterfly. The algorithm can be best described by its signal flow graph, as shown in Figure 8.11 for an 8-point case. From the diagram, it is easy to conclude that it is similar to the DIT_{II} graph, except for changing the position of the twiddle factors from the input to the output of the butterflies. The input time sequence is applied in its normal sequence to give outputs that are bit reversed. The two algorithms are almost identical, especially when considering the computational effort and storage requirements.

In general, when writing a code for either algorithm, it is recommended to precalculate all twiddle factors and store them in a lookup table to speed up calculations. Also, in going through the butterfly diagram it is better to apply the *in-place* calculations technique described earlier. It is worth mentioning here that among possible algorithms,

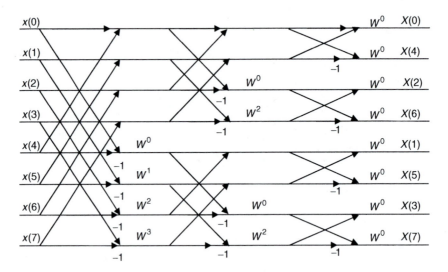

FIGURE 8.11 Signal flow graph for the DIF algorithm.

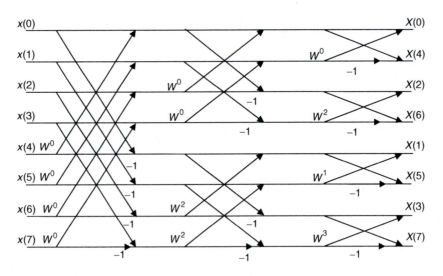

FIGURE 8.12 An alternative signal flow graph for the DIF algorithm.

the radix-2 FFT algorithm discussed here is the most popular one. However, radix-4 and radix-8 algorithms are especially recommended for hardware implementations, due to the possibility of adopting pipelining techniques [17, 20] (see also Chapter 9).

A rather different, yet equivalent, signal flow graph could be derived. As shown in Figure 8.12, it differs from the previous one in that the twiddle factors are located at the input of each butterfly stage and distributed differently as illustrated in the figure. The reader is asked to test the consistency of the results from both graphs.

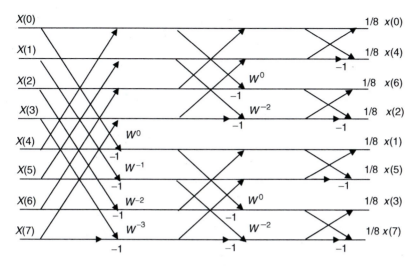

FIGURE 8.13   Butterfly diagram for an 8-point IFFT of the DIT algorithm.

EXERCISE 8.3

Using the DIF–FFT algorithm, recalculate the transform for the sequence given in Example 8.4. Compare the two results.

8.5.3   Inverse Fast Fourier Transform (IFFT)

Given the frequency spectrum of a certain time sequence, the IFFT is defined as

$$x(nT) = \frac{1}{N} \sum_{K=0}^{N-1} X(K\Omega) \cdot W_N^{-Kn} \quad n = 0, 1, 2, 3 \ldots, N-1 \qquad (8.12)$$

which is similar to Equation 8.9 of the forward transform, except for the negative power of the twiddle factors. A butterfly diagram for the inverse transform is illustrated in Figure 8.13. It is clear that it is almost similar to the one used to represent the DIF forward algorithm except for the sign change of the power of the twiddle factors and the division by N at the output. It is also equivalent to the butterfly diagram of the DIT algorithm, with the direction of the signal flow reversed. It was found that if

$$y = \sum_{i=1}^{m} x_i$$

then

$$y^* = \sum_{i=1}^{m} x_i^*$$

where the sign * stands for the conjugate; that is, the conjugate of a sum is the sum of the conjugates.

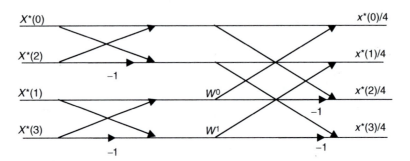

FIGURE 8.14   Butterfly diagram for Example 8.6.

Applying this rule to Equation 8.12, we get

$$[x(nT)]^* = \frac{1}{N}\left[\sum_{K=0}^{N-1} X(K\Omega) \cdot W_N^{-Kn}\right]^*$$

(8.13)

$$x^*(nT) = \frac{1}{N}\sum_{K=0}^{N-1} X^*(K\Omega) \cdot W_N^{Kn}$$

which indicates that the summand represents the DFT of the sequence $X^*(K\Omega)$. Thus from the previous calculations, we can conclude that the IDFT can be obtained as follows:

1. Get the complex conjugate of the input sequence $X(K\Omega)$.
2. Apply the sequence (bit reversed whenever necessary) to any FFT algorithm.
3. Get the complex conjugate of the resulting sequence.
4. Divide each output sample by N.

EXAMPLE 8.6

Find the IDFT for the results obtained in Example 8.3, using the forward FFT algorithm.

Solution

The frequency components obtained in Example 8.3 were $\{3, j1, 1, -j1\}$. Their complex conjugates are $\{3, -j1, 1, j1\}$.

From the butterfly diagram shown in Figure 8.14, we can write directly

$$x(0) = \{X^*(0) + X^*(2) + W^0[X^*(1) + X^*(3)]\}^* / 4 = \{3 + 1 - j1 + j1\}/4 = 1$$

$$x(1) = \{X^*(0) - X^*(2) + W^1[X^*(1) - X^*(3)]\}^* / 4 = \{3 - 1 - j(-j1 - j1\}/4 = 0$$

$$x(2) = \{X^*(0) + X^*(2) - W^0[X^*(1) + X^*(3)]\}^* / 4 = \{3 + 1 - j1 + j1\}/4 = 1$$

$$x(3) = \{X^*(0) - X^*(2) - W^1[X^*(1) - X^*(3)]\}^* / 4 = \{3 - 1 + j[-j1 - j1]\}/4 = 1$$

which is exactly the original input of Example 8.3.

General Properties of the Butterfly Diagrams

From the previous discussion, we can conclude the following properties of the butterfly diagram:

1. All butterflies of width 1×1 either at the output or the input have their input or output samples in bit reversal order.
2. Butterflies having widths of $N/2 \times N/2$ have their inputs or outputs in normal sequence.
3. Butterflies of the DIT have their weights W at the inputs, whereas those of the DIF have their weights at their outputs.
4. Usually twiddle factors and the multiplier -1 are associated with each row in the lower half of each butterfly unit.
5. The butterflies diagram for the IFFT algorithm are similar to those of the DIF diagrams, except that the exponents of the W's are negative.
6. The same forward butterfly diagram can be used to get the inverse transform, with some modifications at both the input and output sequences.

8.6 SOME OTHER TRANSFORMS

The DFT just described is the most commonly used one. Yet it is not the transform for all seasons. There are many other powerful transforms that are more efficient in some specific applications other than spectral analysis, high-speed convolution, and correlation computations. Examples of such transforms are the DCT, the Haar transform, the Walsh, the Walsh–Hadamard transforms, the wavelet transform, and many others. In this section, we highlight the properties of some of these transforms, especially those that are considered essential tools in key digital signal processing applications such as pattern recognition, data compression, and so on. The suitability of a transform for a certain application is decided by many factors that depend on the specific application. In data compression applications, for example, transforms are rated according to their *compression efficiency* η, defined as

$$\eta = \frac{\text{number of input samples} - \text{number of output samples}}{\text{number of input samples}}$$

expressing the degree of achievable compression. Furthermore, transforms are also compared according to the achievable *compression ratio* (CR) defined as

$$CR = 1 - \eta = \frac{\text{number of output samples}}{\text{number of input samples}}$$

However, the deciding factor for a certain transform, besides its compression efficiency, is the existence of a fast computational algorithm. It is considered in general as a crucial criterion in this respect.

8.6.1 Discrete Cosine Transform

The output components of a DFT are complex quantities. They contain real cosinusoidal terms and imaginary sinusoidal terms; that is, they give, in addition to the magnitude spectrum, a phase spectrum for the signal. However, in some applications, phase information is not of great importance, so one can omit it.

Let us consider again Equation 8.5, describing the DFT:

$$X_{FT}(k\Omega) = \sum_{n=0}^{N-1} x(nT) e^{-jkn\Omega T} = \sum_{n=0}^{N-1} x(nT)[\cos nk\Omega T - j \sin nk\Omega T]$$

$$\text{for } K = 0, 1, 2, \ldots, N - 1$$

Neglecting the imaginary part and dividing by N, we get a definition for the DCT as follows:

$$X_{CT}(k\Omega) = \frac{1}{N} \sum_{n=0}^{N-1} x(nT) \cos nK\Omega T \quad \text{for } K = 0, 1, 2, \ldots, N - 1 \qquad (8.14)$$

It is worth mentioning here that dividing each frequency component by N in the forward transform is a common practice in all transforms except the Fourier transform. Another definition for the DCT, due to Ahmed and Rao [21], is

$$X_{CT}(k\Omega) = \frac{1}{N} \sum_{n=0}^{N-1} x(nT) \cos\left(\frac{2\pi nK + K\pi}{2N}\right) \qquad (8.15)$$

which is a modified version of Equation 8.14 with ΩT replaced by $2\pi/N$ with an added angle of $K\pi/N$. The whole argument is then divided by 2. Similar to the idea of FFT algorithm, there are fast DCT algorithms; for example, the one due to Chen et al. (1977) [22].

DCT is used extensively in many signal processing applications, as we shall see in Chapter 10. Good examples are video compression in satellite communications to reduce the bit rates, speaker recognition, and diagnosis of heart and brain diseases. The DCT could be considered a near-ideal transform for the following reasons:

1. It is easy to compute (real terms).
2. It achieves high data CRs at minimum mean square errors.
3. It concentrates the signal energy in the low-frequency components so that high-frequency terms below a certain specified amplitude (threshold) can be neglected. This insures the reconstruction of the original signal with minimum distortion.

EXAMPLE 8.7

Calculate the DCT of the sequence {1, 0, 4, 1}; then estimate the compression efficiency η, assuming a threshold of ±0.8.

Solution

With $n = 4$, we can rewrite Equation 8.14 as

$$X_{CT}(k) = \frac{1}{4}\sum_{n=0}^{3} x(nT) \cdot \cos\left(\frac{\pi n K}{2}\right) \quad \text{for } K = 0, 1, 2, 3$$

$$= \frac{1}{4}\left[x(0) \cdot \cos 0 + x(T) \cdot \cos\frac{\pi K}{2} + x(2T) \cdot \cos \pi K + x(3T) \cdot \cos\frac{3\pi K}{2}\right]$$

$$= \frac{1}{4}\left[1 + 0 + 4 \cdot \cos \pi K + \cos\frac{3\pi K}{2}\right]$$

$$X(0) = \frac{1}{4}[1 + 0 + 4 + 1] = 1.5$$

$$X(1) = \frac{1}{4}[1 + 0 - 4 + 0] = -0.75$$

$$X(2) = \frac{1}{4}[1 + 0 + 4 - 1] = 1$$

$$X(3) = \frac{1}{4}[1 + 0 - 4 + 0] = -0.75$$

With the threshold set to ±0.8, the output sequence becomes {1.5, 0, 1, 0}. The achievable data compression efficiency is then

$$\eta = \frac{4 - 2}{4} \times 100 = 50\%$$

Alternatively, we can construct a butterfly diagram for these calculations, bearing in
mind that the twiddle factors are pure real values. A diagram for a 4-point transform is
illustrated in Figure 8.15. The input sequence is seen to be bit reversed.

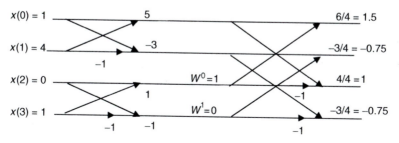

FIGURE 8.15 Butterfly diagram for a 4-point DCT.

8.6.2   DISCRETE WALSH TRANSFORM (DWAT)

Fourier transforms and its related transforms are based on analyzing their input sequences into harmonically related sinusoidal and co-sinusoidal terms. The Walsh transforms, on the other hand, analyze them into pulse-like (rectangular) waveforms that are also in one way or another harmonically related. Such waveforms can assume only two distinct levels—either +1 or −1. Compared to the DFT, it should be faster to compute and need smaller storage sizes. Due to the shape of its waveforms, it is more suitable for data sequences containing discontinuities. Therefore, it lends itself to applications like image processing, due to the inherent abrupt changes at the edges and contours of the picture contents.

Because the kernel (mother) functions are nonperiodic rectangular functions (Walsh functions), the term "frequency" is not applicable here. A rather more indicative term, *sequency*, is used. It gives actually the number of zero crossings (ZC)—that is, the number of sign changes—over the interval of interest. A Walsh function is usually abbreviated as WAL(n, t), where n is the sequency. Inspection of Figure 8.16 reveals that there are equal numbers of odd- and even-indexed functions exactly, as in the case of FT, where we get equal numbers of sin and cos terms. Thus, to make it easy to differentiate between even and odd Walsh functions, the abbreviations CAL(k, t) and SAL(k, t) have been coined to them where the letter C stands for even (co-sinusoidal) terms and the letter S for the odd (sinusoidal) terms. Thus, we can write

$$\text{WAL}(2k, t) = \text{CAL}(i, t) \quad \text{where } i, k = 0, 1, 2, \ldots, (N/2)-1$$

and

$$\text{WAL}(2k+1, t) = \text{SAL}(j, t) \quad \text{where } j = 1, 2, 3, \ldots, N/2$$

The forward discrete Walsh transform (DWAT) is defined as

$$X(k) = \frac{1}{N} \sum_{i=0}^{N-1} x(nT)\,\text{WAL}(k, i) \quad k = 0, 1, 2, \ldots, (N-1) \qquad (8.16)$$

whereas the inverse discrete Walsh transform (IDWAT) is given by

$$x(nT) = \sum_{i=0}^{N-1} X(k)\,\text{WAL}(k, i) \quad i = 0, 1, 2, \ldots, N-1 \qquad (8.17)$$

Recalling that the function WAL(k, i) can assume only the values +1 or −1, we can conclude that both of the forward and inverse transforms are equivalent, except for the constant $1/N$.

Due to their nature, both transforms could be better calculated through matrix multiplications as follows:

$$[X(k)] = \frac{1}{N}[x(iT)] \cdot [W_{ki}] \qquad (8.18)$$

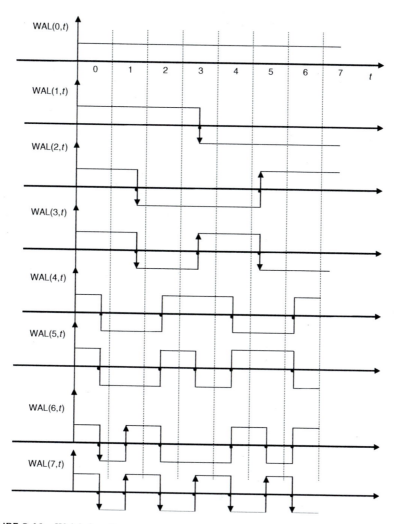

FIGURE 8.16 Walsh functions.

where

$\quad x(iT)$ = input vector (a row matrix)
$\quad W_{ki}$ = Walsh-transform matrix (a square matrix).

It is to be noted here that there exists a fast DWAT algorithm also. The matrix W can be derived from the waveforms given in Figure 8.16. It is given here as well, for the simple case of a 4-point transform:

$$W_4 = \begin{bmatrix} 1 & 1 & 1 & 1 \\ 1 & 1 & -1 & -1 \\ 1 & -1 & -1 & 1 \\ 1 & -1 & 1 & -1 \end{bmatrix}$$

From the waveforms and the matrix W, it is easy to notice that row number 0 has no ZC, because there is no sign change (direct current [DC] component), and row 1 has one ZC (fundamental), row 2 has two ZC's (second harmonic), and row 3 has three ZC's. This conclusion could be helpful when checking the correctness of the matrix—especially for a large number of points.

Alternatively, the Walsh matrix of any dimension can be derived from the basic 2×2 matrix:

$$W_b = \begin{bmatrix} 1 & 1 \\ 1 & -1 \end{bmatrix}$$

For example, to get a 4×4 matrix, we simply use four basic matrices as follows:

$$W_4 = \begin{bmatrix} 1 & 1 & 1 & 1 \\ 1 & -1 & 1 & -1 \\ 1 & 1 & 1 & 1 \\ 1 & -1 & 1 & -1 \end{bmatrix}$$

Changing the signs of the elements inside the hatched square (core matrix), we get

$$W_4 = \begin{bmatrix} 1 & 1 & 1 & 1 \\ 1 & 1 & -1 & -1 \\ 1 & -1 & -1 & 1 \\ 1 & -1 & 1 & -1 \end{bmatrix}$$

which is the Walsh matrix $W4$ for a 4-point transform. The matrix W, being a square matrix, can be decomposed into four square submatrices—namely W_0, W_{R2}, W_{C2}, and W_D—as follows:

$$W_4 = \begin{bmatrix} W_0 & W_{R2} \\ W_{C2} & W_D \end{bmatrix}$$

where W_0 is a submatrix with *no (zero) negative ones*, W_{R2} is a submatrix whose second row (R_2) is made of negative ones, W_{C2} is a submatrix with its second column

(C_2) containing negative ones, and finally W_D is a submatrix that has negative ones at its *principal diagonal* (D). In general, for any dimension, each of the submatrices can be subdivided into four smaller square matrices, as shown here. The process is repeated until the basic matrix (2×2) is reached. The Walsh functions can be deduced from the waveforms illustrated in Figure 8.16.

To get the inverse discrete Walsh transform (IDWAT), the same matrix W can be used as follows:

$$\text{IDWAT}\{X(k)\} = x(iT) = [X(k)] \cdot [W] \tag{8.19}$$

The following example demonstrates the ease of computations required to get both the forward and inverse Walsh transforms.

EXAMPLE 8.8

Calculate the DWAT for the sequence {1, 0, 4, 1}, and then derive its inverse.

Solution

According to Equation 8.17,

$$X(k) = \frac{1}{4}[1 \ 0 \ 4 \ 1] \cdot \begin{bmatrix} 1 & 1 & 1 & 1 \\ 1 & 1 & -1 & -1 \\ 1 & -1 & -1 & 1 \\ 1 & -1 & 1 & -1 \end{bmatrix}$$

$$= \frac{1}{4}[6 \ -4 \ -2 \ 4] = [1.5 \ -1 \ -0.5 \ 1]$$

The inverse transform can be derived by the same way using the same matrix as follows:

$$x(nT) = [1.5 \ -1 \ -0.5 \ 1]$$

$$\begin{bmatrix} 1 & 1 & 1 & 1 \\ 1 & 1 & -1 & -1 \\ 1 & -1 & -1 & 1 \\ 1 & -1 & 1 & -1 \end{bmatrix}$$

Giving

$$x(nT) = [1 \ 0 \ 4 \ 1]$$

EXAMPLE 8.9

Derive the Walsh matrix W_8 for an 8-point transform.

Solution

We repeat the matrix W_4 four times and change the signs of the elements of the core matrix as before.

$$
{}^8W = \begin{bmatrix}
1 & 1 & 1 & 1 & 1 & 1 & 1 & 1 \\
1 & 1 & -1 & -1 & 1 & 1 & -1 & -1 \\
1 & -1 & 1 & -1 & -1 & 1 & 1 & 1 \\
1 & -1 & -1 & 1 & -1 & 1 & 1 & -1 \\
1 & 1 & -1 & -1 & -1 & -1 & 1 & 1 \\
1 & 1 & 1 & 1 & -1 & -1 & -1 & -1 \\
1 & -1 & 1 & 1 & 1 & -1 & -1 & 1 \\
1 & -1 & 1 & -1 & 1 & -1 & 1 & -1
\end{bmatrix}
$$

8.6.3 DISCRETE WALSH–HADAMARD TRANSFORM (DWHT)

The discrete Walsh–Hadamard transform is a modified version of Walsh transform, in which the rows of the transform matrix are reordered. The basic (2×2) matrix ${}^2H$ has the form

$$
{}^2H = \begin{bmatrix} 1 & 1 \\ 1 & -1 \end{bmatrix}
$$

Thus, a matrix of $2N \times 2N$ is expressed as

$$
{}^{2N}H = \begin{bmatrix} {}^NH & {}^NH \\ {}^NH & -{}^NH \end{bmatrix}
$$

EXAMPLE 8.10

Deduce the WH matrix for the case of a 4-point transform. Compare it with a similar matrix of the Walsh transform.

Solution

Following the above rule, we can write

$$
{}^4H = \begin{bmatrix} {}^2H & {}^2H \\ {}^2H & -{}^2H \end{bmatrix} = \begin{bmatrix} 1 & 1 & 1 & 1 \\ 1 & -1 & 1 & -1 \\ 1 & 1 & -1 & -1 \\ 1 & -1 & -1 & 1 \end{bmatrix}
$$

Comparing the above matrix with that of the Walsh transform, we can conclude that it includes the same elements, except that rows are reordered as shown in Table 8.2.

TABLE 8.2
Relation between Walsh and Walsh–Hadamard Matrices

Walsh Matrix	Walsh–Hadamard
Row 0	Row 0
Row 1	Row 3
Row 2	Row 1
Row 3	Row 2

TABLE 8.3
Reordering the Walsh Matrix to Construct the Walsh–Hadamard Matrix

Row	Binary-Coded	Bit Reversed Binary	Gray-Coded Order	New Order
0	0 0 0	0 0 0	0 0 0	0
1	0 0 1	1 0 0	1 1 0	6
2	0 1 0	0 1 0	0 1 1	3
3	0 1 1	1 1 0	1 0 1	5
4	1 0 0	0 0 1	0 0 1	1
5	1 0 1	1 0 1	1 1 1	7
6	1 1 0	0 1 1	0 1 0	2
7	1 1 1	1 1 1	1 0 0	4

A rule can then be derived that gives the WH matrix from the corresponding W matrix of any order as follows:

1. Express the rows indices of the W matrix in binary representation.
2. Bit reverse the binary number (exchange the LSB's with the MSB's).
3. Convert the resulting bit reversed binary numbers into corresponding gray-coded ones.
4. Get the equivalent decimal number again.

To get the gray-coded equivalent of binary-coded indices, one should follow the next steps:

1. Keep the MSB unchanged.
2. EX-OR each two adjacent.

The reordering procedure of the Walsh matrix is illustrated for the case of an 8-point transform in Table 8.3.

Just like the case of DFT and the DCT transforms, we can draw a signal flow graph (butterfly diagram) describing the calculations. As an example, we consider again the 4-point sequence shown in Figure 8.17. The result of the Butterfly diagram is

$$X_{WH}(0) = X_W(0) = x(0) + x(1) + x(2) + x(3)$$
$$X_{WH}(1) = X_W(3) = x(0) + x(1) - x(2) - x(3)$$
$$X_{WH}(2) = X_W(1) = x(0) - x(1) - x(2) + x(3)$$
$$X_{WH}(3) = X_W(2) = x(0) - x(1) + x(2) - x(3)$$

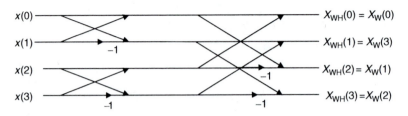

FIGURE 8.17 Butterfly diagram for a 4-point DWHT.

From these results, it is easy to conclude that the Walsh and the Walsh–Hadamard transforms yield the same results, but differently ordered. It is also evident that like FFT, there exists also a fast Walsh transform and a fast Walsh–Hadamard transform as well. The only difference is that the Walsh–Hadamard transform could be calculated a little bit faster than the Walsh transform, due to the reordering of calculations and much faster than the FFT [4], because it deals with pure real data.

EXAMPLE 8.11

Deduce the Walsh–Hadamard transform matrix for an 8-point case.

Solution

The matrix can be constructed as follows:

$$
{}^{8}W_{H} =
\begin{bmatrix}
1 & 1 & 1 & 1 & 1 & 1 & 1 & 1 \\
1 & -1 & 1 & -1 & 1 & -1 & 1 & -1 \\
1 & 1 & -1 & -1 & 1 & 1 & -1 & -1 \\
1 & -1 & -1 & 1 & 1 & -1 & -1 & 1 \\
1 & 1 & 1 & 1 & -1 & -1 & -1 & -1 \\
1 & 1 & 1 & 1 & -1 & 1 & -1 & 1 \\
1 & 1 & 1 & 1 & -1 & -1 & 1 & 1 \\
1 & -1 & -1 & 1 & 1 & 1 & 1 & -1
\end{bmatrix}
$$

EXAMPLE 8.12

Given the sequence {1, 2, 0, 3}, calculate:

1. The DFT
2. The DCT
3. The DWAT
4. The discrete Walsh–Hadamard transform (DWHT)

Calculate also the power spectral components for each case.

Solution

Using these rules, we get the following results:

DFT { 6, 1+j, −4, 1−j}, power spectral components are {36, 2, 16, 2}
DCT {1.5, 0, −1, 0.5}, power spectral components are {2.25, 0, 1, 0.25}
DWAT {1.5, 0, 0.5, −1}, power spectral components are {2.25, 0, 0.25, 1}
DWHT {1.5, −1, 0, 0.5}, power spectral components are {2.25, 1, 0, 0.25}

8.6.4 TIME–FREQUENCY TRANSFORM

The so far considered transforms decompose the signals, as shown before, into either sinusoidal, co-sinusoidal, or rectangular waveforms. The FT, for example, is obtained by the inner product of the signal with sinusoidal functions of infinite duration. Being time-independent, such transforms deliver pure frequency-domain descriptions of the signal they represent. They deliver no information about the time such frequencies occurred. This means that any localized time variation (an abrupt change) in the signals could not be identified; it is rather spread out over the whole frequency axis. Therefore, such transforms are suitable for transforming stationary signals only; that is, signals whose statistical properties do not change with time [23]. In this section, a new class of transforms—namely, time–frequency transforms— are considered where the resulting spectrum contains information about the time a certain frequency of the spectrum occurs. A three-dimensional representation is depicted in Figure 8.18.

8.6.5 SHORT-TIME FOURIER TRANSFORM (STFT)

To get time–frequency representation of a certain sequence of data that contains sudden and abrupt changes, a new tool is needed that can deal with nonstationary signals by transforming them into a joint time-frequency domain. This could be

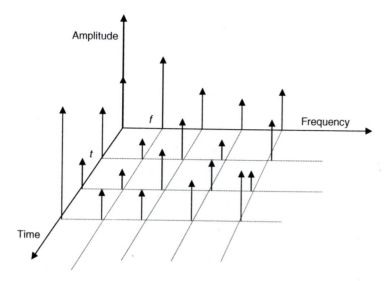

FIGURE 8.18   Illustrating the time–frequency relation.

achieved by introducing time dependency in the conventional Fourier analysis while maintaining linearity. The transform will look at the signal through a time window that should be short enough that the signal could be considered approximately stationary. One example of such a tool is the so-called *short-time Fourier transform* (STFT), which is defined as

$$STFT(\tau, \omega) = \int f(t) \cdot g(t - \tau)e^{-j\omega t} \cdot dt \qquad (8.20)$$

It is actually a modified form of the conventional FT where $g(t)$ is a finite and constant duration sliding window function with τ representing the time shift. Scanning the signal $f(t)$ by shifting this window on the time axis provides successive overlapped transforms that describe the spectrum as a function of time, thereby giving a three-dimensional (3D) representation of the spectrum. However, as said before, this can be valid only if the signal remains stationary over the duration of the window. To ensure this condition, the duration of the window should be as short as possible. Thus the sinusoidal basis functions should accordingly be modified to be more concentrated in time, giving rise to the STFT kernel. Such basis functions are in essence windowed sinusoids having the same duration, different frequency ω_i and time τ_i shift.

Because we are dealing with time–frequency description of signals, we have to define the terms *time resolution*, Δt, *frequency resolution*, Δf. The time resolution Δt of a transform is given by the length (duration) of its window. A short window comprises fewer time samples; the obtainable number of discrete frequencies are proportionately small, resulting in a bad *frequency resolution* Δf. According to the definition of time resolution, two bursts (pulses) in a function $f(t)$ can be detected by the transform if they are more than Δt apart. Also, two frequency components could be discriminated if they are separated by more than Δf, on the frequency scale. It follows from this rule that one has to trade off either the frequency resolution Δf for time resolution Δt or vice versa. Both resolutions are related by the uncertainty (Heisenberg) principle:

$$\Delta f \cdot \Delta t \geq \frac{1}{4\pi} \qquad (8.21)$$

This inequality states that it is impossible to know what frequencies exist at what times. All we can conclude are the time intervals during which a certain band of frequencies do exist. In such a multiresolution analysis, once the resolutions are determined for a certain analysis, they remain unaltered throughout the process, giving rise to uniform resolution cells in the time–frequency plane shown in Figure 8.19. The basis functions (kernel) of the STFT are windowed sinusoids.

8.7 WAVELET TRANSFORM

The STFT just described fails to provide good time and frequency resolutions simultaneously, as it uses fixed-length windows. A new transform has recently emerged [24]; namely, the *wavelet transform*. It employs basis functions (windows) having variable lengths; that is, they could be stretched (dilated) or compressed. This property allows the selection of a window duration that fits the frequency range of

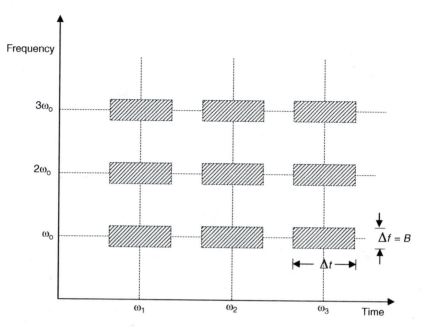

FIGURE 8.19 Time–frequency representation of the STFT.

interest. In other words, one can change the time and frequency resolutions. This means that for low frequencies, the window should be wide enough to accommodate a large number of samples (improved frequency resolution). On the other hand, at high frequencies, the windows should be narrow enough to have good time resolution. This means that such multiresolution analyses are designed to provide good time resolution and modest frequency resolution for high frequencies, while at low frequencies the opposite is valid; that is, they show good frequency resolution and poor time resolution. The situation is made clear in Figure 8.20. The "windowed" basis functions are called *wavelets*. A family of wavelets can be derived from a single "mother" wavelet by dilation/compression and translation; that is, scaling and shifting. Any signal can then be constructed by performing a weighted sum of translated/scaled wavelets. Wavelet transform thus provides time–frequency representation of a given signal, as it gives simultaneously time and frequency information about the signal it represents. It is now considered to be one of the most powerful tools in multiresolution analysis for nonstationary signals.

Though time analysis is performed employing a contracted high-frequency version of a basis function, frequency analysis can be obtained using a dilated, low-frequency version of the same wavelet. Wavelet analysis of signals can then be done in terms of wavelet coefficients as linear combinations of wavelet functions. A prototype wavelet can be seen as a band pass filter with a constant Q factor; that is,

$$Q = \frac{\omega_i}{BW_i} = \text{constant}$$

where BW_i is the bandwidth and ω_i is the center frequency.

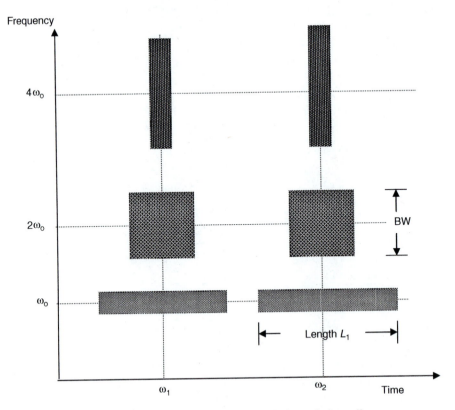

FIGURE 8.20 Time–frequency plane illustrating wavelet's resolution cells.

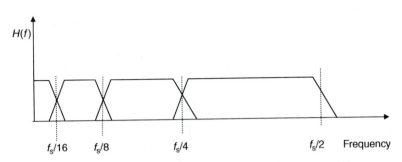

FIGURE 8.21 Frequency responses of constant-Q filters.

Thus to ensure constancy of Q, one should allow Δf and Δt to vary simultaneously, giving rise to a multiresolution analysis. This could be demonstrated by the shifted frequency response of the equivalent analysis filters shown in Figure 8.21.

As the central frequency increases, so does the bandwidth proportionally, to keep a constant Q and hence higher time resolution. In wavelet theory, the term "scale" is used in place of "frequency," such that the transform is said to be time–scale rather than time–frequency representation. The symbols "a" and "b" will be used from now on to replace the frequency ω and the time shift τ, respectively.

8.7.1 Wavelet Kernel

The kernel of a wavelet transform is represented by

$$K_{a,b}(t) = \frac{1}{\sqrt{a}} g^* \left(\frac{t - b}{a} \right) \tag{8.22}$$

where $g^*(t)$ is a zero-mean band–pass function that is scaled by a and time-shifted (translated) by b, and the term $1/\sqrt{a}$ is used to keep the energy content of the wavelet constant, irrespective of the scale. The question now is: What kind of functions can be used as mother wavelets? Actually, any oscillatory function that decays rapidly towards zero and has a zero average (zero DC component) is an "admissible" function as a mother wavelet. However, there are a limited number of mother wavelets that have proven to be useful. The most important of them will be discussed later in this chapter.

8.7.2 Continuous Wavelet Transform

The continuous wavelet transform (CWT) is defined as

$$\text{CWT}(a, b) = \frac{1}{\sqrt{a}} \int g^* \left(\frac{t - b}{a} \right) \cdot f(t)\, dt \tag{8.23}$$

from which one can conclude that the wavelet transform represents signals by decomposing them into many components that are derived from a single mother wavelet through shifting and scaling. Figure 8.22 depicts some popular wavelet functions such as the Morlet, Coiflet, Haar, Mexican Hat, and Daubechies. Compressed and dilated versions of the Morlet wavelet (a Gaussian-weighted function) are illustrated in Figure 8.23.

In contrast to the STFT basis functions, which have equal durations (constant length window) and a variable frequency, the mother wavelets of the wavelet transform, due to scaling, do have different durations that decrease as the frequency increases. Such an adaptive-length property improves greatly the performance of the transform, thereby making it ideal for analyzing signals having short duration spikes as well as low-frequency components lasting for long periods.

8.7.3 Applications and Implementations

The wavelet transform is now being used in a multitude of applications, such as data compression, image processing, pattern recognition, de-noising of signals, biomedical signal analysis for diagnosis, and so on. The CWT is usually implemented in discrete time, giving rise to the discrete-time wavelet transform (DWT). The band pass property of the mother wavelets is implemented by parallel connection of a low-pass and a high-pass filter whose frequency responses are complementary, as depicted in

Figure 8.24, They are usually designated as *quadrature mirror filters* (QMF). The following relationships should be valid:

$$G\overline{H} = H\overline{G} = 0$$

$$G\overline{G} = H\overline{H} = I$$

The filters, being designed as QMF, split the input signal equally into low-frequency components describing a coarse *approximation* and high-frequency components giving *detailed* information. A bit rate reduction (decimation) by a factor of 2 is followed, aiming at reducing the redundant data of each stream without losing any significant information.

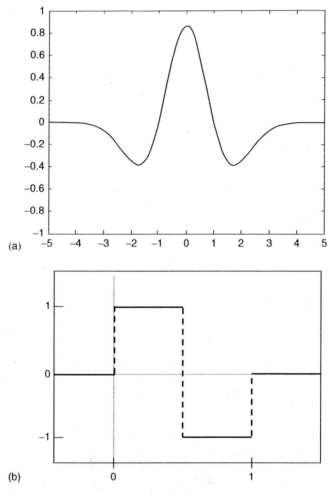

FIGURE 8.22 Some important mother wavelet functions.

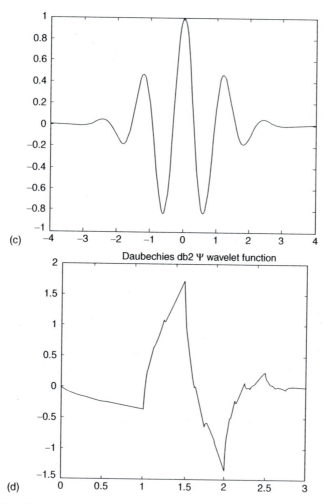

(c)

(d)

FIGURE 8.22   (Continued)

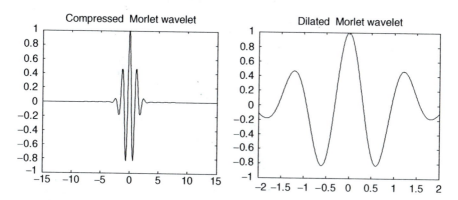

FIGURE 8.23   Compressed and dilated versions of Morlet wavelet.

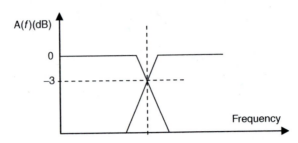

FIGURE 8.24 Illustrating the frequency responses of QMF.

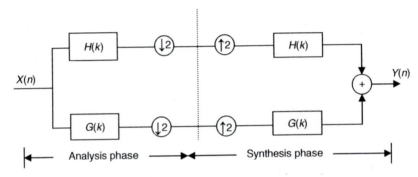

FIGURE 8.25 Wavelet analysis and synthesis.

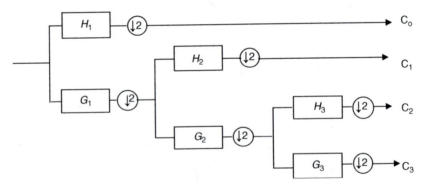

FIGURE 8.26 Malat tree structure for the analysis, illustrating successive dilation.

A direct implementation of the transform is shown in Figure 8.25 and is detailed in Figures 8.26 and 8.27.

8.7.4 COMPUTATION OF THE DWT

The octave-band-based analysis due to decimation could be made finer by subdividing the octaves into several "voices." This is possible because there are no constraints on the scale parameters a. The necessary computations for calculating the transform

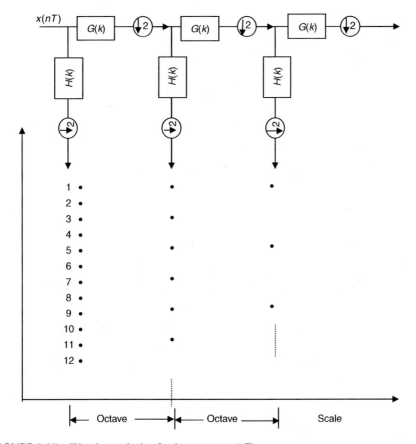

FIGURE 8.27   Wavelet analysis of a data stream $x(nT)$.

in a tree structure, as shown in Figure 8.26, are great. Ordinary filter operations and dilation realized through decimation result in nearly

$$4N\left(1 - \frac{1}{2^K}\right)\text{operation/point}$$

where
   N = number of samples (points)
   K = number of octaves

Using a grid of M voices/octave needs approximately $4.N.M.K$ operations/point.

   An efficient means to reducing this extensive amount of computation is possible via FFT. Through FFT, convolution operations required to calculate the DWT are turned into simple multiplication, as follows:

$$\text{FFT}\{\text{CWT}(a,b)\} = \frac{1}{\sqrt{a}} \cdot G^*\left(\frac{\omega}{a}\right) \cdot F(\omega)$$

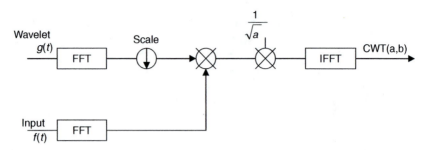

FIGURE 8.28   Fast wavelet transform algorithm.

in which $G^*(\omega)$ and $F(\omega)$ are the FT's of the wavelet function $g(t)$ and the input signal $f(t)$, respectively. The idea of the algorithm is illustrated in Figure 8.28.

   This FFT-based wavelet transform clearly requires a reduced number of calculations, so it could be called "fast wavelet transform."

8.7.5   WAVELET PACKETS

Decomposing both sides of the Malat tree, given in Figure 8.26—that is, zooming on both the low-frequency and the high-frequency components in a DWT—results in what are called *wavelet packets*. This kind of wavelet transform now finds many applications.

8.8   SUMMARY

Transforming time domain signals into any frequency domain and vice versa is an important, useful, and sometimes indispensable tool in many DSP applications. This chapter presents a comprehensive treatment of almost all popular transforms. After introducing the problem, some typical applications are discussed that incorporate, as an integral part, one or more transform. The traditional discrete Fourier transform (DFT), being the most popular transform, is presented together with some definitions and basic properties. The fast Fourier transform (FFT) as an efficient computational algorithm for the DFT is then described using signal flow graphs. Two versions of the algorithm are introduced; namely, the decimation-in-time and the decimation-in-frequency algorithms. The inverse transform is then presented, where a technique that allows the use of the forward transform for getting the inverse transform is described. The widely used discrete cosine transform (DCT) as a special case of the DFT is next presented. Other transforms, such as the Walsh, Walsh-Hadamard, and so on, are described together with their unique properties. The short time Fourier transform is then described as an elementary multiresolution transform. The emerging "wavelet transform" is presented as an important tool for multiresolution analysis for nonstationary signals where some basic definitions are provided. The concept of mother wavelets is presented alongside several examples of some prominent mother wavelets. Methods of implementation of the transform and some of its applications are then described. A hint about wavelet packets finally concludes the chapter.

8.9   REVIEW QUESTIONS

1. What are the advantages of transforming time functions into the frequency domain?
2. Why is it instrumental to subject a time series representing functions to tapered functions (windows) before transforming to the frequency domain?
3. What are the main criteria that are taken into consideration when comparing different transforms?
4. What are the main differences between the STFT and the wavelet transform?

8.10   PROBLEMS

1. Find the DFT for the sequence {1 0 1}. Sketch the spectrum (magnitude and phase), assuming a sampling frequency of 10 kHz.
2. Find the DFT for the sequence {1 0 0 0 0 0 0 1}, using the DIT–FFT algorithm, and assuming a sampling frequency of 8 kHz. Verify your results using MATLAB.
3. Calculate the DFT for the series {2 2 0 0}, using the DIT–FFT algorithm.
4. Sketch the butterfly diagram for a 4-point FFT algorithm, and then use it to find the time sequence whose spectrum is [4, $-2 + j2$, 0, $-2 - j2$].
5. Using the DIT–FFT algorithm, find the spectrum of the sequence {1 0 1 0 1 0 1 0}. Verify your results using MATLAB.
6. Using the forward DIT–FFT algorithm, find the IDFT for the shown signal spectrum.

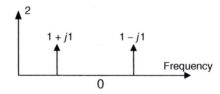

7. Find the DFT for the sequence {1 1 1 1 0 0 0 0}, using the DIT–FFT algorithm. Sketch the corresponding signal flow graph illustrating the twiddle factors on the butterfly diagrams.
8. Draw a butterfly diagram for an 8-point DIT–FFT algorithm, then use it to get the components $X(0)$ and $X(6)$ in the sequence {0 1 2 3 3 2 1 0}.
9. Find the DFT for the sequence {1 0 0 0 0 0 0 1}, using the DIF–FFT algorithm, assuming a sampling frequency of 8 kHz.
10. Calculate the DFT for the series {1 1 0 0}, using the DIT–FFT algorithm.
11. Sketch the butterfly diagram for a 4-point IFFT algorithm, and then use it to find the time sequence whose spectrum is [5, $0 - j1$, -1, $0 + j1$].
12. Using the DIT–FFT algorithm, find the spectrum of the sequence {1 1 0 0 1 1 0}, assuming a sampling frequency of 16 kHz.
13. Deduce an 8-point Walsh transform matrix from the basic 2 × 2 matrix, then sketch it.
14. From the result in Problem 13, using the reordering rules, deduce an equivalent 8-point Walsh–Hadamard transform matrix and its related waveforms.

8.11 MATLAB® PROJECT

Consider an image of your choice from the MATLAB database. Apply the following two-dimensional (2D) transforms:

1. DFT
2. DCT
3. DWT

Select a proper threshold, then calculate the obtainable percentage of compression for each case.

9 Digital Signal Processors

9.1 INTRODUCTION

Conventional processors are optimized for general-purpose computations. Their speed capabilities cannot cope with the high speed requirements of most signal processing operations. Therefore, special high-speed hardware has been developed to match signal processing needs. Almost all mathematical operations that are employed in signal processing applications are limited to multiplication and addition. Other operations, like logical and memory management tasks, have modest speed requirements. Specially tailored processors for DSP applications naturally aim at real-time processing. Low-level languages like Assembly or higher-level languages such as C or C++ are usually used.

Data processing is performed either as:

1. *Stream (batch) processing*, in which one sample at a time is processed, as in digital filters, for example.
2. *Block processing*, in which strings of data are processed simultaneously, as in FFT calculations.

DSP processors are classified into general-purpose processors and special-purpose processors.

9.1.1 GENERAL-PURPOSE PROCESSORS

Unlike ordinary processors found in most PCs, laptops, desktops, and so on, general-purpose digital signal processors are tailored to perform almost all DSP operations, yet at moderate speeds. Examples of existing general-purpose DSP processors are the TMS320C25 from Texas Instruments and the ADSP56000 from Motorola.

9.1.2 SPECIAL-PURPOSE PROCESSORS

This breed of processors are specially designed and optimized to perform a single specific DSP job at the maximum possible computational efficiency. They are classified into:

1. *Algorithm-specific processors.* Tailored to perform specific DSP algorithms, such as FFT.
2. *Application-specific processors.* Designed for very specific DSP applications.

9.2   BASIC ARCHITECTURE OF DSP PROCESSORS

Conventional Von-Neumann processor architecture, depicted in Figure 9.1, is the most commonly used architecture in almost all processor systems. It relies on sequential operation of the individual blocks, causing some components to remain idle while others are active. Therefore, to enhance the speed, faster components should be employed. However, there is certainly a limit imposed by the available current technology as well as the cost.

On the other hand, processors designed to perform DSP tasks should have an optimized architecture that matches real-time operation. They should feature:

- Separate buses for data and addresses
- Separate memory spaces for data and programs
- An input/output port containing interfacing devices such as ADCs and DACs for data exchange
- Possibility of direct memory access (DMA) that provides fast transfer of data blocks by bypassing the processor
- A computational unit for arithmetic and logical operations (ALU) and a hardware multiplier to enhance the speed.

As the name implies, data memory is required to store input and output data, to hold intermediate data values, to store coefficients of repeated algorithms, and so on. Program memory, on the other hand, holds the instruction sets for performing the DSP algorithm. Almost all DSP algorithms include repetitive arithmetic operations like multiplication and addition.

Therefore, most DSP processors contain, as an integral part, a MAC, which performs all mathematical operations. Thus, to get a DSP processor with an optimized speed performance, one has to optimize not only its architecture but its instruction set as well. A typical architecture of a DSP processor is illustrated in Figure 9.2.

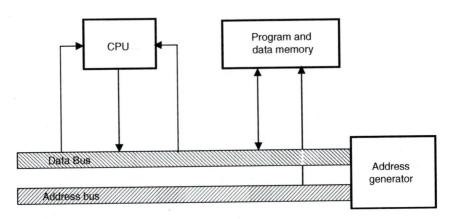

FIGURE 9.1   Von-Neumann processor architecture.

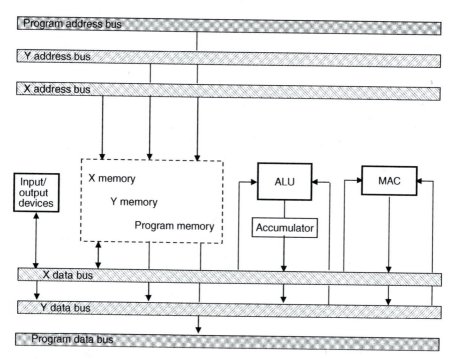

FIGURE 9.2 Basic processor architecture for signal processing.

9.3 FEATURES OF A DIGITAL SIGNAL PROCESSOR

Digital signal processors should be designed to allow the highest possible degree of parallelism (Table 9.1). To do this, these features (detailed in the following sections) should be available [4]:

1. Harvard architecture
2. Pipelining
3. Hardware MAC
4. Replication
5. Special instruction set
6. On-chip (cache) memory

9.3.1 HARVARD PROCESSOR ARCHITECTURE

Harvard architecture, shown in Figure 9.3, represents a milestone in the design of modern DSP processors. It features separate bus structures for data and programs as well as separate spaces for program and data memories. This design allows all components to operate in parallel rather than sequential; in contrast to the Von-Neumann architecture, all components are kept busy simultaneously, such that the system can operate at its best possible speed.

TABLE 9.1
A Summary of Important Features of Some DSP Processors

Manufacturer	Fixed/Floating-Point	Accum. Word Length (bit)	On-Chip Memory/ RAM (words)	On-Chip Memory/ ROM (words)	Cache Memory (words)	Bit Reversal Capability	Cycle Time (ns)
Texas Instruments							
TMS320C10	Fixed	32	144	1.5 k	0	No	114
TMS320C25	Fixed	32	544	4 k	0	Yes	78
TMS320C30	Floating	40	2 k	4 k	64	Yes	50
Motorola							
DSP5600	Fixed	24	1.5 k	544	0	Yes	60
DSP96002	Floating	32	2 k	1088	0	Yes	50
Analog Devices							
ADSP2100	Fixed	40	0	0	16	Yes	60
ADSP21000	Floating	80	0	0	2	Yes	40

Source: Ifeachor, E. and B. Gervis, *Digital Signal Processing*, 2nd Ed., Prentice Hall, New York, 2002.

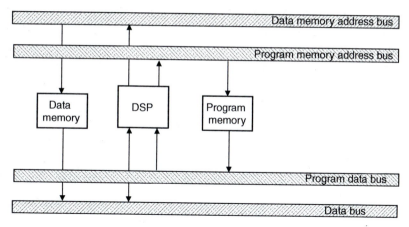

FIGURE 9.3 Simplified Harvard architecture. (Adapted from Ifeachor E. and B. Jervis, *Digital Signal Processing*, 2nd Ed., Prentice Hall, New York, 2002.)

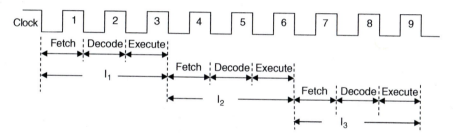

FIGURE 9.4 Sequential execution of instructions in a non-Harvard architecture.

To illustrate the speed gap between sequential and parallel operation, let us first consider the timing diagram depicted in Figure 9.4. Three instructions, I_1, I_2, and I_3, are required to be executed. Each one should be performed in three successive steps; that is, in three clock cycles: *fetch*, *decode*, and *execute*. In ordinary processors, the three instructions are executed sequentially, so that they need nine clock cycles.

Parallel operation, on the other hand, allows optimal utilization of the available units by performing several tasks simultaneously; for example, executing instruction I_1, and decoding instruction I_2 while fetching instruction I_3. The technique is known as *pipelining*.

9.3.2 PIPELINING

Theoretically speaking, Harvard architecture allows full overlap of the several steps, so that all components can be kept busy simultaneously. As illustrated in Figure 9.5, for a three-stage pipeline, it is possible while fetching one instruction, to decode

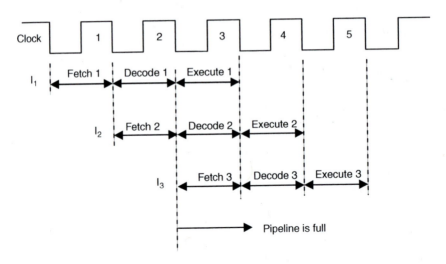

FIGURE 9.5 Pipeline operation.

another preceding instruction, while a third one is being executed. If this is the case, the pipeline is said to be "full" and the speed is tripled.

In the three-stage pipeline shown in Figure 9.5, the stages should be synchronized and adjusted according to the delay of the slowest stage. Thus speed improvement, assuming full overlap (ideal conditions), can be calculated from the relation

$$\text{speed gain} = \frac{\text{instruction time (nonpipelined operation)}}{\text{instruction time (pipeline operation)}}$$

In fact, full overlap of all instructions is a practical impossibility, due to the overheads of the pipeline registers usually added to enable pipelining. However, optimized designs can achieve near-ideal performance; the speed gain made approximately equals the number of stages in the pipeline. The following example highlights this fact.

EXAMPLE 9.1

The fetch, decode, and execute times in a nonpipelined processor are 45, 35, and 50 ns, respectively. Calculate the speed gain if pipeline operation is introduced, provided that a pipeline overhead of 5 ns should be added to each stage.

Solution

The processor should have a fixed machine cycle that should be adjusted to the slowest stage, 50 ns. This means that an instruction without pipeline operation would need

$$T_{\text{nonpipeline}} = 50 \times 3 = 150 \text{ ns}$$

while with pipeline operation, the time becomes

$$T_{\text{pipeline}} = 50 + 5 = 55 \text{ ns}$$

The speed gain is thus

$$\text{speed gain} = \frac{150}{55} = 2.72$$

which is slightly less than 3 (the ideal gain if the pipeline overhead were neglected).

9.3.3   HARDWARE MULTIPLY AND ACCUMULATE

DSP applications usually imply intensive use of the operations of multiplying, adding, and memory access. Software multiplication and floating-point addition are by nature time-consuming tasks. To enable real-time operation, it is thus necessary to implement such functions in hardware as in MACs. Two hardware realizations for a MAC are depicted in Figure 9.6.

The realization shown in Figure 9.6b is equipped with temporary storage elements (the so-called "pipeline registers") to allow for pipeline operation. The added delay (overhead) due to such registers, as shown in Example 9.1, is not serious and hence could be neglected.

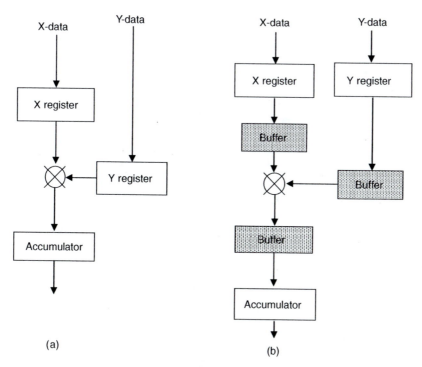

FIGURE 9.6   Hardware implementation of a MAC unit. (a) Nonpipelined and (b) pipelined.

9.3.4 REPLICATION

The availability of more than one fundamental unit such as an ALU, a MAC, or even a memory unit in one processor greatly boosts its speed performance by allowing parallel operation. Furthermore, it is the trend now to have more than one DSP processor operating—in parallel—on a single task [4].

9.3.5 SPECIAL INSTRUCTION SET

A very instrumental factor in determining the popularity of a certain processor is its associated instruction set. A compact set ensures better speed performance, in addition to saving memory space.

9.3.6 ON-CHIP MEMORY/CACHE

Sometimes, the memory unit in a certain processor becomes the speed-limiting factor if it cannot cope with the processor's speed. Usually in such a case, wait states have to be added to slow down the processor. Moreover, communicating with a remote memory unit (one not on the processor) may add some delay. Better speed performance can be achieved, however, if a high-speed memory (Erasable Programmable Read-Only Memory, or EPROM) is built onto the processor chip's "cache memory." It is employed to hold repeated instructions so as to reduce the number of memory fetch cycles. Almost all modern DSP processors have several kb of built-in cache.

9.4　HARDWARE IMPLEMENTATION OF SOME BASIC DSP ALGORITHMS

9.4.1　FIR FILTERS

To implement an FIR filter (Figure 9.7), the hardware components shown in Figure 9.8 are needed. Beside the MAC unit, ROM for the coefficients and RAM

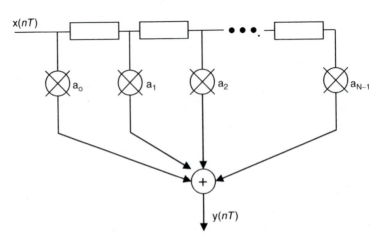

FIGURE 9.7　An FIR filter.

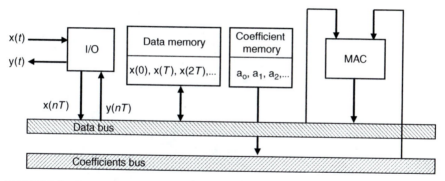

FIGURE 9.8   Hardware implementation of an FIR filter.

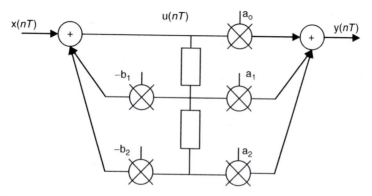

FIGURE 9.9   A second-order filter.

for the data are needed. This is in addition to the ADC at the input and the DAC at the output.

9.4.2   IIR FILTERS

A second-order canonic IIR filter section is shown in Figure 9.9. The corresponding hardware realization illustrated in Figure 9.10 is similar to that of an FIR, except for the inclusion of a register for the auxiliary variable $u(nT)$.

9.4.3   FFT ALGORITHM

A hardware implementation of a butterfly diagram is shown in Figure 9.11. An implementation of butterfly arithmetic is shown in Figure 9.12. The bit reversed data are applied to the input of the processor through buses A and B, and the outputs are available on buses C and D.

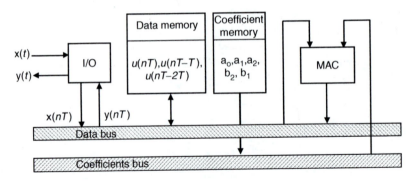

FIGURE 9.10 Hardware implementation of an IIR.

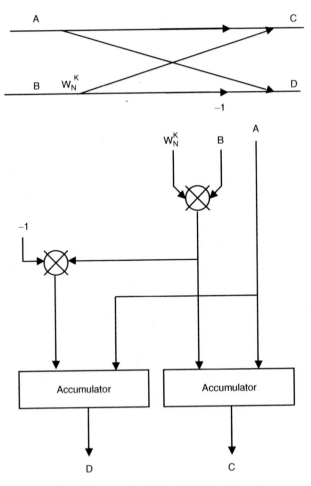

FIGURE 9.11 A butterfly implementation for an FFT algorithm.

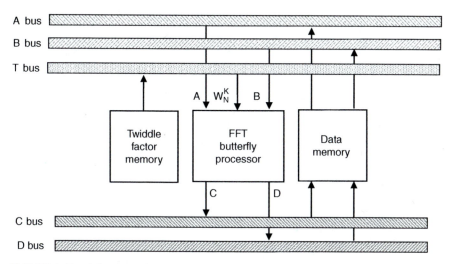

FIGURE 9.12 A hardware implementation of an FFT algorithm.

9.5 EXAMPLES OF PRACTICAL DSP PROCESSORS

As said before, DSP processors are in fact microprocessors that are optimized to perform DSP operations at the maximum possible speed and computational efficiency. To achieve this, most of the processors on the market implement some or all of the following features:

1. Replication
2. Harvard architecture
3. Parallelism
4. Pipelining

Due to fast advancement in integrated circuits technology in the last three decades, it was possible to introduce continuous improvements on the architectures the fact that have led to the availability of several generations. In the following, examples of some popular DSP processors are discussed.

9.5.1 TMS320XX FAMILY FROM TEXAS INSTRUMENTS

As a representative of this family, the TMS320C25 is considered here. It was presented in 1986 as a successor of the TMS32020. It features a 16-bit wide bus, three on-chip memory blocks, 4 k program memory, two data memories (each 2 k), and an extra 4 k module that can be configured as data or program memory. Figure 9.13 illustrates its simplified architecture. The series includes successors like TMS320C30, TMS320C50, and TMS320C40, supporting both fixed and floating-point arithmetic.

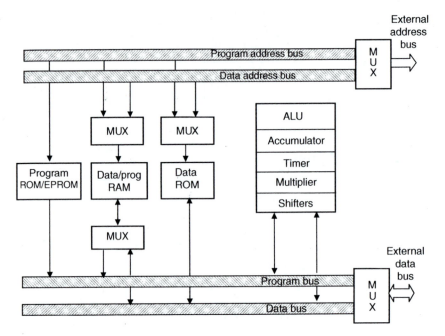

FIGURE 9.13 Simplified architecture of the Texas Instruments processor TMS320C25 (N.B.: all buses are 16 bits wide). (Adapted from Ifeachor E. and B. Jervis, *Digital Signal Processing*, 1st Ed., Addison-Wesley, Reading, MA, 1993.)

9.5.2 MOTOROLA DSP5600 FAMILY

The Motorola DSP5600 series of processors is a family of high-precision fixed-point processors. Figure 9.14 depicts a simplified architecture for a member of the DSP family, the DSP56002 chip. The availability of two separate data memory units and program ROM/RAM adds flexibility and improves speed performance. The buses are all 24 bits wide. The MAC unit thus accepts 24 bits and delivers 56-bit output to the 56-bit accumulators. The processor allows the execution of some special functions, such as bit reversal algorithms, for reordering the input data of FFT and output data of IFFT.

9.5.3 ANALOG DEVICES OF ADSP2100 FAMILY

The main feature of this family is the absence of on-chip memory; it has two external memory modules, one for data and a second for both data and programs. High-speed memory is thus required for full-speed operation. An on-chip program memory cache is added to eliminate probable bottlenecks. It holds the last 16 executed instructions to avoid repeated instruction fetches upon executing loops. The MAC unit, the ALU, and the shifter are all connected via a bus called the *result bus*, which enables immediate communications between the three units. A simplified block diagram of the processor is shown in Figure 9.15. It is worth mentioning that new members of this family are provided with on-chip memory.

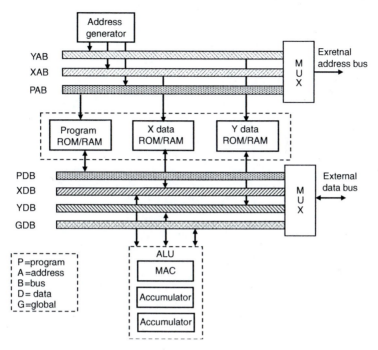

FIGURE 9.14 Basic architecture of the Motorola processor DSP56002. (From Motorola, Inc. DSP56002 Data Sheet, 1996.)

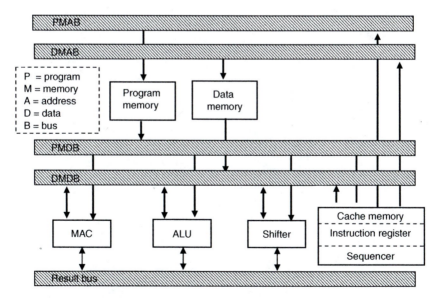

FIGURE 9.15 Functional block diagram of Analog processor ADSP2100. Table 9.1 summarizes the important characteristics of some DSP processors. (From Analog Devices, ADSP-2100 Family, Data Sheet.)

9.6 SUMMARY

This chapter dealt with hardware architectures that are specially designed for DSP applications. After introducing the problem, the basic Harvard architecture was described, with an illustration of its distinguished features. Dedicated hardware for some basic DSP operations like filtering and FFT were then presented. Three practical examples of popular DSP processors—namely, those from Texas Instruments, Motorola, and Analog devices—were detailed. The chapter closed with some review questions.

9.7 REVIEW QUESTIONS

1. Why it is necessary to have dedicated hardware for signal processing operations?
2. What are the main modifications that should be introduced to the conventional Von-Neumann architecture to suit signal processing applications?
3. Draw a block diagram for a Harvard processor, stating its main feature.
4. What is meant by "pipelining"? How does it affect the performance of a processor? Explain the speed gain.

10 Digital Signal Processing Systems

10.1 INTRODUCTION

Due to its many advantages, practical applications of DSP are increasing ever more quickly over time. The following are just examples of such applications:

1. Telecommunications (cellular phones, digital TV, digital cameras, satellite communications, and so on)
2. Biomedical engineering (CT scans, MRI, enhancement, and storage of X-ray images)
3. Entertainment (HDTV, MP3, DVD's … etc.)
4. Instrumentation and control (Data Acquisition Systems [DAS], telemetry, … etc.)

This chapter is devoted to highlighting some representative DSP techniques and applications.

10.2 DATA ACQUISITION SYSTEMS

There are several industrial applications where hundreds of variables must be monitored, recorded, and/or kept continuously under control 24 hours a day, 7 days a week, as in nuclear reactors, some manufacturing processes, and ICUs. In such applications, it is mandatory to monitor, store, and analyze a huge amount of data collected every day. Any deviation in the nominal value of any variable should be accompanied by an instant corrective action through a feedback loop. To achieve this, an efficient acquisition system that is capable of collecting, storing, and processing such data is required. The computer-aided measurement and control (CAMAC) system used mostly in monitoring nuclear reactors is an example of such systems. A block diagram of a typical data acquisition system is shown in Figure 10.1. Signals representing all variables, produced through suitable transducers, are first low-pass-filtered before being multiplexed. The analog output is then converted to digital after being sampled and held. The digitized data is then processed and/or stored. One of the major processing steps is to compare the obtained values with their nominal values. An error in a certain variable that exceeds a set threshold is fed back to an actuator to force the specific variable to resume its nominal value and/or release an alarm signal.

Several types of transducers—for example, strain gauges, tachometers, thermocouples, and piezoelectric crystals—are needed to measure the many physical quantities involved in different engineering systems. Based on the specific application, suitable

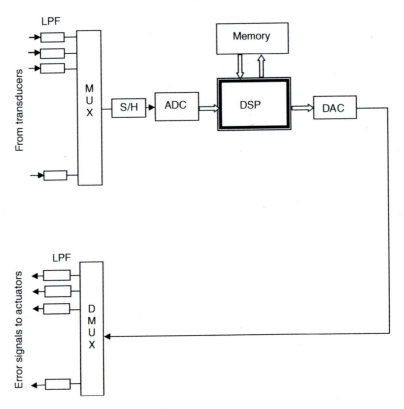

FIGURE 10.1   A typical data acquisition system.

transducers are selected according to the nature of the quantities to be measured considering their cost, size, response time, and—more importantly—required accuracy. Signals developed in the transducers are usually preprocessed in signal conditioners where they are, for example, amplified, filtered, and de-noised. Calibration facilities should be provided in those systems wherever the precision of measurements is critical. The used ADC and DAC are the main determining factors of the system's accuracy. They should be fast and precise enough to enable online processing. Both the multiplexer (MUX) and demultiplexer (DMUX) should be synchronized. The nominal values of the different variables are stored in the memory together with their thresholds.

10.3   TELEMETRY

Telemetry is the technique of measuring at a distance (remotely). The word is derived from the Greek roots *tele*, which means "remote," and *metron*, which means "to measure." Weather satellites are good examples of telemetry systems. While orbiting the earth in the upper layers of the sky, they continuously measure several physical quantities—for example, the atmospheric pressure, temperature, and wind speed—and send them to ground logging stations. The received data is then displayed, analyzed, and/or stored. Modern telemetry systems are very complicated, as they deal with huge amount of data. Therefore, they comprise data acquisition systems as integral parts. The measured data are grouped, digitized, and processed (for example, compressed) to be suitable for transmission over a channel and/or storage. At the receiving end (ground stations), the data stream is then reprocessed to assume their original format (decompressed, regrouped, converted from digital to analog, and so on). Telemetry makes it possible to safely measure data from dangerous locations; for example, cores of atomic reactors, high-temperature furnaces, test aircraft, and chemical reactors.

Another example of telemetry systems are data acquisition systems that are used in automated ICUs. Patients should be monitored around the clock. Several parameters (such as heart rate, blood pressure, temperature, respiration rate, and EKG measurements) are measured and transmitted to a data logger that is located either in a nearby room or at a distant location. Doctors can diagnose, prescribe treatment based on an available long record of monitored data, and trace the effect of medicines on patients as well as the accompanying changes in the patient's measurements. Recently, the branch of *telemedicine* has been introduced, where outpatients who are at risk and cannot be hospitalized are monitored remotely by their doctors. To achieve this, transducers and transmitting devices are either attached or implanted in the patient. Any alarming conditions can be immediately detected by doctors who can promptly advise the patient or the patient's family how to react.

A rather interesting application of telemetry is the study of wildlife, where members of an endangered species are observed through video cameras and global positioning system (GPS)* transceivers that are mounted on their bodies to provide updated information about their position and behavior.

10.4 PATTERN RECOGNITION

Pattern recognition is a branch of a well-established science called *artificial intelligence* (AI). It is concerned with the description and classifications of objects that can be represented in form of signals. Examples of such signals are speech, images, or video signals that are developed in different transducers (sensors) such as microphones, video cameras, and strain gauges. Pattern recognition techniques are applied in many fields, such as:

- Speech/speaker recognition and optical character recognition (OCR).
- Biometrics, where biological parameters of humans (such as fingerprints, voice prints, signature, and iris patterns), are captured, recorded, analyzed, and used to identify persons. Such techniques are necessary in some applications, as in forensic science, where criminals have to be identified, in electronic banking, passport control, and so on, as will be discussed later.
- Biomedical applications, as in automated diagnosis of diseases.
- Computer vision, as in robotics.

---

* A satellite-based system used for navigation.

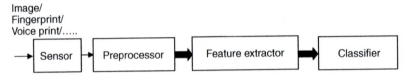

Image/
Fingerprint/
Voice print/.....

FIGURE 10.2   A functional block diagram for a typical pattern recognition system.

In pattern recognition, two distinct steps are performed:

1. Features extraction
2. Classification

The block diagram in Figure 10.2 illustrates a typical pattern recognition system. It consists of a sensor, preprocessor, feature extractor, and classifier. Captured signals by the sensor are improved in quality in the preprocessor, where they are amplified to enhance their signal-to-noise ratio (SNR), then digitized before being transformed to the frequency domain, for example. Distinct features of the pattern are extracted, then classified into a limited number of classes for recognition.

10.4.1   FEATURE EXTRACTION

Features of an object are those distinctive measurements, attributes, or descriptors that characterize exclusively and unambiguously that object. Examples of useful features of an object are color, volume, contours, shape, lines, and area. Features selected for classification should be robust; that is, they should not vary with time or any sort of processing. The effectiveness of a pattern recognition system is mainly based on the selected set of features that might lead to a successful classification or more precisely to high recognition rates (number of successful recognitions divided by the total number of recognition trials). Let us consider, for example, a ball and a cone. One straightforward and characteristic feature is their shape. Moreover, a cone and a pyramid could also be classified according to the shape of their bases, as one is circular and the other is squared. To make things a little bit difficult, consider two pyramids that are similar in every aspect except their heights. In such a case, they should be classified according to their heights. To be able to compare several features in practical problems, they should be projected in an m-dimensional feature space, where they are represented by their *descriptors*. Table 10.1 gives examples of some features and their favorite descriptors.

In image recognition applications, for example, Fourier descriptors are basically employed. These are in fact Fourier coefficients of the image containing the object to be detected. They have the advantage of being independent of translation, rotation of the object, change of origin, and/or scale. Usually the first few coefficients are sufficient to allow a user to distinguish between two different objects. However, the number of relevant coefficients is usually application-dependent. It has been proved that the first coefficient represents the position of the center of gravity of the object; hence, it is irrelevant to its shape, while the second term (the first frequency component) gives the size of the object. The color histogram, on the other hand, is a

TABLE 10.1
Some Feature Descriptors

Features		Descriptors
Image	Shape	Geometric descriptors (such as perimeter and area), Fourier descriptors
	Color	Color histogram, color moments
Texture		DCT coefficients, Gabor wavelet coefficients
Speech		LPC coefficients, mel cepstral coefficients

Note: The mel-frequency cepstrum uses a logarithmic frequency scale (mel scale) that approximates the response of the human auditory system better than the linear one.

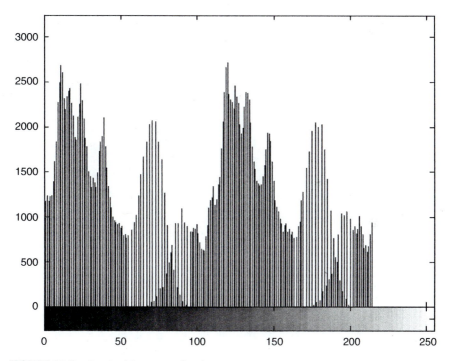

FIGURE 10.3   A color histogram of an image.

plot of the frequency of occurrence of the different gray levels in a picture. Because it is robust with regard to image size and orientation, it is the most commonly used color feature. In such cases, a histogram is plotted for each basic color (R, G, and B) of a colored image. Due to probable errors resulting from color quantization, color moments could be used instead. Nevertheless, only the first few moments are of interest, as most of the information is concentrated in them. Usually the first moment (the mean) and the next two moments are considered. Figure 10.3 illustrates a color histogram of an image.

The texture of an image results from regular and irregular variation in its gray levels or color. Recently, most feature-extraction techniques of an image texture are based on the statistics (such as the mean and variance) of a wavelet sub-band's coefficients (wavelet transform was described in Chapter 8). Of special interest in this respect are the Gabor wavelets [27]. Representative features are usually obtained using lengthy and complicated procedures comprising a sequence of mathematical operations.

10.4.2   CLASSIFICATION

The obtained features of a certain object or a collection of objects are grouped into sets or domains of similar features. Each set of features is a pattern. The job of a *pattern classifier* is to find out to which set of the defined sets an input pattern belongs. To do this task, these steps should be followed:

1. The classifier should first be *trained* using the features of all possible patterns.
2. A training set is constructed using all available data.
3. In the simplest pattern matching case, the nearest-neighbor rule is used, in which the feature set of an input pattern (an *m*-element feature vector) is compared with the feature sets stored in the database to find the best match, a process that is usually called *template matching*.
4. Such comparison is performed in an *N*-dimensional feature space in which, for example, the *Euclidian distance* between the input feature set and all data base sets is measured. The nearest feature set in the database to the feature set of the input pattern (the one that has the minimum Euclidean distance) is the best match. The Euclidean distance E between two points p and q in an N-dimensional feature space is

$$E = \sqrt{\left(\sum_{i=1}^{N}(p_i - q_i)^2\right)} \qquad (10.1)$$

Many other types of classifiers are known, such as the Gaussian-mixture models (GMMs), the neural network models, and the hidden Markov models (HMMs). These last classifiers are very successful in classifying sequences of feature vectors, as in speech or bioinformatics [28].

10.4.3   ARTIFICIAL NEURAL NETWORK CLASSIFIERS

Among several known classifiers, *artificial neural networks* [29] are the most popular and widely used classifiers. These are networks consisting of several numbers of highly interconnected computing elements called *neurons*. As the name implies, they emulate the human brain cell's architecture. One important property of the human brain that an artificial neural network cannot achieve is its extremely high speed, due to its unique parallel processing capability. A neuron by itself, as mentioned earlier, is a computing element. It comprises an adder that sums up all inputs and a threshold element that "fires"—that is, issues a binary 1—if the sum of all inputs exceeds a

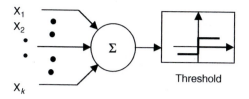

FIGURE 10.4 Structure of an artificial neuron.

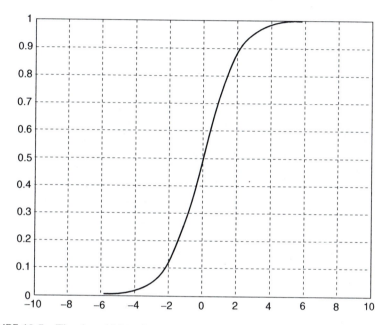

FIGURE 10.5 The sigmoid function.

specified threshold. A schematic diagram for the structure of an artificial neuron is given in Figure 10.4.

The block labeled "threshold" is usually based on the sigmoid function S, plotted in Figure 10.5. It is expressed as

$$S(x) = \frac{1}{1 + e^{-x}} \tag{10.2}$$

There are different known topologies for neural networks. However, the feed-forward back-propagation topology is the most popular one. Figure 10.6 illustrates what is called *a multilayer feed-forward network*. It consists of three layers: an input layer, one or more intermediate (hidden) layers, and an output layer. The lines connecting each two neurons include a multiplier "weight." The number of neurons in the input layer should equal the number of extracted features, and the number of output neurons should equal the number of classes (k). Generally, there is no rule for

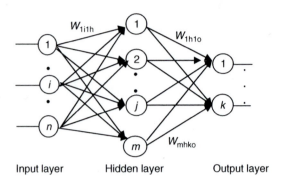

FIGURE 10.6   A feed-forward neural network.

determining how many hidden layers are necessary. However, one or two layers are found to be sufficient in most cases. The number of neurons (*m*) in each hidden layer depends on the complexity of the problem. The network is connected such that each neuron in a layer is connected to all neurons in the next layer [29].

The process of classification is performed in two phases:

1. The learning or training phase, where sets of extracted features of known patterns (classes) are employed to train the network. Once the corresponding output for each input is known, the weight for each connection is then determined. The starting values of the weights are arbitrary and randomly chosen. One of the most popular training algorithms is the *back-propagation algorithm*. The weights are determined backward; that is, from the output side to the input side, iteratively. Learning is completed when values at the input side are reached that are nearest to the original inputs with an allowed error. The network undergoes several iterations until a minimum error is attained. In such optimization problems, there should be an upper limit for the number of iterations (*epochs*) to avoid trapping in local minima.
2. The testing phase, where the network response is tested using input feature sets—preferably different from those used for training. The number of feature sets used for training should be large enough to achieve the highest possible correct recognition rates. However, a very large number of training sets are not always advantageous, as the network could reach a state of *overtraining*, which might lead to incorrect recognition results.

The result of the training phase is a set of weights that leads to a minimum error between the desired output and the actual "attainable" output. It is evident that the process of determining these weights is an optimization problem that should be achieved in a reasonable time; that is, a finite number of epochs. The steepest ascent method is usually the optimization tool of choice. Figure 10.6 depicts a neural network that has *n* input nodes, *m* neurons in the hidden layer, and *k* output neurons; that is, it has an *n*–*m*–*k* topology. The symbol W_{1i1h} stands for the weight of the link connecting the input node one "1i" and hidden node one "1h"; similarly, W_{1h1o} stands for the weight of the junction between hidden node one "1h" and output node one "1o."

Neural networks are used not only as classifiers and feature extractors, but also as universal predictors, as they offer arbitrary nonlinear predictors. This type of predictors can be used successfully in predicting time series data with highly nonlinear relations, such as a financial time series [30]. As feature extractors, neural networks have succeeded in extracting nonlinear principal components, which remove linear and nonlinear correlations from extracted features [31].

10.5 DATA COMPRESSION

10.5.1 INTRODUCTION

Data compression techniques are employed to bring large multimedia files (such as text, images, and video) into compact forms. To imagine the extent of what data compression can do, consider an encyclopedia—say, the *Encyclopedia Britannica*, which is available in 24 printed volumes. It is now possible to fit such a huge amount of data on a single DVD! (Until recently, this would have been considered science fiction.) However, it has been made possible only through well-developed data compression techniques. The main idea of data compression is to remove redundant data during recording and then restore them when played back. There are several well-developed and efficient techniques for performing compression. Examples of some popular techniques are run-length coding, entropy coding (Huffman and Lempel-Ziv-Welch [LZW]), delta encoding, JPEG, and MPEG [32].

Compression techniques can be classified into the following categories:

- *Lossless techniques.* In this type of compression, the restored data after decompression is nearly identical with the input data. Such a property is instrumental in several applications in which any minor loss of data can cause large errors. The obtainable compression ratios are, however, modest.
- *Lossy techniques.* In this type of compression, some errors in the restored data are allowed. In return, much higher compression ratios are achievable. Such techniques are suitable for those applications in which the input data contains redundancy and/or noise. Therefore, it is to be expected that the higher the achieved compression ratios are, the larger the incurred errors/noise.

Table 10.2 gives a classification of some known compression techniques. They will be described later.

TABLE 10.2
Some Important Compression Techniques

Classification	Technique
Lossy	LPC, JPEG, MPEG
Lossless	Run-length, Huffman, delta, LZW

Note: LZW is a lossless data compression algorithm from A. Lempel, J. Ziv, and T. Welch.

10.5.2 Lossless Compression Techniques

10.5.2.1 Run-Length Coding

This compression technique is based on minimizing data redundancy common in several applications as in data loggers, images, and so on, in which digits and/or characters are repeated several times. One of the most popular techniques is based on replacing each run of zeros by two digits (a zero followed by the number of zeros in the run). To illustrate the idea, let us consider the input stream:

$$\{1\ 0\ 0\ 0\ 5\ 0\ 0\ 0\ 0\ 2\ 1\ 6\ 0\ 0\ 0\ 0\ 0\ 8\ 3\ 2\}\quad \text{containing 20 elements}$$

which when encoded, reduces to

$$\{1\ 0\ \underline{3}\ 5\ 0\ \underline{4}\ 2\ 1\ 6\ 0\ \underline{5}\ 8\ 3\ 2\}\quad \text{containing 14 elements}$$

giving rise to a compression ratio of

$$\frac{20-14}{20}\times 100 = 30\%$$

It is clear that the more the number of zeros in the input string is, the higher the achievable compression ratio.

A rather general scheme, the pack bits, is based on writing the repeated digit once followed by the number of times it is repeated, headed by a minus sign. The following example describes the technique. Consider the following string of data, containing 20 elements:

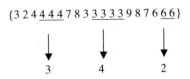

$$\{3\ 2\ 4\ \underline{4\ 4\ 4}\ 7\ 8\ 3\ \underline{3\ 3\ 3\ 3}\ 9\ 8\ 7\ 6\ \underline{6\ 6}\}$$

that can be reduced to contain only 14 elements, as follows:

$$\{3\ 2\ 4\ -\mathbf{3}\ 7\ 8\ 3\ -\mathbf{4}\ 9\ 8\ 7\ 6\ -\mathbf{2}\ 1\}$$

The negative numbers (written in boldface) indicate the number of times the preceding digit is repeated. In the example string, the number 4 is repeated three times, number 3 is repeated four times, and the number 6 is repeated two times. The resulting record length is thus reduced to 14 elements. The obtainable compression ratio is therefore:

$$\frac{20-14}{20}\times 100 = 30\%$$

TABLE 10.3

Probability Distribution of Some English Characters and Their Given Huffman Code

Letter	Probability	Assigned Huffman code
A	0.154	1
B	0.110	01
C	0.072	0010
D	0.063	0011
E	0.059	0001
F	0.015	000010
G	0.011	000011

Source: Smith, S.W., *The Scientist and Engineer's Guide to Digital Signal Processing*, California Technical Publishing, San Diego, CA, 2006.

10.5.2.2 Huffman Coding (Variable-Length Coding)

In this entropy-coding method, a statistical analysis of the raw data to be coded should first be performed to estimate the frequency of occurrence of each character. The compression power of the technique relies on the idea of assigning shorter codes (a small number of bits) to more frequent characters and longer codes to less frequent ones. Table 10.3 gives an example of the probability distribution of some English language characters and their assigned Huffman codes.

To illustrate the idea, consider the following string of characters:

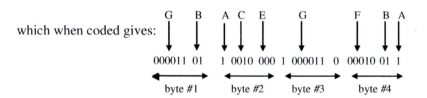

which when coded gives:

It is obvious from this example that the coded text is then handled as a continuous stream of bytes (8 bits). The decoding algorithm differentiates between the different characters by considering that each code consists of two subcodes: a number of zeros and a two-digit code that is binary 1 or binary 3.

10.5.2.3 Delta Coding

In this coding technique, only the difference between each two successive samples, rather than the sample, itself is stored/transmitted. To illustrate the idea, consider the following stream of data:

$$\{10 \ 15 \ 17 \ 16 \ 8 \ 8 \ 8 \ 13 \ 17 \ 18 \ 18 \ 19 \ 20\}$$

which when encoded, reduces to

$$\{10\ 5\ 2\ -1\ -8\ 0\ 0\ 5\ 4\ 1\ 0\ 1\ 1\}$$

To achieve a high compression efficiency, especially for a slowly varying data stream, delta encoding is usually followed by another type of coding, such as run-length coding or Huffman coding.

10.5.3   LOSSY COMPRESSION TECHNIQUES

10.5.3.1   Linear Predictive Coding

Linear predictive coding (LPC) is a powerful technique, especially for encoding speech signals. Due to the good quality of the reproduced sound, even at low bit rates, it is being widely used in several applications, such as GSM systems (a cellular "mobile" phone system), secure wireless communications, and electronic music. It is based on the assumption that a speech signal is produced by a buzzer at the end of a tube. The vocal cords generate the buzz that is characterized by its frequency and intensity. The vocal tract (mouth and nasal cavity) acts as the tube. It is characterized by its resonances (formants).

   An LPC algorithm analyzes (encodes) a speech signal by first dividing it into frames of relatively short periods (30–50 frames/s), detecting the formants, canceling their effects from the speech signal (inverse filtering), and estimating the intensity and frequency of the residual buzz. Analysis is performed employing a linear predictor (for example, the standard system LPC-10 uses a tenth-order FIR filter). It is obvious that the longer the length of the predictor is, the better the compression ratio will be. Compression is achieved because only the numbers describing both the formants (coefficients of the filter) and the residue for each frame are transmitted and/or stored. Speech is synthesized (decoded or reconstructed) by reversing the process; that is, by using the residue to create a source signal, the formants to construct a filter, and allowing the signal to pass through the filter. Because speech is a time-varying quantity, analysis and synthesis are usually done on short time frames.

   Analysis. Speech is produced either by the vibration of the vocal cords (voiced), or due to a random hiss produced by air coming out of the lungs. For voiced segments of speech, the source is assumed to produce a train of impulses. The pitch period describes the spacing between the impulses. For unvoiced segments, a random noise generator is used as the source.

   Synthesis. The synthesis step is basically the inverse of the analysis step. The 54 bits due to one frame (22.5 ms) are read in and used to determine the parameters of the filter: voicing, pitch period, gain, and so on. One thing that is different between the analysis and synthesis is that the synthesizer attempts to smooth its output. It stores one old frame in its buffers and uses it to smooth the output signal. This causes a delay of the output signal by one frame.

   Encoding the residue. There are several known techniques for efficient encoding of the residue signal so as to get better speech quality compared to LPC-10e, while keeping the bit rate almost unchanged. In one approach, the use of a *codebook* (a table set by the system's designer describing typical residue signals) is suggested. During operation, the residue is compared with all entries in the codebook to select the closest match, which is then sent. At the other end, the synthesizer retrieves the corresponding residue from a similar codebook, where it is used to "excite" the formant filter. This type of coding is thus called *code-excited linear prediction* (CELP). An efficient CELP system should have a codebook that is big enough to include all the possible residues. However, if it is too big, the search time might be excessive, while requiring large codes to be assigned to the large list of residues. To overcome this, two small codebooks could be employed, rather than one: a fixed one, containing enough code words to represent one pitch period of residue, and a second one that starts empty and is then filled through operation by previous residues.

Advantages and drawbacks. The main advantage of LPC-10 is its extremely low bit rate. It is therefore suitable for low-bandwidth applications. For example, more than 20 simultaneous LPC coded conversations could take place at 2400 BPS using a single 56 k modem! However, LPC-10 has some drawbacks, in that it is more a vocoder rather than a waveform encoder—it is inappropriate for all signals (for example, music) except voice. The algorithm is computationally intensive, and therefore requires a powerful processor or even a digital signal processor. The noise performance of the system is not optimum.

10.5.3.2 Joint Photographic Expert Group

The Joint Photographic Expert Group (JPEG) standard is a technique for lossy image compression. The JPEG commission established the standard in 1992. The technique and its successor, JPEG2000, are widely used now for image compression in several applications like remote sensing, satellite communications, and medical images. The block diagram in Figure 10.7 illustrates the steps encountered in the process of encoding. The image is first divided into 8 × 8 pixel tiles; each tile is then DCT-transformed. The obtained coefficients are quantized (rounded), leaving more than half the coefficients at zero. The tiles are then zigzag-scanned in a manner illustrated in Figure 10.8, starting with the low-frequency coefficients (upper-left corner). The nonzero coefficients are then encoded using a variable-length code (VLC) or

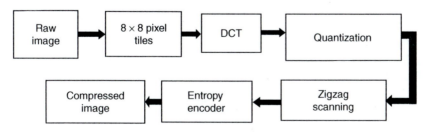

FIGURE 10.7   Sequence of operations involved in JPEG encoding.

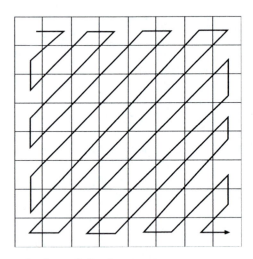

FIGURE 10.8 **(See color insert following page 262.)** Zigzag scanning.

entropy code; for example, Huffman code. After encoding all tiles of an image, an end-of-file marker is then produced. The technique provides high compression ratios, due to the following facts:

- Most of the image energy is concentrated in the low-frequency coefficients of the DCT transform.
- DCT coefficients are real; thus they can be represented by small number of bits.
- Due to quantization, the number of coefficients is greatly reduced.
- Entropy encoding results in an optimized size of the data, as it assigns shorter word lengths to frequent codes and vice versa.

In case of colored images, the three color components (RGB or Y, I, and Q) are treated each as a grayscale image, that is, each component undergoes the same steps as depicted in Figure 10.9. Recently, a modified and improved version of JPEG, JPEG2000, has been released. It employs DWT instead of DCT, in which the Daubechies 9/7 mother wavelet is used in a three-level analysis. The resulting image quality is far better than the conventional JPEG.

10.5.3.3 Moving Pictures Expert Group (MPEG)

10.5.3.3.1 Introduction
Video compression is achieved if irrelevant and redundant data in a video sequence is discarded, leaving a manageable bit rate. Each event is then assigned a VLC, such that frequent events are assigned short codes, and rare events are given lengthy codes. Four tools are usually used in video compression standards: DCT, vector quantization (VQ), fractal compression, and DWT [32,33].

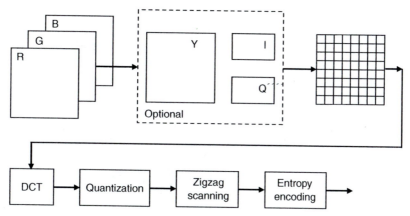

FIGURE 10.9   JPEG encoding of a colored image.

MPEG is a compression standard set by the Moving Pictures Expert Group" for digital TV transmission. There are now five MPEG standards:

- *MPEG-1.* This standard is employed mainly for compression of video and audio signals up to 1.5 Mbit/s; for example, in video CD-ROM, video on the Internet, and MP3s for compression of digital audio, as we shall see later.
- *MPEG-2.* This standard is based on MPEG-1, and is employed for data rates between 1.5 and 15 Mbit/s. It is used mainly for compression of digital broadcast TV signals, DVDs, and HDTV.
- *MPEG-4.* This standard employs object-based compression and is designed for multimedia and web compression. It deals efficiently with both low and very high data rates. Moreover, it allows a user to independently control certain objects in a scene.
- *MPEG-7 and MPEG-11.* These standards are concerned with multimedia applications.

10.5.3.3.2   Some Basic Definitions

MPEG-1 is a data reduction system that operates on redundant information in video signals. On the sending end, the encoder removes redundant data, which are then reinserted by the decoder at the receiving end. Two distinct types of redundancy can be distinguished in video signals:

- *Spatial and temporal redundancy.* Adjacent pixel values are not independent; they are more or less correlated with each other within the same frame (intraframe redundancy). The same applies to corresponding pixels in a sequence of frames (interframe redundancy). This makes the value of a pixel predictable, given the values of neighboring pixels.
- *Psychovisual redundancy.* The human eye is less sensitive to fine spatial detail (changes in a picture quality due to data reduction can be tolerated by the human eye). Thus each MPEG *codec* (coder-decoder) employs two techniques for data reduction: DCT for intraframe coding and motion-compensation for interframe prediction.

10.5.3.3.3   Intraframe DCT coding

A 2D DCT is performed on blocks of 8×8 pixels of a picture to produce an equal number of DCT coefficients. The values of the DCT coefficients in a block are a measure of the contribution of horizontal and vertical spatial frequencies to the respective picture block. One of the main characteristics of a DCT transform, as said before, is its ability to concentrate the energy in the low-frequency coefficients, leaving many other coefficients either at zero or near zero. Bit rate reduction can be achieved by discarding all coefficients below a certain threshold, while quantizing and coding the rest. Variable quantization is used, so that high-frequency coefficients are quantized differently than low-frequency ones. It is to be noted here that the noise introduced by quantization is irreversible; therefore, this technique is a lossy one.

The quantized DCT coefficients are then scanned in a diagonal zigzag fashion, starting from the northwest corner. The scanned coefficients are then entropy-coded, making use of the property of clustering of energy in the low-frequency coefficients and the many other zero-valued coefficients, using a VLC.

10.5.3.3.4   Motion-Compensated Interframe Prediction

Interframe coding makes use of the temporal redundancy by predicting a frame based on a previous "reference" one. The simplest interframe prediction of a block is that which repeats the corresponding blocks of the reference picture (co-sited). This should render an exact prediction for almost stationary regions of the image and a poor one for moving areas. Therefore, a rather complicated technique called *motion-compensated interframe prediction* is employed to offset any translational motion between frames. To estimate the shift between a current block and the corresponding block in the reference frame, the "block-matching" search is used, in which a large number of offsets are tried and tested by the encoder employing the luminance component of the picture. The offset that produces the minimum error between the block and the prediction is a best match. Motion-compensated prediction (MCP) should add a bit rate overhead, due to the motion vectors. As an example, MPEG-2, when used to compress standard-definition video to 6 Mbit/s, would have a motion vector overhead of about 2 Mbit/s in a picture containing many moving objects.

10.5.3.3.5   MPEG-2 Codec Structure

The encoder in MPEG-2 systems combines DCT and motion-compensated interframe, as shown in Figure 10.10. The MCP is subtracted from the source picture, giving what is called a "prediction error" picture. This prediction error is DCT-transformed; the coefficients are then quantized and coded using a VLC. The coded prediction errors due to luminance and chrominance signals are then combined with some "side information," that is needed by the decoder, such as motion vectors and synchronization information to form a bit stream ready for transmission [34].

At the decoder, the DCT coefficients are restored and inverse-transformed to provide the prediction error. This error is then added to the MCP, due to a previous picture giving the decoded output.

In MPEG-2 codecs, the motion-compensated predictor can "forwardly predict" a block from a previous picture, "backwardly predict" it from a future picture, or "bidirectionally predict" it by finding the average of a forward and backward predictions. The used prediction method varies from one block to the next. Moreover, two

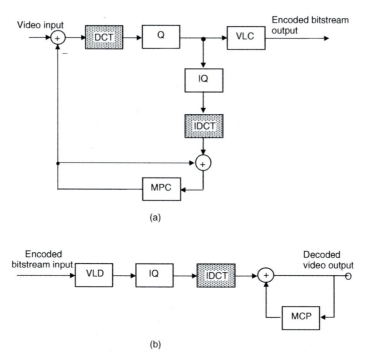

FIGURE 10.10   Motion-compensated DCT codec. (a) Encoder, (b) decoder, IDCT; VLC; MCP; Q, quantizer; IQ, inverse Q.

fields within a block may be predicted jointly, using a common motion vector, or separately, using their own motion vector.

10.5.3.3.6   Picture Types
Three "picture types" are known in MPEG-2; each type defines the prediction modes that it uses for coding:

- *Intrapictures (I-pictures)*. These are coded without reference to other pictures. Because only spatial redundancy, not temporal redundancy, is reduced, they cause moderate compression. They are used periodically to give access points in the bit stream for decoding to begin.
- *Predictive pictures (P-pictures)*. This type makes use of the previous I- or P-pictures for motion compensation and hence can be used as a reference for further prediction. Blocks in such pictures can either be intracoded or predicted. Due to their associated reduction of spatial and temporal redundancy, P-pictures offer more compression as compared to I-pictures.
- *Bidirectionally predictive pictures (B-pictures)*. Motion compensation is achieved here by using the previous and next I- or P-pictures. They therefore produce the best compression ratio among I- and P-pictures. Any block here could be forward, backward, or bidirectionally predicted or intracoded. To allow backward prediction from a future frame, the coder should reorganize the pictures from natural sequence to "bit stream" order.

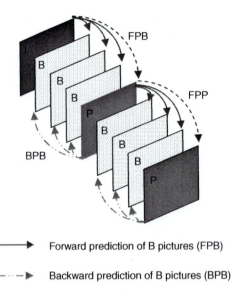

⟶   Forward prediction of B pictures (FPB)

–··–▶   Backward prediction of B pictures (BPB)

– – –▶   Forward prediction of P pictures (FPP)

FIGURE 10.11   **(See color insert following page 262.)** A sequence of MPEG frames.

The different types of pictures, as depicted in Figure 10.11, usually occur in a repeated "group of pictures" (GOP). An example of a natural stream and its corresponding bitstream is

$$B_1\ B_2\ I_3\ B_4\ B_5\ B_6\ P_7\ B_8\ B_9\ B_{10}\ P_{11}\ B_{12}\ B_{13}\ B_{14}\ P_{15}$$

The corresponding bitstream order is

$$I_3\ B_1\ B_2\ P_7\ B_4\ B_5\ B_6\ P_{11}\ B_8\ P_9\ B_{10}\ P_{15}\ B_{12}\ B_{13}\ B_{14}$$

A GOP is usually described with two parameters; namely, N, the number of elements in the GOP and M, the spacing between P-pictures. The example given here has $N = 15$ and $M = 4$. The MPEG-2 standard does not imply a regular GOP composition.

10.5.3.3.7 Buffer Memory

It is practically impossible to compress all frames in a video sequence by the same degree, as some parts may have low spatial redundancy (such as complex picture content), while others show low temporal redundancy due to fast-moving objects. Therefore, the output of the encoder has a variable bit rate, depending on the nature of the source picture and the quality of motion compensation prediction. Because a fixed bit rate is usually required on the transmitting channel, a buffer memory is used at the encoder output, such that it accepts variable rates and then supplies the channel with data at constant rates. The required buffer size is defined by the MPEG-2 standard.

One way to control the transmission rates is to arrange the data in a buffer in a descending manner with respect to its detail. Compression can be made effective

by discarding some irrelevant information. The overall picture quality can be pre-
served, in spite of compression, if detailed information are removed leaving the less
detailed content.

10.5.3.4 MPEG-1 Layer III

MP3 is an abbreviation for MPEG-1 Layer III encoding technology. It is nowa-
days widely used for audio compression, where it provides up to 12:1 compression
with nearly no noticeable degradation of quality. It was originally developed by the
German University of Erlangen and the Frauenhofer Institute. Later, the MPEG
set a series of standards containing different techniques for both audio and video
compression. The audio standards included three "layers" of different complexity
and performance. The third layer, Layer III, is capable of compressing high-quality
music from 1.4 Mbit/s down to 128 kbit/s without any degradation [35].

According to the MPEG-1 standard, it is possible to transmit/store audio/video
signals at a bit rate of 1–2 Mbit/s. There are three layers of compression and complexity: Layer I, Layer II, and Layer III. They differ in complexity, quality, and more
importantly, the bandwidth occupied. Table 10.4 lists the characteristics of each layer
as compared to uncompressed PCM CD quality [35].

MP3 is basically a perceptual codec, as it relies on the psychoacoustic properties
of the human auditory system (HAS)—for example, masking. For every tone in an
audio signal, there is a masking threshold. Another tone that lies below this threshold will be masked and thus inaudible. Inaudible components in an audio signal
are imperceptible to the HAS and thus could be safely eliminated by the encoder.
Layer III part of the MPEG-1 standard applies Huffman coding, cyclic redundancy
coding (CRC), FFT, and a modified DCT (MDCT). It defines bit rates between 8
kbit/s and 320 kbit/s, however the default rate is usually 128 kbit/s. A high bit rate
means that the samples are evaluated precisely, giving a better audio resolution. Two
bit rates are specified by the standard: *constant bit rate* (CBR) and *variable bit rate*
(VBR). Encoding using CBR (the default) implies that the whole work is encoded
using the same number of bits. However, because the dynamics differ from one work
to another, audio streams are encoded using VBR. Based on the required quality,
a threshold is set using a proper bit rate. MP3 files are divided into *frames*, of a
duration of 26 ms at a rate of 38 frame/s. Each frame stores 1152 audio samples. As
illustrated in Figure 10.12, a frame consists of five segments: a header, CRC, side
information, main data, and ancillary data.

TABLE 10.4
Comparing MPEG Layers

Technique	Compression Ratio	Bit Rate
CD quality	1:1	1.4 Mbps
Layer I	4:1	384 kbps
Layer II	8:1	192 kbps
Layer III	12:1	128 kbps

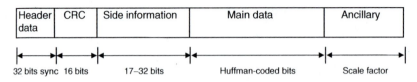

Header data	CRC	Side information	Main data	Ancillary

32 bits sync   16 bits   17–32 bits   Huffman-coded bits   Scale factor

FIGURE 10.12 MP3 frame structure.

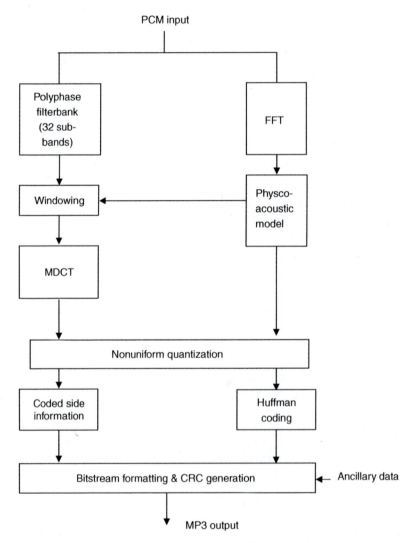

FIGURE 10.13 A simplified block diagram for an MP3 encoder.

A simplified block diagram for an MP3 system is shown in Figure 10.13 [35]. A sequence of 1152 PCM samples are filtered in a bank of 32 filters equally spaced by 689 Hz. The MDCT is then applied to each frame of the subband samples after being windowed. Simultaneously, as the signal is processed by the polyphase filterbank, it is also FFT-transformed. A psychoacoustic model retrieves the input data from the FFT output and applies it to algorithms that will model the human sound perception to classify audio signals into audible and nonaudible. This classification is important in selecting the window and the proper method for nonuniform quantization.

To decide upon a window type prior to the MDCT block, three FFT spectra are compared: the present one and the two previous spectra. If differences are detected, switching to short windows is requested; otherwise, the MDCT block should continue using long (normal) windows. The samples are then nonuniformly quantized and Huffman-coded. This explains why MPEG-1 Layer III retains a high quality at low bit rates. Finally, in the last block, all data—including, among others, a frame header, CRC, and Huffman-coded spectra—are assembled in frames of size 1152 samples.

10.6 BIOMETRICS

Biometrics are a set of unique physiological or behavioral characteristics of an individual that can be used to identifying that individual. A biometric system is thus a foolproof pattern recognition system that can be used to prove the authenticity of a certain identity. Examples of such characteristics are fingerprints, voice prints, iris textures, and signatures.

Use of biometrics is really old; they date back to the ancient Egyptians, where people were identified employing their measurements. Behavioral characteristics such as signature, voice print, and keystroke are usually emotional and physiological status–dependent and may vary over time. On the other hand, intrapersonal physical characteristic changes are minimal compared to behavioral ones. Recently, great interest has been focused on biometric systems, due to the increased need for reliable security measures and/or authentication techniques that are necessary in, for example, airports, restricted military and civilian areas, and e-banking. State-of-the-art "smart" ID cards can accommodate a person's fingerprint, voice print, and/or any other biometric stored on a small chip introduced into the card.

A typical biometric system consists usually of:

- An enrollment module, in which individuals are exposed to the system to generate a template
- An identification module that is used to identify individuals at the point of access

Biometric systems are required to be precise enough to detect intruders and allow (authenticate) only authorized persons to deal with the systems they are monitoring. The accuracy of a biometric system is measured by:

- False rejection rate (FRR), which measures the rate of rejection of an authorized (*genuine*) person
- False acceptance rate (FAR), which gives the rate of acceptance of intruders

A genuine acceptance rate (GAR) is defined as

$$GAR = 1 - FRR$$

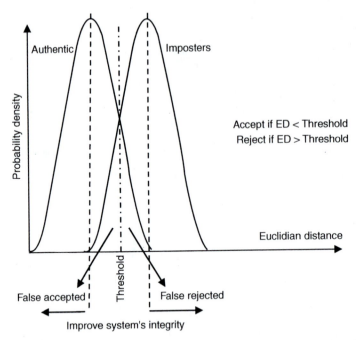

FIGURE 10.14 Statistical distributions of genuine and imposters in a typical biometric system.

The curves shown in Figure 10.14 illustrate the statistical distributions of genuines and imposters in an ideal biometric system. The separation between the peaks is considered a figure of merit of the system. The wider the separation is, the better the system. The receiver operating characteristics (ROC) curve, shown in Figure 10.15, is a plot of the GAR versus the FAR. It is considered to be a measure of system's performance. A steep rising curve indicates a nearly perfect system.

In the following sections, some popular biometrics will be discussed briefly, together with examples of their potential applications.

10.6.1 FINGERPRINT

The unique pattern of ridges and furrows found at the fingertips of each individual has been used successfully for a long time in personal identification. Such a pattern is best described by a set of "minutiae points." As illustrated in Figure 10.16, a *minutiae pattern* is a map representing those points where a ridge bifurcates or ends.

A fingerprint should first undergo an enhancement procedure to be able to precisely determine the distribution of the minutiae. Fingerprint matching techniques can be classified into the following categories:

- Minutiae-based
- Correlation based

Minutiae-based techniques detect minutiae points before mapping their relative location on the fingerprint. This implies, however, that the fingerprint is of reasonable

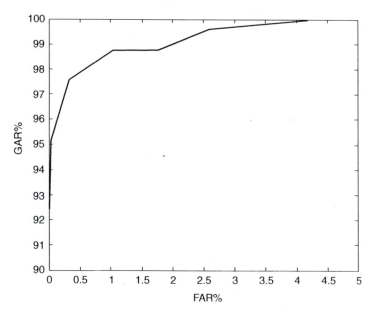

FIGURE 10.15   Receiver operating characteristics.

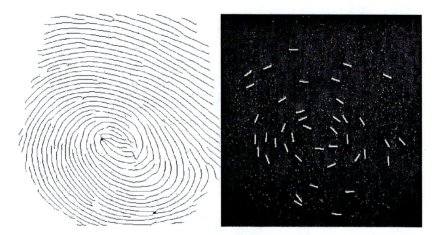

FIGURE 10.16   A fingerprint and its minutiae pattern.

quality. On the other hand, the correlation-based method does perform well, yet it requires precise localization of registration points and is sensitive to image translation and rotation.

An efficient commercial fingerprint identification system should be accurate and robust to noise in fingerprint images. The ROC, shown in Figure 10.15, usually measures accuracy. The system should also have a very low FRR for a given FAR, as mentioned earlier.

10.6.1.1 Fingerprint Classification

A huge number of fingerprints are collected and stored every day in several applications; for example, forensics and access control in restricted areas. An automated fingerprint-based recognition system should be able to find out of the database the best match to the presented fingerprint. Thus, to accelerate the process, it is mandatory to classify fingerprints accurately so that the input image is first matched with a subset of the fingerprints in the database rather than the whole set of data. Fingerprint classification can be considered as a preliminary or a coarse level matching step, where each print is assigned a class out of five known classes; namely, whorl, right loop, left loop, arch, and tented arch. Examples of such classes are given in Figure 10.17. A finer classification level then follows, where the input image is matched with the fingerprints of such subset. One published algorithm allocates the number of ridges among four directions (0°, 45°, 90°, and 135°) by filtering the central part of a fingerprint in a bank of Gabor filters. The result is then quantized giving a feature vector—"the FingerCode"—that is then used for classification. Different classification techniques are possible.

10.6.2 IRIS TEXTURE

The iris is the colored portion of the eye, which is bounded by the pupil and the sclera. A photograph of an iris is shown in Figure 10.18. In spite of the apparent visual similarity of iris patterns, either for the same person or for genetically identical twins, experiments have revealed that their textural details are uncorrelated and

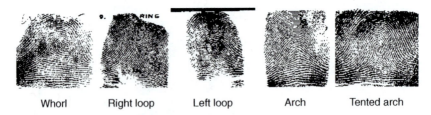

Whorl   Right loop   Left loop   Arch   Tented arch

FIGURE 10.17   Examples of fingerprint subsets.

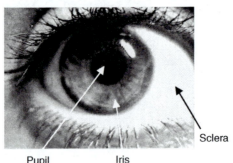

Pupil   Iris   Sclera

FIGURE 10.18   A photograph of an iris.

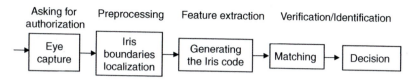

FIGURE 10.19   Stages of an iris recognition system.

independent. An iris print is superior to fingerprints (the most famous and reliable biometrics) in that it is highly stable, impossible to surgically modify its texture, inherently protected, isolated from the physical environment, and also it is easy to monitor its physiological response to light. It is also user-friendly, in that it can be captured without physical contact. It is therefore considered to be a reliable means for automatically verifying and/or identifying persons. A sequence of steps should be carried out to classify iris patterns:

1. Capturing an image for the eye, using specified illumination and standard distance.
2. Preprocessing the captured image, where the iris image is extracted from the rest of the eye image by locating its inner and outer boundaries.
3. Extracting distinctive features for the iris texture (iris code), using transforms like Gabor wavelet transform [27], DWT, and DCT.
4. Matching the extracted features with those stored in the database to verify or identify a person.

The block diagram shown in Figure 10.19 gives the components of an iris recognition system.

Several classification techniques have been published, and some of them have become industry standards. The Daugman technique [27], which is Gabor filter–based, is an example of such techniques.

10.6.3   Voice Print

The human voice is a useful means to prove the identity of a certain person. The features of a person's voice are mainly determined by the shape and size of the vocal tract, mouth, nasal cavities, and lips; that is, it is a characteristic of each individual. Figure 10.20 illustrates an oscillogram of a spoken letter.

10.6.3.1   Features of a Human Voice

Sound is, in general, a highly dynamic (nonstationary) phenomenon. Therefore, short-time analysis is a must. This is achieved by dividing the speech into frames of durations of 20–30 ms each, using special windows. To avoid loss of information, there should be an overlap of about 50%, as shown in Figure 10.21. LPC coefficients (mentioned before), have been the favorite features of classical speech-recognition and speaker-verification systems. However, they proved to be unsuitable for real-time applications, due to the need of lengthy and hence time-consuming calculations.

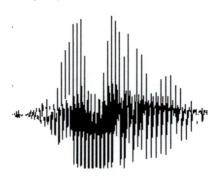

FIGURE 10.20   An oscillogram of a spoken letter.

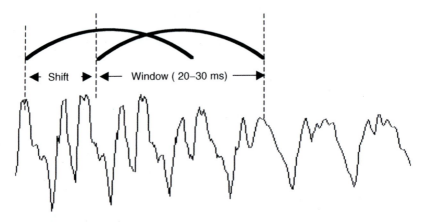

FIGURE 10.21   Framing of speech.

The cepstrum, defined as the inverse Fourier transform of the logarithm of the magnitude spectrum, has been recently suggested. It allows the use of the Euclidean distance for matching two sets of cepstral coefficients. Better matching scores could be achieved if the LPC-derived cepstral coefficients were used instead.

Speaker recognition approaches are divided into speaker identification and speaker verification. These in turn are either text-dependent or text-independent. In text-dependent speaker identification/verification, the speaker is asked to utter a certain word or phrase. In the text-independent case, on the other hand, the speaker is free to choose the words selects to be used for authentication.

10.6.3.2   Speaker Modeling

It has been observed that there are differences in the features derived from an utterance repeated by the same speaker. A usable speaker model should therefore be able to capture such variations in the extracted set of features. In this respect, stochastic models are the best tool. A stochastic model treats the speech production process as a parametric random process. The hidden Markov models (HMM)—specially, the

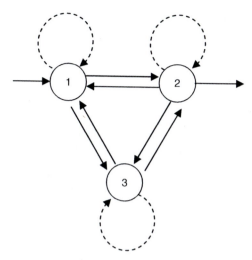

FIGURE 10.22　A three-state HMM.

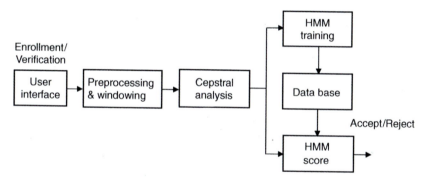

FIGURE 10.23　A typical speaker verification system.

left/right model—is a widely used tool for modeling the speech production process. A fully connected three-state HMM is depicted in Figure 10.22.

10.6.3.3　Pattern Matching

Input feature vectors are compared with the speaker model of the claimed identity. A matching score is then produced to express the probability that the given set matches that set generated by the model as shown in Figure 10.23.

10.6.4　FACE RECOGNITION

Face recognition is an ideal tool for personal identification, as it is fast, reliable, and noninvasive—that is, it is user-friendly. An efficient face recognition system should tolerate changes due to variations in illumination, pose, and expression. The reliability of the system depends on the selection of a proper feature vector.

Several techniques for the extraction of facial features have been suggested [36]. They include:

- Geometrical techniques, where use is made of parameters such as ratios of dimensions (for example, height/width), distances between basic features such as eyes and ears, nose width and length, lips, and so on.
- Statistical techniques, where algebraic tools such as the Principal Component Analysis (PCA) [37], the Independent Component Analysis (ICA), and the Linear Discriminant Analysis (LDA) are used to find distinctive features.

To construct an efficient database that helps in achieving high recognition rates, it is recommended to have M grayscale images having the same size for each person. Each face image is first converted into a vector $F(n)$ of length N (N = image width $\times$ height). An average face $\overline{F}$ is then calculated as follows:

$$\overline{F} = \frac{1}{M}\sum_{n=1}^{M} F(n)$$

The difference $D(n)$ between the average image and each face image in the database is then estimated from:

$$D(n) = F(n) - \overline{F}$$

The covariance matrix C that measures the degree of correlation between two data sets can be evaluated as follows:

$$C = \frac{1}{M}\sum_{n=1}^{N} D(n) \cdot D(n)^{T}$$

The eigenvectors of matrix C are used to generate M eigenfaces for each person in a set of k individuals. The required lengthy computations can be reduced through PCA by considering only the highest eigenvalues. The resulting $M \times k$ eigenfaces are employed to construct a useful knowledge base.

Any face image can be roughly reconstructed by combining some weighted eigenfaces. In this respect, it is similar to Fourier analysis, where a sum of weighted sinusoids at different frequencies are summed to perfectly reconstruct a signal. The eigenvectors are ordered to represent different amounts of variation, respectively, among the faces. Each face can then be represented exactly using only the best eigenvectors with the largest eigenvalues. The best M eigenfaces construct an M-dimensional space called the "face space."

PCA has proven to be an effective tool in face recognition, as faces have a lot in common—such as similar general structure. However, it suffers from two limitations:

1. Poor discriminatory ability, because the measured similarity between two images of the same face or two different faces is very high
2. Heavy computational load associated with eigenvector calculation for each face image (eigenface)

PCA analysis is usually preceded by a transform like DFT, DCT, DWT, and so on.

10.7 DIGITAL WATERMARKING

10.7.1 INTRODUCTION

As a result of the huge amount of digital information that has been currently made available on the diversity of media, it has become necessary to protect intellectual property (IP) rights of copyright owners against any form of illegal duplication for commercial purposes, manipulation, and/or plagiarism. To achieve this purpose, digital watermarking has been introduced. A copyright information in the form of a code, mostly invisible, is embedded in a work (host, cover, or carrier signal) to prove its authenticity and/or integrity, exactly like the watermark that can barely be seen on paper currency. Watermarks are designed to be inseparable and/or difficult to alter without causing damage to the carrier signal (tamper-proofing). Although the inclusion of such watermarks does not prevent theft or manipulation, it helps prove the claimed authenticity and ownership of a piece of information; for example, a multimedia signal. Digital watermarks are usually made invisible (imperceptible); therefore, they are detectable only if the carrier signal is subjected to a special detector known only by the owner or creator using a special key. Ideally, there should be no clear differences between the watermarked and the original signal. The term *just-noticeable difference* (JND) is used in this respect to identify the maximum change that could be introduced through the embedded watermark without being observable. A digital watermark in a product, besides proving authenticity, can provide information about its creator, date of production, and so on.

Watermarking is a branch of a broader field known as *steganography* (covered writing), in which messages or data are hidden in cover information. In this context, they differ from those in the field of *cryptography*, in which a message is encoded (ciphered) such that it cannot be read by any one except the one who has the key used in encrypting the message. Watermarks are introduced in several types of host signals like images, texts, graphs, and audio and video signals. They can be classified into the following categories:

- *Robust watermark* (resilient). This type withstands many kinds of processing (attacks), such as change of content, removal, rescaling, rotation, resampling, compression, and filtering of additive noise. The watermark is usually embedded into a perceptually significant part of the host, thus any trial to remove the watermark will result in destroying the work itself. This type is therefore suitable for copyright protection.

- *Fragile watermark* (vapor watermark). Breaks easily after any modifications in the host signal. The watermark information (data stream) is used to replace the least significant bits of the pixels or a perceptually insignificant part of the host. Any lossy compression, if performed on the watermarked content, could easily be detected, because it affects the integrity of the watermark. Fragile watermarks are therefore used for tamperproofing and for proving the authenticity of a work.
- *Semifragile watermark* (marginally robust). Allows simple operations; for example, lossy compression that exceeds an owner-specified threshold. (A zero threshold renders a fragile watermark.) On the other hand, it should be sensitive to other attacks that may affect the integrity of the host signal. It serves mainly for authentication purposes and tamper detection.

Watermarks can also be divided according to the extraction method into:

- *Blind* (*public*) *watermark.* To be detected, it is mandatory to have the key used to generate the watermark. The cover (host) signal is not needed.
- *Nonblind* (*private*) *watermark.* Can be detected only if the original (nonwatermarked) signal and the extraction key are available.
- *Semiblind watermark.* Needs the extraction key and the watermarked signal for detection.

10.7.2 Applications

Current applications of watermarking are many. To mention a few [38]:

- Copyright protection for authors, composers, photographers, manufacturers of video tapes, DVDs, and multimedia products
- Archiving of medical images where patients data are embedded in their digitized X-ray images, CAT scans, and MRI images, … etc., so as to avoid possible mixing of patients' images
- Verifying the authenticity of documents, banknotes, and checks
- E-banking and e-commerce
- Forensic applications, as in watermarked passports
- Secure communications

Digital watermarks should fulfill the following requirements:

- *Security.* The embedding algorithm should be based on a secret key that is difficult to guess by any attacker.
- *Invisibility.* In some applications, the watermark should be perceptually invisible or inaudible. In such cases, the characteristics of the human visual system (HVS) and the HAS should be considered.

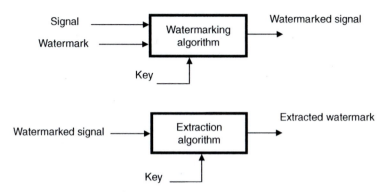

FIGURE 10.24   Embedding and extraction of a watermark.

10.7.3   WATERMARKS EMBEDDING AND EXTRACTING TECHNIQUES

There are several known techniques for embedding a watermark in a host signal; most of them use a special key that is either made available to all (public key) or to authorized people only (secret key). Embedding a watermark in the host signal is achieved through a special algorithm and the key, as depicted in Figure 10.24. To extract the watermark, the reverse process is performed using the same key. Watermarks can be embedded either in the spatial or the transform domain. Spatial domain techniques are straightforward, but are generally not robust. In the transform domain, as in the DCT and the DWT, for example, both the signal to be watermarked and the watermark itself are first transformed, then merged.

10.7.4   IMAGE WATERMARKING

There are two techniques for watermarking of images:

- *Visible watermarks* that are used for authenticity, as well as advertisement purposes
- *Invisible watermarks* that are usually used for copyright protection, copy control, authentication, and tamper detection

10.7.4.1   Embedding Techniques of Invisible Watermarks

The following sections describe the most common watermark embedding techniques.

10.7.4.1.1   The m-Sequence Technique

A watermark, in form of an *m*-sequence (pseudorandom sequence), can be added to the LSB of either the pixel values (gray level) in the spatial domain or to the coefficient of a frequency transform; for example, the DCT or the DWT. Detection of the watermark is easy in this case and is done by correlating the watermarked signal with the *m*-sequence used in watermarking where a peak is detected. Any other sequence yields either no peak or a small one. However, this method is highly sensitive to most signal processing operations; for example, ADC and DAC.

10.7.4.1.2   The Blue Component Technique

The HVS is less sensitive to blue color; hence a watermark, if embedded into the blue component of an RGB image, will be undetectable. Furthermore, it is robust to filtering of additive noise (such as median filtering). However, it is slightly sensitive to lossy compressions, such as JPEG.

10.7.4.1.3   Texture-Rich Watermark

A watermark that has a similar texture to that of the host image if embedded would be difficult to detect. The best extraction tool in this case is correlation.

10.7.4.1.4   Block-Based DCT Watermarking

A DCT is commonly performed on 8×8 blocks of the image. The watermark, in the form of a pseudorandom number (PN) sequence, is added to the coefficients of a chosen range of frequency. The algorithm can be described by

$$S_w = S[1 + \alpha W]$$

where

S = host signal (image, music, etc.)
S_w = watermarked signal
W = watermark
α = factor used to provide compromise between robustness and impercepti-
bility, which varies between 0.1 and 1

There is no definite range of frequency that is optimal. Although embedding in the high-frequency coefficients does not greatly affect the image quality, it renders an image that is sensitive to LP filtering attacks. Embedding in the low-frequency coefficients, on the other hand, degrades the image quality while yielding a robust image against common image processing attacks. Therefore, embedding watermarks in the mid-frequency bands coefficients provides a trade-off. Genetic algorithms (GA) are now being used to determine the optimal bands for embedding watermarks into the DCT domain. This simultaneously improves security, image quality, and robustness.

10.7.4.1.5   Patchwork Watermarking

It is known that luminance rather than chrominance is a robust medium for embedding image watermarks. Within the luminance signal itself, areas of high gradients (such as edges) are the best locations. The image to be watermarked is split into two groups, X and Y, where they are subjected to changes in luminance in opposite directions. While the brightness of part X is incremented slightly, that of Y is decremented by the same amount. The degree of change is usually chosen such that no change in intensity can be perceived. The location of the groups (considered to be the key) is usually not public. Watermark extraction is simple and is achieved by averaging the difference between the corresponding pixel intensities in the two groups. For N pixels in each group, the sum of differences should be $2N$ for watermarked images and zero if there is no watermark.

10.7.4.1.6   Wavelet Transform–Based Watermarking

Wavelet transform can be efficiently used to embed and extract watermarks as follows:

1. The image is first wavelet-transformed using any mother wavelet.
2. A set of *m*-sequences is generated and wavelet-transformed [38].
3. Only one transformed sequence is chosen to be the watermark.
4. The selected *m*-sequence can be embedded in the host signal in different ways. It is a common practice, for example, to replace those coefficients in the approximation (LL) whose values are below a certain selected threshold of the watermark coefficients. Other strategies do the same for the LH, HL, or the HH subbands' coefficients.*
5. To extract the watermark, the image is wavelet-transformed. All coefficients of the encoded subbands that are below the applied threshold are selected and inverse wavelet–transformed.
6. The inverse wavelet–transformed coefficients are then correlated with the generated set of *m*-sequences. Only one sequence out of the set will show a global maximum. It represents the watermark.

10.7.4.2   Embedding Techniques for Visible Watermarks

One of the most common methods for embedding visible watermarks is the DWT [28].

10.7.5   TEXT WATERMARKING

Checking the authenticity of documents is an important legal issue. Accords, contracts, agreements, classified reports, and so on should be protected against manipulation and/or unauthorized copying. Different techniques have been successfully used to embed a hidden message in a text. The following are some possibilities:

- *Line shifting*, where a certain specified line is slightly shifted either up or down by, for example, 0.1 mm.
- *Word shifting*, where individual words or strings of words are slightly shifted horizontally, except the last words in the line.
- *Feature coding*, where the geometry of the character is slightly distorted or modified, for example, elongated in either direction.

Extraction of the watermark from a printed or photocopied document is usually preceded by a preprocessing step that includes skew corrections and noise removal. The line shifts, word shifts, and/or feature changes are detected to form the extracted watermark.

10.7.6   AUDIO WATERMARKING

To check the originality and authenticity of audio records that are presented as evidence to courts, or to protect composed music and other audible information from

---

* Where L = low frequency and H = high frequency.

illegal copying, duplications, and manipulations, they should be watermarked. A perfect audio watermark should satisfy the following conditions:

- Inaudibility (psychoacoustically inaudible)
- Robustness to any kind of processing—for example, compression, filtering, and resampling
- Immunity to unauthorized extraction or removal
- Illegal copying

Among other usable audio watermarking techniques only two will be discussed here namely:

- *Spread spectrum technique.* A binary code word is modulated by the audio signal to be watermarked and a PN-sequence using biphase shift keying is generated. It is then added to the audio signal, causing a detectable noise. To satisfy the condition of inaudibility, such noise should be kept to a minimum.
- *Phase coding.* The phase of each frequency component in a STFT signal is modified. Because the HAS is by nature insensitive to phase distortions, nothing will be detected.

10.7.7 WATERMARKING A WATERMARKED IMAGE

To guarantee a high degree of security of a certain work, multiple (dual) watermarking is done. This is achieved either by watermarking the host twice with two different watermarks or embedding a watermarked watermark. In the second case, a primary watermark is used to watermark a secondary watermark before being embedded in the host. The embedding technique could be different in the two stages; that is, the first watermark could be embedded in the spatial domain, while the other is in the transform domain [39]. Also, the two watermarks could be of different types; that is, one is robust while the other is fragile.

10.8 ACTIVE NOISE CONTROL

10.8.1 INTRODUCTION

Noise is an environmentally unfriendly agent. It could be as dangerous as toxic materials. Continuous exposures to high noise levels can cause hearing deterioration that might lead to total hearing loss. Therefore, there is a need for systems that can combat such a harmful environmental pollutant. Acoustical noise is classified into two categories:

- *Broadband noise.* This is usually a random noise that is evenly distributed along a wide band of frequencies. A good example of this class is the noise due to a jet plane.
- *Narrowband noise.* This is a periodic noise with its energy concentrated around a specific frequency; for example, sounds produced due to the rotational motion of compressors, refrigerators, and internal combustion engines.

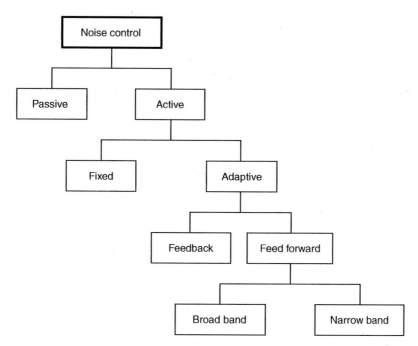

FIGURE 10.25   Classifications of noise cancellation schemes.

10.8.2   NOISE CONTROL

Noise level reduction and/or cancellation can be categorized into passive and active noise control (ANC), as illustrated in the tree structure shown in Figure 10.25.

10.8.2.1   Passive Noise Control

This type of noise control reduces noise by introducing sound absorbents, such as barriers, perforated panels, and earmuffs. Thus, its contribution to the noise reduction is modest; it does not eliminate it.

10.8.2.2   Active Noise Control

In ANC, noise reduction or cancellation is achieved by generating an antinoise (secondary noise); that is, noise that resembles the disturbing (primary) noise in all respects but the phase. For perfect cancellation, it should be in antiphase with the primary noise. ANC is thus a superposition process where primary and secondary noises are combined acoustically. Being out of phase, the difference is then the residue whose amplitude depends on how perfect the secondary noise is generated to be an exact replica of the original primary noise. ANC is divided, in turn, into fixed and adaptive noise control.

In fixed noise control (synthesis method), one or more noise cycles are sampled and stored to be used in the generation of an antinoise. A constant amount is then subtracted from the original noise, thereby limiting its efficiency in noise

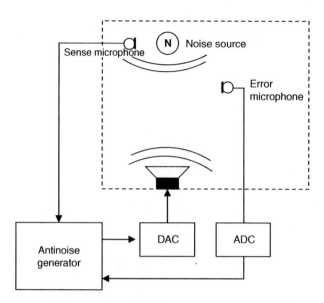

FIGURE 10.26   Block diagram of an ANC system.

cancellation to a few decibels (dBs). The situation is much better in case of periodic noise, in which a system like the one shown in Figure 10.26 is used. A sense microphone picks up samples of the noise that is then used in the generation of an antinoise. An error signal is then used to adjust the antinoise generator through another microphone (error microphone) based on the difference between the secondary noise, propagated through the shown loudspeaker, and the primary noise. DSP in conjunction with field-programmable gate arrays (FPGAs) can be used in the implementation of such a system.

Adaptive noise control systems, on the other hand, detect noise through one or more sense microphones and generate an antinoise that is then radiated through one or more loudspeakers. The difference signal is then sensed by another microphone (error microphone), as before, to present an adaptation signal to the built-in adaptive filer. Such an adaptation signal is used to vary the filters' coefficients in a direction to minimize the error. A well-designed adaptive noise control system can achieve up to 60 dB reduction in the noise level. Furthermore, adaptive systems are usually designed to be selective, such that only unwanted noise is reduced/eliminated, leaving other sounds unaltered. Figure 10.27 depicts a block diagram of an adaptive noise cancellation system.

The performance of an ANC system depends mainly on the following:

- How similar is the antinoise waveform to that of the primary noise, and how far can the system maintain exactly a phase shift of 180° between them? Failure to fulfill any of these conditions may result in a second noise rather than cancellation of the original one.

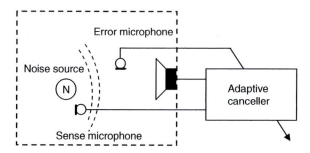

FIGURE 10.27   An adaptive noise cancellation system.

- The location of the noise source. For an ANC system to be effective, the noise source should be stationary relative to the speaker producing the antinoise, while keeping closest to the noise source to avoid any delay differences due to their physical distances.
- The ability of the system to predict noise from its past history or how closely it can adapt the response to the incoming signal. Prediction of periodic noise (most noises in practice are periodic) is made using the past few cycles of noise.

10.8.3   APPLICATIONS

1. Engine noise cancellation inside cars, vans, trucks, and military vehicles. Many automobile manufacturers have already implemented ANC systems inside their products.
2. Reduction of noises due to home appliances such as air conditioning, refrigerators, washing machines, vacuum cleaners, food mixers, and hair dryers.
3. Industrial applications, as in textile factories, air compressors, pumps, blowers, engines, and operating enclosures of large equipment such as bulldozers and heavy cranes.
4. Noise reduction systems inside the cockpits and cabins of jet planes, boats, helicopters, diesel locomotives, and so on.
5. Helicopter pilots can have now their headphones equipped with ANC to attenuate engine noise as much as possible while clearly receiving warnings, instructions, and so on.

10.8.4   ADAPTIVE INTERFERENCE CANCELLATION

Engine noise and street noise interfere inside a vehicle with received telephone calls, played piece of music, and the like, thereby impairing the quality of the signal. The system depicted in Figure 10.28 illustrates how interference is eliminated/minimized employing an adaptive filter $H(Z)$. Usually an optimization algorithm—for example, the least mean square (LMS)—is used in the adaptation process of the filter.

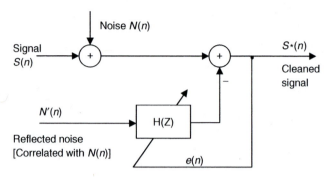

FIGURE 10.28   An adaptive noise canceller.

The quality of the useful signal $S * (n)$ at the output of the system is measured by the *S/N* ratio before and after the canceller.

10.9   SUMMARY

In this chapter, many representative examples of DSP systems were selected and described, together with their practical applications. Industrial systems like data acquisition and telemetry systems were first introduced in terms of their components and applications. Pattern recognition, being a pivotal DSP application, was thoroughly discussed and terms like feature extraction and feature classification are highlighted. New fields such as telemedicine and personal identification through biometrics were discussed. Neural network classifiers, especially feed-forward backpropagation nets, were described. Data compression through reduction of redundant information was then presented, where terms like "lossy" and "lossless" compression were introduced, together with some relevant techniques. Biometrics, an emerging modern tool for identity verification in secured locations such as airports, military zones, different forensic applications, and banking, was presented. Specific features that have proven to be unique for each person—such as iris pattern, voice print, fingerprint, facial thermograms, and retinal patterns—were described.

Watermarking as a highly secure means for intellectual property protection against intruders was then discussed in some detail. Efficient embedding and extraction algorithms of watermarks for images, text, and audio records were presented. Finally, the important issue of active noise reduction was introduced. After presenting the problem, different techniques of noise cancellation were discussed together with their applications. The chapter closed with some review questions.

10.10   REVIEW QUESTIONS

1. Draw a functional block diagram for a data acquisition system, stating the function of each block. Modify the system to be usable for monitoring an ICU in a hospital.
2. What is meant by "telemetry"? Give some practical applications.
3. Draw a block diagram for a typical pattern recognition system, giving the function of each block.

4. Artificial neural networks are used as classifiers in pattern recognition systems. Sketch a feed-forward back-propagation network that has the topology 3–4–2. Sketch the used threshold function, giving its describing equation.
5. What is meant by "lossy" and "lossless" compression? Give examples for each type.
6. Describe how the LPC coding technique is performed.
7. Draw a block diagram for the JPEG technique, illustrating the function of each block.
8. Describe how MP3 achieves high compression ratios for the music, while keeping such a high quality of sound.
9. Suggest some biometrics to be used in a car, both for the security (antitheft measures) and comfort of the driver. Describe how they would function.
10. Suggest a design for an efficient ANC system that can be implemented to reduce the noise in a factory.

Appendix A1

Butterworth Active LP/HP Values

Order N	C_1/C or R/R_1	C_2/C or R/R_2	C_3/C or R/R_3
2	1.414	0.7071	
3	3.546	1.392	0.2024
4	1.082	0.9241	
	2.613	0.3825	
5	1.753	1.354	0.4214
	3.235	0.3090	
6	1.035	0.9660	
	1.414	0.7071	
	3.863	0.2588	
7	1.531	1.336	0.4885
	1.604	0.6235	
	4.493	0.2225	
8	1.020	0.9809	
	1.202	0.8313	
	1.800	0.5557	
	5.125	0.1950	
9	1.455	1.327	0.5170
	1.305	0.7661	
	2.000	0.5000	
	5.758	0.1736	
10	1.012	0.9874	
	1.122	0.8908	
	1.414	0.7071	
	2.202	0.4540	
	6.390	0.1563	

0.01 dB Chebyshev Active LP/HP Values

Order N	C_1/C or R/R_1	C_2/C or R/R_2
2	1.4826	0.7042
4	1.4874	1.1228
	3.5920	0.2985
6	1.8900	1.5249
	2.5820	0.5953
	7.0522	0.1486
8	2.3652	1.9493
	2.7894	0.8197
	4.1754	0.3197
	11.8920	0.08672

0.1 dB Chebyshev Active LP/HP Values

Order N	C_1/C or R/R_1	C_2/C or R/R_2	C_3/C or R/R_3
2	1.638	0.6955	
3	6.653	1.825	0.1345
4	1.900	1.241	
	4.592	0.2410	
5	4.446	2.520	0.3804
	6.810	0.1580	
6	2.553	1.776	
	3.487	0.4917	
	9.531	0.1110	
7	5.175	3.322	0.5693
	4.546	0.3331	
	12.73	0.08194	
8	3.270	2.323	
	3.857	0.6890	
	5.773	0.2398	
	16.44	0.06292	
9	6.194	4.161	0.7483
	4.678	0.4655	
	7.170	0.1812	
	20.64	0.04980	
10	4.011	2.877	
	4.447	0.8756	
	5.603	0.3353	
	8.727	0.1419	
	25.32	0.04037	

0.25 dB Chebyshev Active LP/HP Values

Order N	C_1/C or R/R_1	C_2/C or R/R_2	C_3/C or R/R_3
2	1.778	0.6789	
3	8.551	2.018	0.1109
4	2.221	1.285	
	5.363	0.2084	
5	5.542	2.898	0.3425
	8.061	0.1341	
6	3.044	1.875	
	4.159	0.4296	
	11.36	0.09323	
7	6.471	3.876	0.5223
	5.448	0.2839	
	15.26	0.06844	

0.25 dB Chebyshev Active LP/HP Values (continued)

Order N	C_1/C or R/R_1	C_2/C or R/R_2	C_3/C or R/R_3
8	3.932	2.474	
	4.638	0.6062	
	6.942	0.2019	
	19.76	0.05234	
9	7.766	4.891	0.6919
	5.637	0.3983	
	8.639	0.1514	
	24.87	0.04131	
10	4.843	3.075	
	5.368	0.7725	
	6.766	0.2830	
	10.53	0.1181	
	30.57	0.03344	

0.5 dB Chebyshev Active LP/HP Values

Order N	C_1/C or R/R_1	C_2/C or R/R_2	C_3/C or R/R_3
2	1.950	0.6533	
3	11.23	2.250	0.0895
4	2.582	1.300	
	6.233	0.1802	
5	6.842	3.317	0.3033
	9.462	0.1144	
6	3.592	1.921	
	4.907	0.3743	
	13.40	0.07902	
7	7.973	4.483	0.4700
	6.446	0.2429	
	18.07	0.05778	
8	4.665	2.547	
	5.502	0.5303	
	8.237	0.1714	
	23.45	0.04409	
9	9.563	5.680	0.6260
	6.697	0.3419	
	10.26	0.1279	
	29.54	0.03475	
10	5.760	3.175	
	6.383	0.6773	
	8.048	0.2406	
	12.53	0.09952	
	36.36	0.02810	

1 dB Chebyshev Active LP/HP Values

Order N	C_1/C or R/R_1	C_2/C or R/R_2	C_3/C or R/R_3
2	2.218	0.6061	
3	16.18	2.567	0.06428
4	3.125	1.269	
	7.546	0.1489	
5	8.884	3.935	0.2540
	11.55	0.09355	
6	4.410	1.904	
	6.024	0.3117	
	16.46	0.06425	
7	10.29	5.382	0.4012
	7.941	0.1993	
	22.25	0.04684	
8	5.756	2.538	
	6.702	0.4435	
	10.15	0.1395	
	28.94	0.03568	
9	12.33	6.853	0.5382
	8.281	0.2813	
	12.68	0.1038	
	36.51	0.02808	
10	7.125	3.170	
	7.897	0.5630	
	9.952	0.1962	
	15.50	0.08054	
	44.98	0.02269	

Appendix A2

Bessel Function (Zero-Order First Kind)

x	$J_o(x)$	x	$J_o(x)$	x	$J_o(x)$	x	$J_o(x)$
0.0	1.0000	2.5	−0.0484	5.0	−0.1776	7.5	0.2663
0.1	0.9975	2.6	−0.0968	5.1	−0.1443	7.6	0.2516
0.2	0.9900	2.7	−0.1424	5.2	−0.1103	7.7	0.2346
0.3	0.9776	2.8	−0.1850	5.3	−0.0758	7.8	0.2154
0.4	0.9604	2.9	−0.2243	5.4	−0.0412	7.9	0.1944
0.5	0.9385	3.0	−0.2601	5.5	−0.0068	8.0	0.1717
0.6	0.9120	3.1	−0.2921	5.6	0.0270	8.1	0.1475
0.7	0.8812	3.2	−0.3202	5.7	0.0599	8.2	0.1222
0.8	0.8463	3.3	−0.3443	5.8	0.0917	8.3	0.0960
0.9	0.8075	3.4	−0.3643	5.9	0.1220	8.4	0.0692
1.0	0.7652	3.5	−0.3801	6.0	0.1506	8.5	0.0419
1.1	0.7196	3.6	−0.3918	6.1	0.1773	8.6	0.0146
1.2	0.6711	3.7	−0.3992	6.2	0.2017	8.7	−0.0125
1.3	0.6201	3.8	−0.4026	6.3	0.2238	8.8	−0.0392
1.4	0.5669	3.9	−0.4018	6.4	0.2433	8.9	−0.0653
1.5	0.5118	4.0	−0.3971	6.5	0.2601	9.0	−0.0903
1.6	0.4554	4.1	−0.3887	6.6	0.2740	9.1	−0.1142
1.7	0.3980	4.2	−0.3766	6.7	0.2851	9.2	−0.1367
1.8	0.3400	4.3	−0.3610	6.8	0.2931	9.3	−0.1577
1.9	0.2818	4.4	−0.3423	6.9	0.2981	9.4	−0.1768
2.0	0.2239	4.5	−0.3205	7.0	0.3001	9.5	−0.1939
2.1	0.1666	4.6	−0.2961	7.1	0.2991	9.6	−0.2090
2.2	0.1104	4.7	−0.2693	7.2	0.2951	9.7	−0.2218
2.3	0.0555	4.8	−0.2404	7.3	0.2882	9.8	−0.2323
2.4	0.0025	4.9	−0.2097	7.4	0.2786	9.9	−0.2403

Appendix A3

PARTIAL FRACTION EXPANSION

In some signal processing applications, it is sometimes necessary to partition a given transfer function into the sum of several first-order terms. An example of such an application is getting the inverse Z-transform of a certain function or in the parallel realization of a filter. The technique is best illustrated through some examples.

EXAMPLE A3.I

Find the PFE for the following function:

$$H(Z) = \frac{Z^{-1}}{3 - 4Z^{-1} + Z^{-2}}$$

Solution

Multiplying by Z^2 gives

$$H(Z) = \frac{Z}{3Z^2 - 4Z + 1}$$

$$\frac{H(Z)}{Z} = \frac{1}{3[(Z - 1)(Z - (1/3))]}$$

which could be written as

$$\frac{H(Z)}{Z} = \frac{A}{(Z - 1)} + \frac{B}{(Z - (1/3))} \qquad (A3.1)$$

The "residues" A and B can be obtained as follows. To get A, we first multiply $H(Z)$ by $(Z - 1)$:

$$H(Z) \cdot \frac{(Z - 1)}{Z} = \frac{1}{3(Z - (1/3))} = A$$

Then, substituting for $Z = 1$, we get

$$A = \frac{1}{2}$$

Similarly, to get B, we multiply $H(Z)$ by $(Z - (1/3))$, and substituting for $Z = 1/3$ gives

$$B = -\frac{1}{2}$$

Substituting in Equation A3.1, we get

$$\frac{H(Z)}{Z} = \frac{1/2}{(Z - 1)} - \frac{1/2}{(Z - (1/3))}$$

Thus

$$H(Z) = \frac{(1/2)Z}{(Z-1)} - \frac{1/2}{(Z-(1/3))}$$

EXAMPLE A3.2

Find the PFE of the function

$$H(Z) = \frac{1 + Z^{-1}}{3 - 4Z^{-1} + Z^{-2}}$$

Solution

Multiplying by the highest power gives

$$H(Z) = \frac{Z+1}{3Z^2 - 4Z + 1}$$

Dividing both sides by Z, we get

$$\frac{H(Z)}{Z} = \frac{Z+1}{Z(3Z^2 - 4Z + 1)} = \frac{Z+1}{3 \cdot Z(Z-1)(Z-(1/3))}$$

$$= \frac{A}{Z} + \frac{B}{(Z-1)} + \frac{C}{(Z-(1/3))}$$

To get the residue A, we multiply by Z, and substituting for $Z = 0$ yields

$$H(0) = A = 1$$

For B, we multiply by $(Z-1)$ and make the substitution $Z = 1$, giving

$$RHS = A \times 0 + B + C \times 0 = \frac{2}{3 \times \dfrac{2}{3}} = 1$$

$$B = 1$$

Similarly, to obtain C, we multiply by $(Z-(1/3))$ and substitute for $Z = 1/3$, giving

$$C = \frac{(1/3)+1}{3.1/3((1/3)-1)} = \frac{4}{3} \times \frac{-3}{2} = -2$$

then

$$\frac{H(Z)}{Z} = \frac{1}{Z} + \frac{1}{(Z-1)} - \frac{2}{(Z-(1/3))}$$

or

$$H(Z) = 1 + \frac{Z}{(Z-1)} - \frac{2 \cdot Z}{(Z-(1/3))}$$

Appendix A4

Important Characteristics of Some Common Windows

Window	N	A_p/dB	A_r/dB	Function
Rectangular	$0.9/\Delta\omega$	0.7416	21	1
Von Hann	$3.1/\Delta\omega$	0.0546	44	$= 0.5 - 0.5\cos\left(\dfrac{2\pi n}{N-1}\right)$
Hamming	$3.3/\Delta\omega$	0.0194	53	$= 0.54 - 0.46\cos\left(\dfrac{2\pi n}{N-1}\right)$
Blackman	$5.5/\Delta\omega$	0.0017	74	$= 0.42 - 0.5\cos\left(\dfrac{2\pi n}{N-1}\right) + 0.08\cos\left(\dfrac{4\pi n}{N-1}\right)$
Kaiser	$\dfrac{A_r - 7.95}{14.36\,\Delta\omega}$	0.0274	50	$= \dfrac{J_o(\beta)}{J_o(\alpha)}, \alpha = 0.1102(A_r - 8.7)$
		0.00275	70	$\beta(n, \alpha) = \alpha\sqrt{1 - \left(\dfrac{2n'}{N-1}\right)^2}, \quad n' = n - \dfrac{N-1}{2}$
		0.000275	90	

Characteristics of Some Windows

Window	N	A_p/dB	A_r/dB	Function
Rectangular	$0.9/\Delta\omega$	0.7416	21	1
Von Hann	$3.1/\Delta\omega$	0.0546	44	$= 0.5 - 0.5\cos\left(\dfrac{2\pi n}{N-1}\right)$
Hamming	$3.3/\Delta\omega$	0.0194	53	$= 0.54 - 0.46\cos\left(\dfrac{2\pi n}{N-1}\right)$
Blackman	$5.5/\Delta\omega$	0.0017	74	$= 0.42 - 0.5\cos\left(\dfrac{2\pi n}{N-1}\right) + 0.08\cos\left(\dfrac{4\pi n}{N-1}\right)$
Kaiser	$\dfrac{A_r - 7.95}{14.36\,\Delta\omega}$	0.0274	50	$= \dfrac{J_o(\beta)}{J_o(\alpha)}, \alpha = 0.1102(A_r - 8.7)$
		0.00275	70	$\beta(n, \alpha) = \alpha\sqrt{1 - \left(\dfrac{2n'}{N-1}\right)^2}, \quad n' = n - \dfrac{N-1}{2}$
		0.000275	90	

References

1. Sedra, A. and K. Smith, *Microelectronic Circuits*. Oxford University Press, U.K., 1998.
2. Savant, C., M. Rohden and G. Carpenter, *Electronic Design*. Addison Wesley, Reading, MA, 1990.
3. Bell, D., *Solid State Pulse Circuits*. Prentice Hall, Englewood Cliffs, NJ, 1999.
4. Ifeachor, E. and B. Jervis, *Digital Signal Processing: A Practical Approach*. 2nd Edition. Prentice Hall, Englewood Cliffs, NJ, 2002.
5. Temes, G.S. and J. Lapatra, *Introduction to Circuit Synthesis and Design*. McGraw Hill, New York, 1977.
6. Hamdy, N., Verbund Transistoren und Y- Gyratoren. Ph.D. Thesis, Erlangen-Nuerenberg University, Germany, 1979.
7. Telegan, B., The Gyrator a New Electric Network Element. *Philips Research Report 3*, 1948.
8. Bruton, L.T., Non Ideal Performance of Two-Amplifier Positive Impedance Converters. *IEEE Transactions on Circuit Theory*, CT-17, 541–549, November 1970.
9. Sallen, R.P. and E.L. Key, A Practical Method of Designing RC Active Filters. *IRE Transactions Circuit Theory*, CT-2, 74, March 1955.
10. Soliman, H., A. Hamad and N. Hamdy, A Video-Sped Switched Resistor A/D Converter Architecture. *Proceedings of the 43rd IEEE Midwest Symposium on Circuits and Systems*, Michigan, IL, August 2000.
11. Hamad, A., Testing of High Speed A/D Converters. M.S. Thesis, Mansoura University, Egypt, 1998.
12. Benetaazzo, A.L., C. Narduzzi, C. Offelli and D. Petri, A/D Converter Performance Analysis by a Frequency Domain Approach. *IEEE Transactions on Instrumentation and Measurement*, 41(December), 834–839, 1992.
13. Tietze, U. and Ch. Schenk, *Electronic Circuits, Design and Applications*. Springer-Verlag, Berlin, Heidelberg, 1991.
14. Texas Instruments, TMS320 Second Generation Devices. Data Sheet, 1987/1990.
15. Ludeman, L., *Fundamentals of Digital Signal Processing*. Harper& Row, New York, 1986.
16. Rabiner, L.R. and B. Gold, *Theory and Applications of Digital Signal Processing*. Prentice Hall, Englewood Cliffs, NJ, 1975.
17. Rabiner, L.R. and R.E. Crochier, A Novel Implementation for Narrow-Band FIR Digital Filters. *IEEE Transactions on Acoustics, Speech & Signal Processing*, 23(5), 457–464, 1994.
18. Defatta, D., J. Lucas and W. Hodgkiss, *Digital Signal Processing. A System Design Approach*. John Wiley & Sons, Chichester, NY, 1988.
19. Cooly, J.W. and J.W. Tukey, An Algorithm for Machine Calculation of Complex Fourier Series. *Mathematics Computation*, 19, 297–301, 1965.
20. Gold, B. and T. Bailly, Parallelism in Fast Fourier Transform Hardware. *IEEE Transactions on Audio and Electroacoustics*, 21, 5–16, February 1973.
21. Ahmed, N. and K.R. Rao, *Orthogonal Transforms for Digital Signal Processing*. Springer-Verlag New York, Inc., Secaucus, NJ, 1975.
22. Hsiung, W., S. Chen and S. Frlick, A Fast Computational Algorithm for the Discrete Cosine Transform. *IEEE Transactions on Communications*, 25(9), 1004–1009, 1977.
23. Bently, P. and J. McDonnell, Wavelet Transforms, An Introduction. *Electronics & Communications Engineering Journal*, 6, 175–186, August 1994.

24. Soliman, H. and N. Hamdy, A Flash-Like Cyclic A/D Converter Architecture for High Resolution Applications. *Proceedings of the 40th Midwest Symposium on Circuits and Systems*, Iowa, USA, August 1997.

25. Motorola, INC. DSP56002 Data Sheet, 1996.

26. ANALOG DEVICES, ADSP-2100 Family. Data Sheet, 1996.

27. Sharkas, M., Iris Recognition for Personal Identification Using Artificial Neural Networks and Different Transforms. Ph.D. Thesis, Alexandria University, Egypt, 2002.

28. Pan, J.-S., H.-C. Huang and L.C. Jain, *Intelligent Watermarking Techniques*. World Scientific Pub., New Jersey, London, Shanghai, 2004.

29. Haykin, S., *Neural Networks, A Comprehensive Foundation*. Prentice Hall, Englewood Cliffs, NJ, 1999.

30. Rabiner, L.R. and B.H. Juang, *Fundamentals of Speech Recognition*. Prentice Hall, Englewood Cliffs, NJ, 1993.

31. Kung, S.Y., *Digital Neural Networks*. PTR Prentice Hall, Englewood Cliffs, NJ, 1993.

32. Tudor, P.N., Tutorial on MPEG-2. *Electronics and Communication Engineering Journal*, 7, 257–264, December 1995.

33. http://vsr.informatik.tu-chemnitz.de/~jan/MPEG/HTML/mpeg_tech.html (last accessed on July 2007).

34. http://www.bretl.com/mpeghtml/codecdia1.HTM (last accessed on July 2007).

35. Raissi, R., The Theory Behind MP3. Report. www.mp3-tech.org/programmer/docs/mp3_theory.pdf, 2002.

36. Sirovich, L. and M. Kirby, Low-Dimensional Procedure for the Characterization of Human Faces. *Journal of Optical Society of America Society (A)*, 4(3), 1987.

37. Feng, G.C., P.C. Yuen and D.Q. Dai, Human Face Recognition using PCA on Wavelet Subband. *SPIE Journal of Electronic Imaging*, 09(02), 2000.

38. www.watermarkingworld.org, 2003.

39. Sharkas, M., D. Elshafie and N. Hamdy, A Dual-Digital Image Watermarking Technique. *Proceedings of the 3rd World Enformatika Conference, WEC'05*, Istanbul, Turkey, April 27–29, 2005.

40. Smith, S.W., *The Scientist and Engineer's Guide to Digital Signal Processing*. California Technical Publishing, San Diego, CA, 2006.

41. Johnson, J., *Introduction to Digital Signal Processing*. Prentice Hall, New York, 1989.

42. Antoniou, A., *Digital Filters*. McGraw Hill, New York, 1993.

43. Rabiner, L.R., J.H. McClellan and J.W. Parks, FIR Digital Filter Design Techniques using Weighted Chebyshev Approximation. *Proceedings of the IEEE*, 63(4), 595–610, 1975.

44. Baher, H., *Analog & Digital Signal processing*. John Wiley & Sons, Chichester, NY, 1990.

45. Mitra, S.K., *Digital Signal Processing, A Computer-Based Approach*. McGraw Hill Int. Editions, Singapore, 1998.

46. Oppenheim, A.V., R.W. Scafer and J.R. Buck, *Discrete-Time Signal Processing*. 2nd Edition. Prentice Hall, Englewood Cliffs, NJ, 1999.

47. Seitzer, D., G. Pretzel and N. Hamdy, *Electronic Analog-To-Digital Converters*. John Wiley & Sons, Chichester, NY, 1983.

48. Chiorboli, G., G. Franco and C. Morandi, Analysis of Distortion in A/D Converters by Time-Domain and Code-Density Analysis. *IEEE Transactions on Instrumentation and Measurements*, 45(1), 45–49, 1996.

49. Akansu, A., Wavelets and Filter Banks, A Signal Processing Perspective. *IEEE Circuits & Devices*, 10, 14–18, November 1994.

50. Hamdy, N., H. Soliman and A. Eid, A Vertical Successive-Approximation A/D Converter Architecture for High-Speed Applications. *Proceedings of the 41st Midwest Symposium on Circuits and Systems*, Notre Dame, IN, August 1998.

51. Daubechies, L., The Wavelet Transform, Time-Frequency Localization and Signal Analysis. *IEEE Transactions on Information Theory*, 36(5), 961–1005, 1990.

52. Zewail, R.F., Fingerprint-Based Identity Verification Employing Biometrics Fusion. M.S. Thesis, AAST, Alexandria, Egypt, 2004.

53. Soliman, H. and N. Hamdy, A High-Speed High-Performance Cyclic A/D Converter. *Proceedings of the International Conference on Electronics, Circuits & Systems. ICECS'95*, Amman, Jordan, December 1995.

54. Hamdy, N. and H. Soliman, A Forward Successive Approximation A/D Converter Architecture. *Proceedings of the 39th Midwest Symposium on Circuits & Systems*, Iowa, USA, August 1996.

55. Hamdy, N. and H. Soliman, A High-Speed A/D Converter Architecture for High-Resolution Applications. *Proceedings of the 39th Midwest Symposium on Circuits & Systems.* Iowa, USA, August 1996.

56. Soliman, H. and N. Hamdy, A Flash-Like Cyclic A/D Converter Architecture for High-Resolution Applications. *Proceedings of the 40th Midwest Symposium on Circuits & Systems*, Sacramento, CA, August 1997.

57. Hamdy, N., H. Soliman and A. Eid, A Vertical Successive Approximation A/D Converter Architecture for High-Speed Applications. *Proceedings of the 41st Midwest Symposium on Circuits and Systems*, Notre Dame, IN, August 1998.

58. Rabiner, L.R., A Tutorial on Hidden Markov Models and Selected Applications in Speech Recognition. *Proceedings of the IEEE*, 77(2), 257–286, 1989.

59. Saal, R., *Der Entwurf von Filtern mit Hilfe des Kataloges normierter Tiefpaesse* (in German). Telefunken Gmbh, Backnang, 1963.

60. Zhang, G., et al., Forecasting with Artificial Neural Networks; The State of the Art. *International Journal of Forecasting*, 14(1), 35–62, 1998.

61. http://www.ics.uci.edu/~magda/Presentations/landscape_part1.pdf (last accessed on July 2007).

Index

A

Active noise control, 481–483
Adaptive interference cancellation, noise
    control, 483–484
Adder component, digital filters, 255–256
ADSP2100 analog devices, 444–445
A-laws, companding characteristics, data
    converter, 161–162
Algorithm-specific processing, digital signal
    processors, 433
Aliasing errors, data converter sampling,
    148–152
All-pass filter, design principles, 71–76
All-pole function, Butterworth approximation,
    84–88
AM index, analog signal processing, 27
Amplification, analog signal
    processing, 11–12
Amplitude
    analog signal processing, 7
        linear operations, 11–16
        nonlinear operations, 16–27
    rectangular window, finite impulse response
        filter design, 279–285
    ripple amplitude
        low-pass filter specifications, 78
        multirate signal processing, 381–386
        tolerance structure, 79
Amplitude modulation (AM), analog signal
    processing, 26–29
Analog comparators
    ADC construction, 172–175
    cascaded-flash ADC converter, 199–204
    parallel (flash) analog-to-digital converter,
        188–189
Analog filters
    all-pass filter, 71–76
    approximation techniques, 78
        Bessel approximation, 97–99
        Butterworth approximation, 80–88
        Chebyshev (equiripple) approximations,
            89–95
        elliptic (Cauer) approximation, 96–97
        ideal response, 80–99
    band-pass filter, 65–66
    band-stop filter, 66–67
    basic principles, 55–56
    breadboarding, 78–79
    design procedures, 76–79
    functions, 56–76
    future research issues, 139–140

    high-pass filter, 64–65
    low-pass filter, 62–64
    notch filter, 67–71
        high-pass notch filter, 71
        low-pass filters, 69–71
        symmetrical filters, 68–69
    realization techniques, 78
        filter catalogues, 121–138
        higher-order filters, 111–112
        inductionless realizations, 101–107
        LC realization, 99–101
        pole-zero distribution, 112–114
        RC realization, 107–111
        scaling factors, 114–117
        tolerance structure, 117–121
    second-order filter, 60–62
    simulation techniques, 78
    specifications, 77–80
    switched capacitor filters, 138–139
Analog input/out specifications, data
    converters, 147
Analog signal processing
    amplitude operations, 11–27
    basic principles, 4–5
    frequency spectrum shaping, 27–30
        filtering, 29
        frequency modulation, 29–30
        harmonic generators, 27–28
    linear amplitude operations, 11–16
        amplification/attenuation, 11–12
        differentiation, 14–16
        integration, 12–14
    nonlinear amplitude operations, 16–27
        amplitude modulation, 26–27
        clamping, 16–19
        clipping, 19–24
        rectification, 24–26
    phase spectra operations, 30–31
    physical quantities, 7–8
    practical system examples, 46–48
    RC circuit sample, 8–10
    transducers, 7
    waveform generation, 31–47
        astable multivibrators, 36–37
        bistable multivibrators, 39
        bootstrap sweep generators, 44–46
        crystal oscillators, 34–35
        LC oscillators, 33–34
        linear sweep generator, 41–44
        Miller integrator, 44
        monostable multivibrators, 37–39
        RC phase-shift oscillators, 32–33